Power Systems

Electrical power has been the technological foundation of industrial societies for many years. Although the systems designed to provide and apply electrical energy have reached a high degree of maturity, unforeseen problems are constantly encountered, necessitating the design of more efficient and reliable systems based on novel technologies. The book series Power Systems is aimed at providing detailed, accurate and sound technical information about these new developments in electrical power engineering. It includes topics on power generation, storage and transmission as well as electrical machines. The monographs and advanced textbooks in this series address researchers, lecturers, industrial engineers and senior students in electrical engineering.

Power Systems is indexed in Scopus

Wei Yao · Hongyu Zhou · Yongxin Xiong ·
Jinyu Wen

Active Control of Large-Scale Offshore Wind Farms Connected Via VSC-HVDC

Springer

Wei Yao
Huazhong University of Science and Technology
Wuhan, Hubei, China

Hongyu Zhou
Huazhong University of Science and Technology
Wuhan, Hubei, China

Yongxin Xiong
The Hong Kong Polytechnic University
Hong Kong, China

Jinyu Wen
Huazhong University of Science and Technology
Wuhan, Hubei, China

ISSN 1612-1287 ISSN 1860-4676 (electronic)
Power Systems
ISBN 978-981-97-9345-7 ISBN 978-981-97-9346-4 (eBook)
https://doi.org/10.1007/978-981-97-9346-4

This Springer imprint is published by the registered company Springer Nature Singapore Pte Ltd.
The registered company address is: 152 Beach Road, #21-01/04 Gateway East, Singapore 189721, Singapore

To our families

Foreword

Offshore wind farms (OWFs) represent an evolving advancement in the wind energy sector, utilizing cutting-edge technologies to harness resources that are more plentiful, powerful, and consistently available compared to those of land-based wind farms. As the integration of large-scale OWFs increases, voltage source converter-based high-voltage direct current (VSC-HVDC) connections have emerged as a superior solution for power transmission. However, this advancement introduces significant challenges in the active control and safe operation of power systems. Due to the independent control of mechanical and electrical systems, VSC-HVDC-connected offshore wind farms cannot directly respond to onshore system faults and frequency fluctuations. As the penetration of OWFs continues to rise, addressing system faults and frequency events becomes increasingly challenging. Therefore, more flexible active control schemes are required to ensure the safe operation of power systems, which is the focal point of this monograph.

Upon submitting the manuscript for publication, the authors invited me to write a Foreword for this book that interests me greatly and represents their recent achievements. The topics in the book are very similar to my research expertise in renewable energy, DC transmission and distribution, and power electronic converter control as I currently lead a research group at Cardiff University, UK, in these areas.

This monograph, *Active Control of Large-Scale Offshore Wind Farms Connected Via VSC-HVDC*, serves as an invaluable technical resource. It not only summarizes existing technologies in the field but also highlights the authors' recent technological advancements. The monograph investigates the core technologies of active control in offshore wind farms connected via VSC-HVDC systems, providing readers with a comprehensive review of both classic and contemporary active control technologies in these systems.

This book goes beyond a mere review. Its core objective is to provide new perspectives and methods for active control in VSC-HVDC-connected OWFs, aiming to optimize performance by leveraging available control freedoms. With these active control strategies, VSC-HVDC-connected OWFs can provide valuable support to onshore power systems.

There are three parts in the monograph: Part I introduces an efficient and high-precision modeling method for large-scale offshore wind farms connected via VSC-HVDC, based on vector modeling; Part II analyzes the characteristics of various faults in offshore wind farms connected via VSC-HVDC systems and proposes fault suppression schemes to enhance system fault ride-through; and Part III presents advanced active frequency support methods for offshore wind farms connected via VSC-HVDC systems. I am sure this coverage will benefit academic researchers and industrial engineers not only in China but also worldwide, like in the UK, Europe, and the US, as I currently have wide communication and collaboration with various sectors internationally on these topics, which reflects the significant needs of the knowledge.

Having known Prof. Wei Yao, the lead author of this monograph, for over 10 years, I am thoroughly familiar with the research conducted by him and his group. Professor Wei Yao has dedicated more than a decade to power system stability analysis and control. His contributions, along with other researchers in his team, have earned wide international recognition. Co-author Mr. Hongyu Zhou is currently pursuing his Ph.D., focusing on fault ride-through for MMC-HVDC integrated offshore wind farms and DC grids. Dr. Yongxin Xiong, another co-author, is a Postdoctoral Fellow at the Hong Kong Polytechnic University, specializing in the control and stability analysis of grid integration of offshore wind farms via VSC-HVDC. Additionally, Prof. Jinyu Wen, with extensive experience in practical projects, has accumulated valuable expertise in the control and analysis of HVDC-connected systems. Their work includes international leading journal and conference papers, as well as granted patents. This book integrates the most critical aspects of their work.

For readers seeking an innovative, informative, and applied exploration of active control in large-scale offshore wind farms connected via VSC-HVDC, this monograph is a must-read. I hope you find the content as insightful and rewarding as I do.

August 2024

Prof. Jun Liang
Cardiff University
Cardiff, UK

Preface

Wind power has become one of the most crucial renewable energy generation technologies. Among its various forms, offshore wind power stands out due to its high resource stability, significant development potential, and proximity to load centers. As offshore wind power capacity and transmission distances increase, traditional AC transmission technology becomes less feasible and economical. Consequently, Voltage Source Converter-based High Voltage Direct Current (VSC-HVDC) transmission, which is more cost-effective for long-distance transmission, has been increasingly employed to connect large-scale offshore wind farms to the grid. Currently, numerous offshore wind farms connected via VSC-HVDC have been constructed worldwide, such as the DolWin 1–6 and HelWin 1–4 projects in Germany and the Rudong project in China.

Compared to onshore wind projects, offshore wind farms connected via VSC-HVDC face significant challenges in modeling, fault management, and frequency support due to their scale, environment, and grid isolation. High-precision modeling of offshore farms is complex because of their size and the intricacies of collector lines. Developing detailed models for each turbine increases simulation size and time, hindering efficient research and design support. Additionally, ensuring VSC-HVDC fault ride-through without blocking under complex faults is a challenge. Multi-mode faults on the grid, valve, DC, and wind farm sides are difficult for conventional strategies to manage, risking widespread turbine disconnection. Furthermore, providing active frequency support is complicated by the decoupling effect of HVDC and the replacement of onshore synchronous units, which reduces grid stability and demands innovative solutions for effective integration and operation.

This monograph addresses these challenges by studying the core technologies of active control in offshore wind farms connected via VSC-HVDC systems to improve safety and grid connection. The monograph is divided into three parts. Part I introduces an efficient and high-precision modeling method for large-scale offshore wind farms connected via VSC-HVDC based on vector modeling. Part II analyzes the characteristics of multiple types of faults in offshore wind farms connected via VSC-HVDC systems and proposes fault suppression schemes to achieve system fault ride-through. Part III details advanced active frequency support methods for offshore wind farms connected via VSC-HVDC systems.

Specifically, Part I is organized as follows:

Chapter 1 introduces the system topology and conventional control structure of offshore wind farms connected via VSC-HVDC. It begins with the topologies and application scenarios of two-terminal VSC-HVDC and VSC-based multi-terminal high-voltage direct current (VSC-MTDC) systems. The chapter also details the conventional control methods for onshore converter stations, offshore converter stations, and offshore wind farms, illustrating the characteristics of each control method through simulation studies. This chapter provides a foundational understanding for readers of various backgrounds on offshore wind farms connected via VSC-HVDC.

Chapter 2 introduces a vector-variable-based method for dynamic modeling of large-scale wind farms with collector system dynamics. It simplifies complex models into wind generator and collector system modules, allowing flexible configurations. The accuracy and simulation efficiency of this model are validated against an electromagnetic model. A detailed wind farm with 25 generators demonstrates that this model accurately reflects stability compared to aggregation models. Furthermore, neglecting collector system dynamics can lead to misjudgments of instability modes, especially in sub-synchronous mode analysis.

Part II consists of the following content:

Chapter 3 proposes an active energy control (AEC) scheme for VSCs using Modular Multilevel Converters (MMCs) to enhance AC fault ride-through (ACFRT) capability. It utilizes MMC sub-module capacitors to absorb residual power during grid-side AC faults. The scheme includes a steady-state control scheme and stages for passive and active energy recovery, maintenance, and release between onshore and offshore converter stations. Additionally, a coordinated control scheme for AEC and active power reduction in offshore wind farms is designed.

Chapter 4 proposes a multi-mode matching control (MMMC) approach to enhance fault ride-through capability and voltage support for offshore wind systems during onshore grid AC faults. Various modes are matched to different fault levels. Mode I maximizes surplus power absorption using MMC sub-modules during high-level faults to mitigate DC overvoltage. Mode II optimizes MMC active and reactive current during medium-level faults to support grid voltage and power output. Mode III adjusts energy absorption rates of both WFMMC and GSMMC during mild faults for optimal energy utilization and grid support.

Chapter 5 analyzes single-phase ground (SPG) fault characteristics in onshore converter stations of MMC-HVDC integrated offshore wind farms and proposes suppression methods. It deduces the propagation and impact mechanism of SPG faults based on MMC control principles. An overvoltage suppression strategy using zero-sequence voltage modulation is proposed, and its effect on transient stability is analyzed. The main parameter of this strategy is designed to meet practical engineering requirements.

Chapter 6 proposes a multi-level coordinated control (MLCC) scheme for symmetrical bipolar MMC-HVDC integrated offshore wind farms to suppress DC overcurrent caused by submarine cable breakage. The MLCC scheme includes four levels: the offshore fault pole MMC absorbs power using sub-module capacitors when a cable breaks; the onshore non-faulty pole MMC raises DC voltage to suppress overcurrent; the offshore non-fault pole MMC balances real-time power delivery; and offshore wind farms reduce power to restore non-fault pole submarine cable transmission power and current to permissible levels.

Chapter 7 introduces a multi-stage sequential network energy control (MSNEC) for offshore AC asymmetric fault ride-through (FRT). It reveals the constraint relationship among sequence networks under such faults and explores the energy control potential of WFMMC. MSNEC reduces fault phase current by engaging submodules in the fault phase and adding submodules in the non-faulted phase to prevent overvoltage. It addresses multiple fault stages, combining sequential network control, energy control, and voltage support.

Part III consists of the following content:

Chapter 8 proposes a two-level combined control (TLCC) scheme for VSC-based multi-terminal HVDC (VSC-MTDC) integrated offshore wind farms to provide frequency support for the onshore system. The TLCC includes step start-up and adaptive inertial droop control at the offshore turbine level and communication-free allocation control at the onshore VSC station level. Each turbine uses inertial and droop control with adaptive coefficients and operates in maximum power point tracking (MPPT) mode. To reduce the second frequency drop (SFD), turbines are clustered by rotor speed, and step start-up control provides sequential frequency support.

Chapter 9 introduces an improved communication-free coordinated control (ICFCC) scheme for VSC-MTDC integrated offshore wind farms (OWFs) to provide fast frequency support. It integrates onshore frequency variations into DC voltage deviation, enabling offshore VSC stations to offer rapid frequency support through an estimator. The scheme features adaptive droop coefficient regulation for efficient power allocation, prioritizing stations closer to disturbances. Following frequency support, offshore turbines recover rotor speed using asymptotic control to mitigate the second frequency drop.

Chapter 10 proposes an adaptive dual droop control (ADDC) scheme for VSC-MTDC integrated offshore wind farms (OWFs) to provide fast frequency support to disturbed onshore systems. ADDC addresses issues with conventional droop control, which can cause frequency variations in undisturbed systems. By detecting disturbed and undisturbed systems, ADDC allows undisturbed systems to provide more support while ensuring safety through power limitation and droop coefficient regulation. Offshore stations estimate onshore DC voltage for fast frequency support, and OWFs recover rotor speed with asymptotic control to reduce the second frequency drop.

Chapter 11 presents a distributed cooperative control (DCC) scheme for offshore wind farms using a multi-terminal DC system to provide rapid frequency support. The DCC employs a consensus algorithm to distribute this support among wind turbines, utilizing their kinetic energy efficiently. It adjusts droop control coefficients for security. Following frequency support, a leading turbine initiates an asymptotic recovery scheme for smooth rotor speed restoration, with others following to reduce a second frequency drop.

We would like to thank Mr. Yulong Li, Dr. Siqi Lin, Mr. Haiyu Zhao, Miss Yuying Chen, Miss Kangyi Sun, and all colleagues in the Smart Grid Operation and Control Group, Huazhong University of Science and Technology, for their assistance.

Special thanks also go to Editors from Springer for their professional and efficient editorial work on this monograph.

Wuhan, China — Prof. Wei Yao
Wuhan, China — Mr. Hongyu Zhou
Hong Kong, China — Dr. Yongxin Xiong
Wuhan, China — Prof. Jinyu Wen
August 2024

Acknowledgements

This book summarizes our recent research on active control for offshore wind farms connected via VSC-HVDC systems, focusing on the core technologies that enhance the safety and grid integration of these systems. This research was conducted at the School of Electrical and Electronic Engineering, State Key Laboratory of Advanced Electromagnetic Engineering and Technology, Huazhong University of Science and Technology, Wuhan, China.

We are deeply grateful to many individuals who contributed to this book in various ways. Special thanks go to Mr. Yulong Li, Dr. Siqi Lin, Mr. Haiyu Zhao, Miss Yuying Chen, Miss Kangyi Sun, and all our colleagues at the Smart Grid Operation and Control Center at Huazhong University of Science and Technology, for their valuable contributions, assistance in identifying errors, and thorough review of the manuscript.

We also extend our appreciation to the staff at Springer for their support and assistance in the preparation of this book.

This work was partially supported by the National Natural Science Foundation of China (52022035, 523B2074, 52237004, U22B20111) and the National Key Research and Development Program Project (2022YFB4202304).

Wuhan, China
August 2024

Prof. Wei Yao
Mr. Hongyu Zhou
Dr. Yongxin Xiong
Prof. Jinyu Wen

Contents

Part II Fault Analysis and Active Suppression in Large-Scale Offshore Wind Farms Connected Via VSC-HVDC

Part I
Modeling of Large-Scale Offshore Wind Farms Connected Via VSC-HVDC

In recent years, as offshore wind farms have continued to develop into deeper waters, multi-terminal high-voltage direct current (MTDC) transmission systems, due to their flexible control characteristics, have gradually received attention for their ability to achieve friendly grid integration of offshore wind power. With the continuous increase in the penetration rate of offshore wind power, some synchronous generator units are being replaced, resulting in a continuous decrease in the overall system inertia. Grid codes in several countries and regions, including the European Union, explicitly require wind farms to have the capability of providing active frequency support. To provide active frequency support in offshore wind farms, corresponding control needs to be added to the conventional control structure.

Existing research on active frequency support for long-distance offshore wind farms integrated with multi-terminal flexible DC systems has mainly focused on control strategy development, lacking an analysis of the factors affecting active frequency support in such offshore wind farms. It is indicated that increasing the virtual inertia control coefficient and droop coefficient of wind turbine units can help mitigate primary frequency drop in the system. However, some references point out that increasing these coefficients implies the need for wind turbine units to release more rotor kinetic energy, leading to an increase in unbalanced power during the rotor speed restoration process, which could potentially cause a secondary frequency drop in the system.

Differently, Reference [1] mitigates the secondary frequency drop caused by wind turbine unit speed restoration under different wind power penetration rates by adjusting the control coefficients and selecting appropriate moments for frequency shedding. On the other hand, Reference [2] achieves this by adjusting the virtual inertia control coefficient and droop coefficient of wind turbine units while raising the minimum point of the secondary frequency drop. Furthermore, besides the virtual inertia control coefficients and droop coefficients of wind turbine units, control parameters of the converter stations in multi-terminal flexible DC systems in offshore wind farms, as well as the communication delay between offshore and onshore, will also have an impact on the effectiveness of frequency support. Therefore, a more in-depth analysis of the primary influencing factors of active frequency support is needed.

Chapter 1 first analyzes the system-level control strategy for the integration of remote offshore wind power with multi-terminal flexible DC systems and introduces the system-level control strategy for the converter station and the control strategy for offshore wind farms. Subsequently, based on the MATLAB/Simulink simulation software [3], and using the power system simulation toolbox proposed in Reference [4], the numerical simulation examples of the multi-terminal flexible DC system integration for remote offshore wind power are built using the MATACDC hybrid system power flow calculation method proposed in Reference [5], enabling the rapid addition of new device models and advanced control strategies [6, 7]. Following this, this chapter further constructs a real-time simulation platform for the integration of remote offshore wind power with multi-terminal flexible DC systems based on the RTLab platform, which can provide real-time simulation verification of the corresponding controllers for the integration of remote offshore wind power with multi-terminal flexible DC systems. Finally, this chapter analyzes the factors affecting the active frequency support in the integration of remote offshore wind power with multi-terminal flexible DC systems, identifies the primary influencing factors affecting the effectiveness of active frequency support, and performs simulation verification of different influencing factors in the dynamic simulation model.

The expansion of wind energy has been characterized by rapid advancements across various countries. As the scale of wind power installations increases, there is a growing focus on maintaining system stability and developing synchronized control mechanisms within wind farms [8]. To comprehensively address these challenges, a detailed dynamic representation of the wind farm is imperative, incorporating each turbine within its operational scope [9]. Unlike other types of power stations, wind farms are often composed of numerous turbines dispersed over extensive areas. These turbines are interconnected through a power collection system, which can be configured in various ways [10]. The complexity inherent in having numerous components within a wind farm poses significant challenges in developing accurate models. Constructing a detailed model of the entire system requires substantial effort, and as the model's scale increases, it imposes a considerable computational burden.

To overcome certain limitations inherent in the aggregation model, numerous studies have constructed comprehensive models of entire wind farms. These studies preserve the dynamics of each wind generator within the wind farm and also develop detailed models of the collector system. However, there are notable variations in the approaches taken to model the collector system's topology. For instance, the collector systems modeled in [11] incorporate only parallel lines, which may not accurately reflect real-world scenarios. In contrast, the collector system topologies presented in [12] more closely align with reality but require manual segmentation of the collector system in a specific sequence. This approach also encounters difficulties when handling networks with arbitrary topologies, such as ring layouts [13]. Notably, the accuracy of these simplified models has not been thoroughly compared.

Recent studies [14, 15] have proposed that the dynamics of the network can significantly impact stability analysis results, not just in the context of high-frequency oscillations [16] but also when examining lower-frequency oscillations within the power system. Although these findings are not specific to wind farm systems, they imply that the dynamics of the collector system should not be overlooked, even when studying modes at lower frequencies. Therefore, considering the dynamics of the collector system in wind farms, particularly when exploring sub-synchronous modes, merits further discussion and investigation.

Chapter 2 introduces a method based on vector variables for dynamic modeling of large-scale wind farms, aiming to construct a comprehensive wind farm model. This method offers a scalable strategy for modeling multiple wind generators and collector systems with arbitrary topologies. The developed model can accommodate any quantity of wind generators and any topology of the collection systems, taking dynamics into account. Given that voltage source converter-based high-voltage direct-current (VSC-HVDC) technology is viewed as a significant means for transmitting power from the wind farm, a wind farm integrated with a VSC-HVDC system is constructed using the proposed method. Leveraging the established model, the chapter analyzes the differences between the detailed model and the aggregation model, as well as the impact of excluding collector system dynamics on the dynamics of the wind farm.

References

1. Díaz-González F, Hau M, Sumper A, Gomis-Bellmunt O (2014) Participation of wind power plants in system frequency control: review of grid code requirements and control methods. Renewable Sustainable Energy Rev 34:551–564. 10.1016/j.rser.2014.03.040
2. Pierri E, Binder O, Hemdan NG, Kurrat M (2017) Challenges and opportunities for a european hvdc grid. Renewable Sustainable Energy Rev 70:427–456. 10.1016/j.rser.2016.11.233
3. Chow J, Cheung K (1992) A toolbox for power system dynamics and control engineering education and research. IEEE Trans. Power Syst. 7(4):1559–1564. 10.1109/59.207380
4. Zimmerman RD, Murillo-Sanchez CE, Thomas RJ (2011) Matpower: steady-state operations, planning, and analysis tools for power systems research and education. IEEE Trans. Power Syst. 26(1):12–19. 10.1109/TPWRS.2010.2051168
5. Beerten J, Belmans R (2015) Development of an open source power flow software for high voltage direct current grids and hybrid ac/dc systems: Matacdc. IET Gener. Transm. Distrib. 9(10):966–974. 10.1049/iet-gtd.2014.0545
6. Ekanayake J, Holdsworth L, Jenkins N (2003) Comparison of 5th order and 3rd order machine models for doubly fed induction generator (dfig) wind turbines. Electr. Pow. Syst. Res. 67(3):207–215. 10.1016/S0378-7796(03)00109-3
7. Xiao L, Xu Z, Xiao H, Zhang Z, Wang G, Xu Y (2019) Electro-mechanical transient modeling of mmc based multi-terminal hvdc system with dc faults

considered. Int. J. Electr. Power Energy Syst. 113:1002–1013. 10.1016/j.ijepes.2019.06.003

8. Wang Y, Wu Q, Yang R, Tao G, Liu Z (2018) H∞ current damping control of dfig based wind farm for sub-synchronous control interaction mitigation. Int. J. Electr. Power Energy Syst. 98:509–519. 10.1016/j.ijepes.2017.12.003
9. Xiong Y, Yao W, Wen J, Lin S, Ai X, Fang J, Wen J, Cheng S (2021) Two-level combined control scheme of vsc-mtdc integrated offshore wind farms for onshore system frequency support. IEEE Trans. Power Syst. 36(1):781–792. 10.1109/TPWRS.2020.2998579
10. Quinonez-Varela G, Ault G, Anaya-Lara O, McDonald J (2007) Electrical collector system options for large offshore wind farms. IET Renew. Power Gener. 1(2):107. 10.1049/iet-rpg:20060017
11. Xia S, Bu S, Zhang X, Xu Y, Zhou B, Zhu J (2018) Model reduction strategy of doubly-fed induction generator-based wind farms for power system small-signal rotor angle stability analysis. Appl. Energy 222:608–620. 10.1016/j.apenergy.2018.04.024
12. Shao B, Zhao S, Yang Y, Gao B, Blaabjerg F (2021) Sub-synchronous oscillation characteristics and analysis of direct-drive wind farms with vsc-hvdc systems. IEEE Trans. Sustain. Energy 12(2):1127–1140. 10.1109/TSTE.2020.3035203
13. Du W, Wang Y, Wang HF, Yu J, Xiao X (2021) Collective impact of multiple doubly fed induction generators with similar dynamics on the oscillation stability of a grid-connected wind farm. IEEE Trans. Power Del. 36(5):2942–2954. 10.1109/TPWRD.2020.3030645
14. Grdenić G, Delimar M, Beerten J (2020) Assessment of ac network modeling impact on small-signal stability of ac systems with vsc hvdc converters. Int. J. Electr. Power Energy Syst. 119, 105, 897. 10.1016/j.ijepes.2020.105897
15. Markovic U, Stanojev O, Aristidou P, Vrettos E, Callaway D, Hug G (2021) Understanding small-signal stability of low-inertia systems. IEEE Trans. Power Syst. 36(5):3997–4017. 10.1109/TPWRS.2021.3061434
16. Song Y, Ebrahimzadeh E, Blaabjerg F (2018) Analysis of high-frequency resonance in dfig-based offshore wind farm via long transmission cable. IEEE Trans. Energy Convers. 33(3):1036–1046. 10.1109/TEC.2018.2794367

Chapter 1
Control Structure of Long-Distance Offshore Wind Power Integrated with VSC-MTDC

Abstract In recent years, offshore wind power has emerged as a critical frontier in global wind energy development, driven by its stable resource conditions, high utilization rates, proximity to load centers, and minimal land use. For offshore wind farms located near the onshore grid with smaller capacities, alternating current (AC) integration is typically employed. However, as these wind farms scale up and transmission distances increase, the voltage source converter-based multi-terminal high-voltage direct current (VSC-MTDC) system has gained prominence. This system effectively addresses challenges related to clustering, long-distance transmission, and multi-regional grid integration of wind energy. Despite its advantages, the VSC-MTDC system's characteristic of decoupling offshore and onshore AC systems poses challenges in frequency support. Frequency fluctuations in the onshore system prevent direct support from offshore wind power, potentially compromising overall power system stability and safety. Therefore, comprehensive research on active frequency support control strategies for VSC-MTDC-connected offshore wind farms is essential to ensure the secure and efficient integration of large-scale offshore wind power.

1.1 VSC-HVDC System Topology

Existing research identifies several main grid integration methods for offshore wind power: high-voltage alternating current (HVAC) transmission, high-voltage direct current (HVDC) transmission, and fractional frequency transmission system (FFTS) technology. Offshore wind power integrated through an HVAC system involves injecting offshore wind power into the onshore AC system using power-frequency AC transmission. HVDC integration can be further categorized based on specific technologies into line-commutated converter-based HVDC (LCC-HVDC), voltage source converter-based HVDC (VSC-HVDC), diode rectifier-based HVDC (DR-HVDC), and voltage source converter-based multi-terminal DC (VSC-MTDC). When employing an HVDC transmission system for grid integration, offshore wind power is collected via offshore AC systems and then converted into a DC system for transmission.

W. Yao et al., *Active Control of Large-Scale Offshore Wind Farms Connected Via VSC-HVDC*, Power Systems, https://doi.org/10.1007/978-981-97-9346-4_1

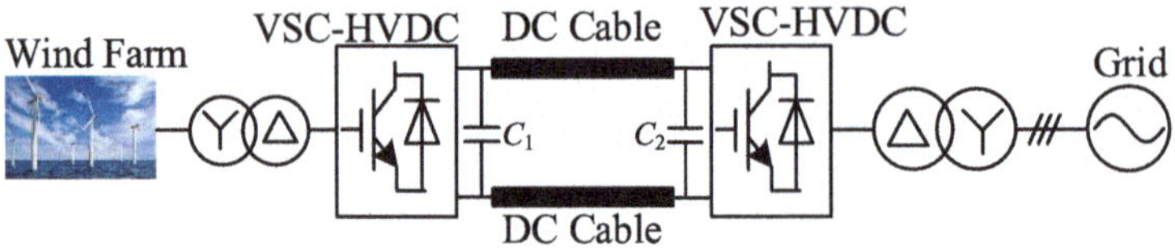

Fig. 1.1 The topology of VSC-HVDC based offshore wind farms

The FFTS for offshore wind power is an emerging grid integration method. It aims to increase the overall transmission capacity of the line by lowering the frequency of the transmission system without raising the voltage level, thereby reducing the number of lines and investment costs.

As the distance between offshore wind farms and the shore increases, longer AC cables exhibit significant capacitive effects, causing AC voltage fluctuations. This necessitates additional investment in reactive power compensation equipment, thereby increasing the transmission cost per unit of power.

For offshore wind farms located far from the shore, integrating them into the grid via voltage source converter-based high-voltage direct current (VSC-HVDC) has significant advantages [1]. The power generated by individual offshore wind turbines is collected and stepped up in voltage, then converted to HVDC at the offshore converter station. It is subsequently transmitted through subsea cables to the onshore converter station, where it is converted back to alternating current (AC) and finally injected into the onshore AC grid (Fig. 1.1).

1.2 VSC-MTDC System Topology

As the capacity, scale, and distribution area of offshore wind farms continue to grow, the integration of wind farms is evolving from single-station access to large-scale multi-station integration. The connection points are also transitioning from a single synchronous AC grid to interconnected multi-AC systems. In this context, using traditional point-to-point DC transmission would require the installation of multiple DC cable lines, significantly increasing costs and maintenance expenses. Therefore, voltage source converter based multi-terminal direct current (VSC-MTDC) grid integration technology has become an effective solution for the cluster-based, long-distance, and multi-region integration of wind power. Utilizing multi-terminal HVDC grid technology can considerably reduce the reactive power control costs and transmission power losses associated with increasing transmission distances, as seen with traditional AC grid technology [2]. Moreover, VSC-MTDC systems can consolidate distributed wind power clusters and transmit power centrally, mitigating the intermittency issues of individual wind farms and enabling the effective distribution of wind power across multiple AC systems.

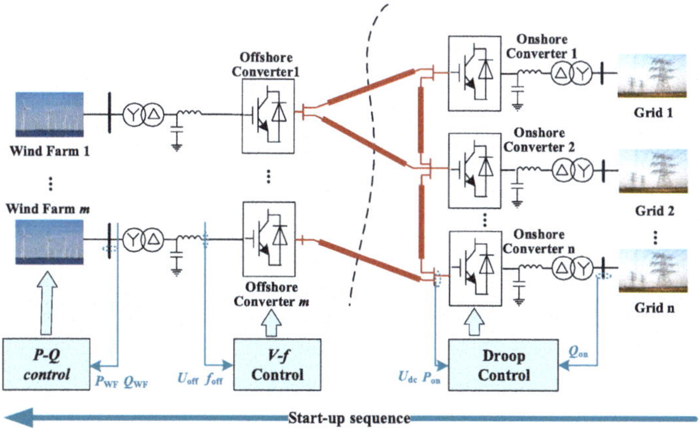

Fig. 1.2 The topology of VSC-MTDC based offshore wind farms

The startup sequence of an offshore wind multi-terminal VSC-HVDC grid follows an onshore-to-offshore order, as shown in Fig. 1.2 [3]. In this multi-terminal VSC-HVDC system, the onshore converter stations typically use dq decoupling control. The d-axis control can be configured with either master-slave or droop control, while the q-axis control is generally used to maintain reactive power at the onshore grid connection bus. When the onshore converter station is activated, it establishes the DC voltage for the entire multi-terminal VSC-HVDC transmission system.

The primary function of the offshore converter station is to aggregate and transmit the wind power generated by the wind farms. Its control objectives include establishing the voltage and frequency of AC collection grid for wind farm, stabilizing the voltage amplitude at the offshore AC collection bus, and maintaining the frequency of the AC system to ensure the full transmission of wind power. Following this, the offshore wind turbine converters use *P-Q* dual-loop control to feed power into the offshore converter station. Once the entire system is fully started, it transitions to normal operation.

1.3 Converter Station System-Level Control

In an offshore wind farm integrated with a multi-terminal VSC-HVDC system, the wind power generated by the farm is transmitted via the offshore converter and then relayed through the multi-terminal VSC-HVDC system to the onshore converter station. At the onshore station, the power is distributed according to various control strategies before being injected into the onshore AC system. The primary function of the offshore converter is to aggregate and transmit the power generated by the offshore wind farm, typically utilizing *V-f* control. The offshore converter station provides the offshore wind farm with frequency and AC voltage amplitude reference values.

Meanwhile, the onshore converter station controls the DC voltage and active power to maintain DC voltage stability and ensure power balance between the offshore and onshore converter stations.

1.3.1 Onshore Converter Station Control

The onshore converter station, serving as the receiving terminal for offshore wind power integrated via a VSC-MTDC system, primarily aims to establish a stable DC voltage, thereby providing the essential conditions for integrating offshore wind farms into the onshore AC grid. For a multi-terminal DC system, the judicious design of control strategies for the onshore converter station can enable the optimal distribution of offshore wind power among various converter stations.

The conventional control strategy for the onshore converter station typically utilizes a phase-locked loop (PLL)-based dq decoupled dual-loop control structure, as illustrated in Fig. 1.3. Through the regulation of the d-axis current, the converter station can manage active power transmission or control the DC voltage. Similarly, by adjusting the q-axis current, the converter station can handle the exchange of reactive power between itself and the AC system, with specific control methods tailored to meet practical needs. The inner-loop current control equations are expressed as follows:

$$\begin{cases} u_{cd} = \left(K_{pI} + \frac{K_{iI}}{s}\right)\Delta i_{sd} - \omega_1 L i_{sq} + u_{sd} \\ u_{cq} = \left(K_{pQ} + \frac{K_{iQ}}{s}\right)\Delta i_{sq} + \omega_1 L i_{sd} + u_{sq} \end{cases} \tag{1.1}$$

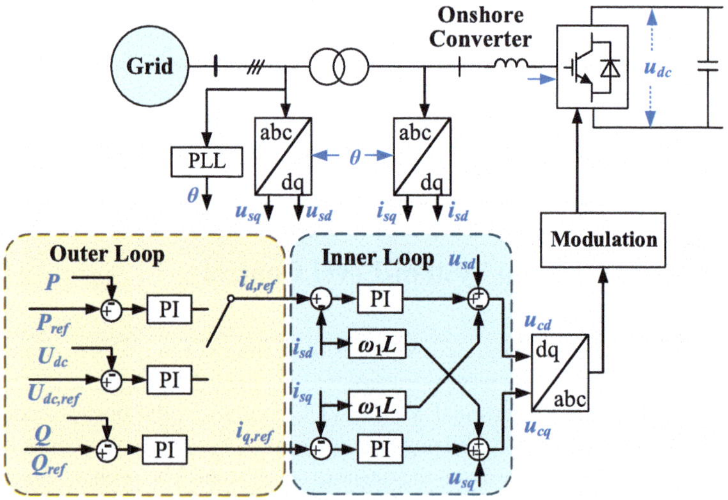

Fig. 1.3 Basic control block diagram of onshore VSC

Here, K_{pI} and K_{iI} represent the proportional and integral coefficients for the d-axis inner-loop control, respectively, while K_{pQ} and K_{iQ} denote the proportional and integral coefficients for the q-axis outer-loop control.

The typical active power outer-loop control strategies for onshore converters can be categorized into constant active power control and constant DC voltage control, forming distinct power outer-loop control structures as illustrated in Fig. 1.3. The output of the outer-loop control serves as the reference value for the inner current loop. The control equations for the d-axis outer loop of a common onshore converter station can be represented as follows:

$$i_{\mathrm{dref}} = \left(K_{pP} + \frac{K_{iP}}{s}\right)(P_{\mathrm{ref}} - P) \tag{1.2}$$

$$i_{\mathrm{dref}} = \left(K_{pU} + \frac{K_{iU}}{s}\right)\left(U_{\mathrm{dc,ref}} - U_{\mathrm{dc}}\right) \tag{1.3}$$

In these equations, K_{pP} and K_{iP} denote the proportional and integral coefficients for the d-axis outer-loop constant power control, respectively, while K_{pU} and K_{iU} represent the proportional and integral coefficients for the d-axis outer-loop constant voltage control.

For onshore converter stations in VSC-MTDC systems, system-level control can be divided into master-slave control and droop control. Master-slave control is a form of single-point voltage regulation. Typically, it involves designating one onshore converter station in the VSC-MTDC system as the master station, with the remaining stations functioning as slave stations. The master station employs constant DC voltage control, serving as the power balancing node, which mitigates power imbalances within the VSC-MTDC system. The slave stations, on the other hand, utilize constant active power control to accurately manage the injection of active power.

The advantages of master-slave control include simplicity of implementation and precise regulation of the master station's DC voltage and the slave stations' DC power. However, its drawbacks include limited control flexibility, as the system's power imbalances can only be mitigated by the master station. When integrating offshore wind farms, the fluctuating wind power introduces significant variability, increasing the burden on the master station. Additionally, if the master station fails and is taken offline, a slave station must be selected to switch roles to master, changing its control mode from constant power to constant DC voltage. This reconfiguration and transition process relies on communication between multiple converter stations. Consequently, if the communication is unreliable or fails, the master station outage could lead to severe repercussions within the system.

To enable each onshore converter station to participate in the absorption of unbalanced wind power and enhance overall system stability, the droop control strategy is commonly employed in VSC-MTDC systems for onshore converter stations. Droop control, also known as multi-point DC voltage control, can be viewed as a combination of constant DC voltage control and constant active power control.

When employing power-voltage droop control, fluctuations in the offshore wind farm's power output are collectively managed by multiple converter stations, distributing the unbalanced power across the multi-terminal VSC-HVDC system. The specific allocation of unbalanced power is determined by the respective droop coefficients of each onshore converter station.

When offshore wind power participates in frequency support for the onshore AC system via a multi-terminal VSC-HVDC system, the converter stations are required to inject DC power into the onshore AC system to provide frequency support. Building upon the existing power-voltage droop control, researchers have further incorporated a frequency droop control structure, thus forming a power-voltage-frequency droop control strategy.

The advantages of the droop control strategy lie in its ability to ensure that each onshore converter station contributes equally to the system by mitigating unbalanced power. The distribution of unbalanced power among the converter stations can be controlled by adjusting their respective droop coefficients, thus achieving a reasonable allocation of power from the offshore wind farm. Additionally, in the event of a fault and subsequent outage at one onshore converter station, the other stations can autonomously redistribute the power without the need for inter-station communication.

However, the droop control strategy does not possess the precise control characteristics of master-slave configurations, leading to the inability to achieve exact control over the DC voltage and active power at the onshore converter stations. Moreover, when using a fixed droop coefficient control strategy, the actual operating conditions of the onshore converter stations and the AC system are not taken into account, which can result in DC voltage and active power limit violations, potentially jeopardizing system stability. Therefore, further design of control parameters is necessary when implementing droop control.

1.3.2 Offshore Converter Station Control

The offshore wind farm, aggregated through AC collection, can be viewed as a weak AC system that requires voltage and frequency support established by the offshore converter station. The offshore converter station can be considered a controllable AC voltage source. By utilizing dq-axis decoupled control, it provides voltage and frequency support to the offshore wind farm, facilitating the full transmission of wind power. The fundamental control structure of this system is illustrated in Fig. 1.4.

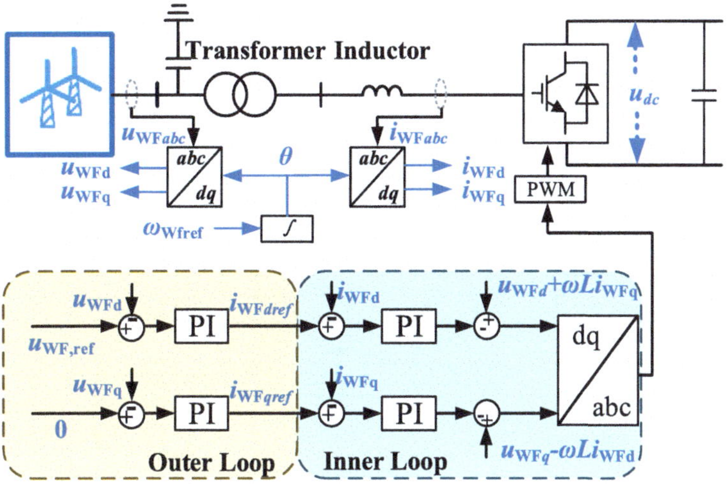

Fig. 1.4 Basic control block diagram of offshore VSC

1.4 Control Strategy for Offshore Wind Farms

1.4.1 Control Structure of Offshore Wind Turbine Generators

A wind turbine generator primarily consists of a wind turbine, shaft control system, and converter. The main function of the wind turbine is to convert wind energy into mechanical energy. In power system-related studies, a simplified modeling approach is often adopted, representing the relationship between the mechanical power P_m output of the wind turbine and the wind speed v as follows:

$$P_{\mathrm{m}} = \frac{1}{2}\rho\pi R^2 v^3 \cdot C_p(\lambda, \beta) \quad \left(\lambda = \frac{\omega_{\mathrm{r}} R}{v}\right) \tag{1.4}$$

Here, ρ represents the air density, R denotes the radius of the wind turbine rotor, and v indicates the real-time wind speed. The term $C_p(\lambda, \beta)$ refers to the power coefficient of the wind turbine, where λ is the tip speed ratio (the ratio of the linear speed of the tip of the blade to the wind speed, with ω_r representing the rotor speed of the wind turbine) and β is the pitch angle. The calculation process for $C_p(\lambda, \beta)$ is given by the following equation. It is evident that C_p is dependent on both the tip speed ratio and the pitch angle. In this study, we utilize the classical formula proposed in [4] for calculating C_p:

$$C_p = 0.5176 \times \left(116\frac{1}{\lambda^*} - 0.4\beta - 5\right) \cdot e^{\frac{21}{\lambda^*}} + 0.0068\lambda \left(\frac{1}{\lambda^*} = \frac{1}{\lambda + 0.08\beta} - \frac{0.035}{\beta^3 + 1}\right) \tag{1.5}$$

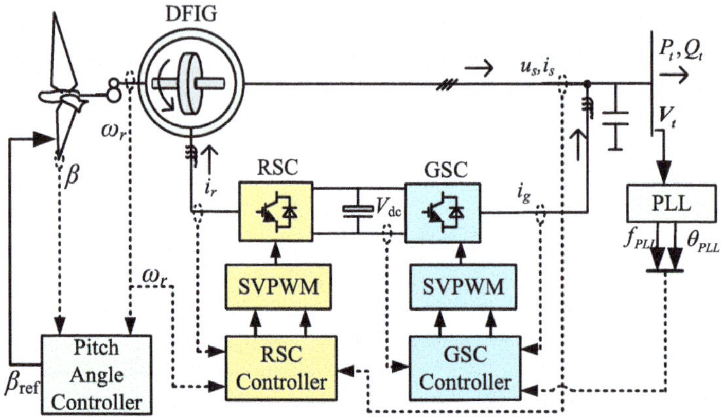

Fig. 1.5 The model diagram of DFIG

This chapter primarily employs an offshore wind farm composed of Doubly Fed Induction Generators (DFIG), with the DFIG structure illustrated in Fig. 1.5 [5].

In this system, the grid-side converter of the wind turbine adopts a dual-loop control structure. Its primary control objective is to provide stable DC voltage for the rotor-side converter. The rotor-side converter utilizes a *P-Q* dual-loop control scheme, with maximum power point tracking (MPPT) employed in the outer d-axis control loop to capture the real-time power of the wind turbine. The specific control strategy can be referred to in [6].

1.4.2 Control Architecture of Offshore Wind Farms

Traditional offshore wind farm power distribution models are based on a centralized control structure. Through this strategy, the active power distribution in an offshore wind farm can be divided into distribution based on individual turbine capacity and distribution based on real-time power [7]. Distribution based on individual turbine capacity entails allocating active power according to the capacity of each wind turbine, as shown in Eq. (1.6).

$$P_{\mathrm{ref},i} = \frac{P_{\max,i}}{\sum_{i=1}^{n} P_{\max,i}} P_{\mathrm{WF}} \tag{1.6}$$

Here, $P_{\mathrm{ref},i}$ and $P_{\max,i}$ represent the active power reference value and the maximum capacity of the wind turbine generator WT_i, respectively, while P_{WF} denotes the active power dispatch command value for the wind farm. Equation (1.6) illustrates that using the allocation method based on individual turbine capacities results in fixed proportions of active power assigned to each wind turbine generator. This method does not consider the varying working conditions due to terrain, wind speed, and other factors, leading to discrepancies in the output of the wind turbine generators. Therefore, some literature proposes distributing the wind farm power command according to the real-time power ratio method, as shown in Eq. (1.7).

$$P_{\mathrm{ref},i} = \frac{P_{\mathrm{ave},i}}{\sum_{i=1}^{n} P_{\mathrm{ave},i}} P_{\mathrm{WF}} \tag{1.7}$$

In this equation, $P_{\mathrm{ave},i}$ represents the predicted output of wind turbine WT_i. By performing real-time predictions of the output from each wind turbine within the wind farm under different wind conditions, the power dispatch instructions for the wind farm can be allocated more accurately. However, as the scale of offshore wind farms continues to increase and the number of wind turbines grows, the communication load on the central controller in a centralized control model increases significantly, demanding higher communication reliability. Should a communication failure occur, the overall stability of the system would be severely impacted. Therefore, further research into offshore wind farm control strategies is necessary, particularly focusing on improving power distribution methods.

References

1. Rouzbehi K, Zhang W, Ignacio Candela J, Luna A, Rodriguez P (2017) Unified reference controller for flexible primary control and inertia sharing in multi-terminal voltage source converter-hvdc grids. IET Gener Transm Distrib 11(3):750–758. https://doi.org/10.1049/iet-gtd.2016.0665
2. Vennelaganti SG, Chaudhuri NR (2018) Selective power routing in MTDC grids for inertial and primary frequency support. IEEE Trans Power Syst 33(6):7020–7030. https://doi.org/10.1109/TPWRS.2018.2854647
3. Attya ABT, Dominguez-Garcia JL (2018) Insights on the provision of frequency support by wind power and the impact on energy systems. IEEE Trans Sustain Energy 9(2):719–728. https://doi.org/10.1109/TSTE.2017.2759503
4. Heier S (2006) Grid integration of wind energy conversion systems, 2nd edn. Wiley, Chichester, England; Hoboken, NJ
5. Xiao L, Xu Z, Xiao H, Zhang Z, Wang G, Xu Y (2019) Electro-mechanical transient modeling of mmc based multi-terminal HYDC system with dc faults considered. Int J Electr Power Energy Syst 113:1002–1013. https://doi.org/10.1016/j.ijepes.2019.06.003
6. Wu Q, Sun Y (eds) (2018) Modeling and modern control of wind power. Wiley, Hoboken, NJ
7. Wu YK, Yang WH, Hu YL, Dzung PQ (2019) Frequency regulation at a wind farm using time-varying inertia and droop controls. IEEE Trans Ind Appl 55(1):213–224. https://doi.org/10.1109/TIA.2018.2868644

Chapter 2
Vector-Variable Based Modeling Method for Large-Scale Wind Farm Integrating with VSC-MTDC

Abstract Unlike conventional power plants, wind farms typically comprise hundreds of wind turbines and intricate collector systems, which significantly increase the complexity and effort required for modeling. This chapter introduces a vector-variable-based approach for dynamic modeling of large-scale wind farms, incorporating the dynamics of the collector system. By employing vector variables, this method simplifies the wind farm's numerous components into a model with two primary submodules: the wind generator module and the collector system module. These submodules allow for flexible modeling of any number of wind generators and various collector system topologies while preserving dynamic characteristics and scalability. The model's accuracy and simulation efficiency are validated by comparisons with an electromagnetic model. Additionally, a detailed wind farm with twenty-five wind generators is modeled using this approach. Comparative analysis with an aggregation model confirms that the proposed model accurately reflects wind farm stability. Moreover, comparison with a model that overlooks collector system dynamics shows that neglecting these dynamics can lead to significant errors and misjudgments of instability modes, especially in sub-synchronous mode analysis.

2.1 Implementation of the Whole Wind Farm Model Based on Vector Variable

An offshore wind farm serves as an illustrative example, with its simplified structural diagram presented in Fig. 2.1. Typically, a significant number of wind generators, ranging from dozens to hundreds, are distributed across the wind farm. The electrical power generated by these wind turbines is collected via a 35 kV network, which subsequently channels the electricity to a transformer station. At this station, the voltage is stepped up to 220 kV and transmitted through a long-distance cable to the VSC station. The topology of the 35 kV network may vary among different wind farms, influenced by factors such as the scale of the farm, reliability requirements, and cost considerations. Consequently, as depicted in Fig. 2.1, the comprehensive model of the wind farm should primarily consist of two components:

W. Yao et al., *Active Control of Large-Scale Offshore Wind Farms Connected Via VSC-HVDC*, Power Systems, https://doi.org/10.1007/978-981-97-9346-4_2

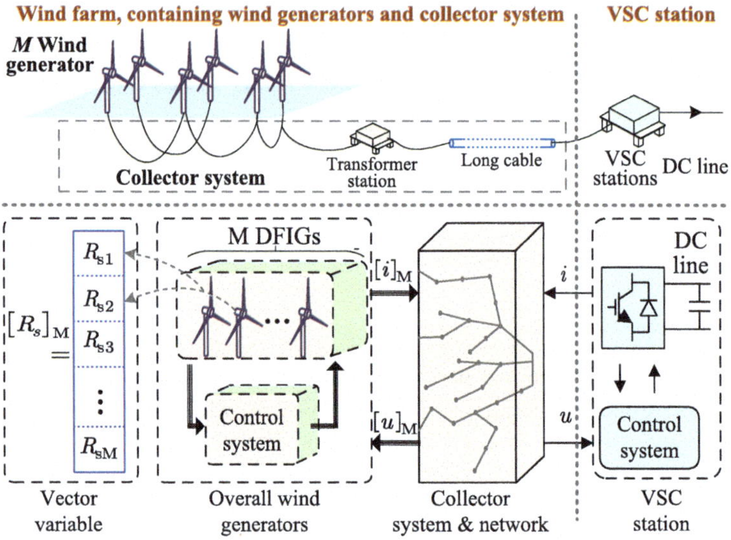

Fig. 2.1 The simplified structure of the wind farm and the main components of the wind farm model

- Wind generator and its controller, the doubly fed induction generator (DFIG) is taken as an example here. The type of wind generator can be changed as needed.
- Collector system, containing 35 kV/220 kV power network, the transform station, and compensator devices like the capacitor.

With these components, a detailed wind farm model can be constructed. Additionally, although it is not essential for modeling the wind farm itself, the inclusion of a VSC station model is necessary to represent the entire system of the wind farm integrated with VSC-HVDC.

Given the large and variable number of wind generators, the model must prioritize flexibility and simulation efficiency. To achieve this, the wind generator model is based on vector variables. A generalized network function is also proposed to manage the collector system model, accounting for dynamics and various topologies. This approach allows for adjustments in the number of wind generators and the topology of the collector system within the standard module, avoiding the need for duplications and manual modifications.

The wind generator and converter control system operates in a two-axis rotating reference frame, known as the dq coordinate system, where the d-axis aligns with the stator voltage and the q-axis leads it. Each device has distinct dq components, denoted by d and q. The devices and collector system share a single, synchronously rotating reference frame, referred to as the xy coordinate system, with variables denoted by x and y.

To develop comprehensive models of DFIGs in a wind farm using vector variables, the first step involves identifying the functions of the key components in a single

DFIG, as illustrated in Fig. 2.2. These components include the wind turbine, induction machine, phase-lock loop (PLL), rotor-side converter (RSC), and grid-side converter (GSC). The models of these components will be briefly summarized in the following section.

Within the modeling framework, the model can be readily adapted as necessary:

$$\frac{d\omega_r}{dt} = \frac{1}{2H}(T_m - T_e) \tag{2.1}$$

where ω_r is the rotor speed of DFIG. H is the inertia constant of DFIG. T_m and T_e are the torque of the wind turbine and generator, respectively.

The functions of the induction machine model comprise the stator and rotor functions, and the functions of the ith DFIG in xy coordinates may be stated as:

$$\begin{cases} \frac{1}{\omega_B}\frac{d\varphi_{sxi}}{dt} = u_{sxi} - R_{si}i_{sxi} + \omega_1\varphi_{syi} \\ \frac{1}{\omega_B}\frac{d\varphi_{syi}}{dt} = u_{syi} - R_{si}i_{syi} - \omega_1\varphi_{sxi} \\ \frac{1}{\omega_B}\frac{d\varphi_{rxi}}{dt} = u_{rxi} - R_{ri}i_{rxi} + s\varphi_{ryi} \\ \frac{1}{\omega_B}\frac{d\varphi_{ryi}}{dt} = u_{ryi} - R_{ri}i_{ryi} - s\varphi_{rxi} \end{cases} \tag{2.2}$$

where φ_s and φ_r are the stator and the rotor flux. i_s and i_r are the stator and the rotor current. R_s, R_r, L_s and L_r are the resistance and the inductance of stator and rotor. s is the slip of DFIG and ω_B is the basic frequency.

A second-order PLL model [1] is used to transform the xy coordinate into the dq coordinate system. The control strategies for the RSC and Grid-Side Converter (GSC) are illustrated in Fig. 2.2. Both RSC and GSC implement a cascaded control structure, consisting of an inner current controller and an outer controller. In the GSC, the d-axis loop is designed to regulate the DC voltage, while in the RSC, the active power reference is determined by the rotor speed, with the reactive power reference set to zero. An anti-windup PI module is used to keep the power and rotor current of

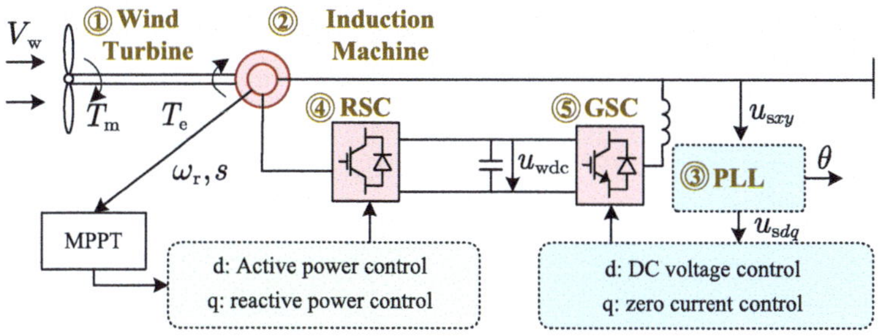

Fig. 2.2 The elements of the single DFIG, including the wind turbine, asynchronous machine, PLL, RSC and GSC

the DFIG within a safe operating range, with current limits set to 1 p.u.. The RSC controller is shown in Fig. 2.3.

While the parameters of wind generators may vary, the control system and structure of the wind generators within a wind farm are typically similar, implying they can be modeled using the same functions. However, as a wind farm usually comprises multiple DFIGs, the parameters of these DFIGs are distributed across these submodules, as depicted in Fig. 2.4. Consequently, the DFIG submodules need to be duplicated numerous times and connected in accordance with the topology to construct the wind farm system model, which increases the modeling efforts, particularly when the wind farm includes hundreds of wind generators. The memory usage by the graphical interface will be substantial, thereby reducing the simulation scale.

A vector-variable approach is proposed to efficiently model any number of wind generators without requiring extensive duplication operations. In a wind farm, the

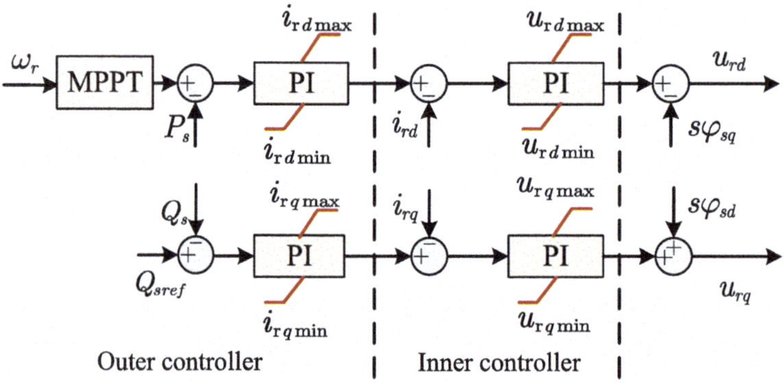

Fig. 2.3 The structure of rotor side controller in DFIG

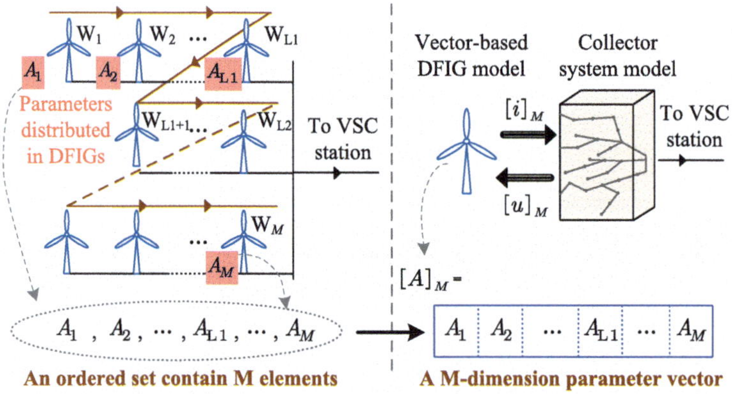

Fig. 2.4 The structure M-dimension vector variable for overall wind generators in the wind farm

wind generators can typically be arranged sequentially, with the parameters associated with different DFIGs organized into an ordered set. Additionally, these parameters of the same type can be expressed as a vector variable, as shown in Eq. (2.3), where each element corresponds to a specific DFIG. This allows the entire set of M-scale DFIGs in the wind farm to be represented by a single DFIG submodule with M dimensions, as illustrated in Fig. 2.4.

$$[A]_M = \left[A_1\ A_2 \cdots A_M\right] \tag{2.3}$$

where $[A]_M$ is one of the parameter vector of the DFIGs, containing M elements. The structure and construction method of the vector variable can be seen in Fig. 2.4.

To illustrate the functions of the stator and rotor, we refer to Eq. (2.2), which characterizes the behavior of the i-th Doubly-Fed Induction Generator (DFIG) within the wind farm. By employing Eq. (2.3), the collective functions of all DFIGs in the wind farm can be represented as follows:

$$\left\{\begin{aligned}
\frac{1}{\omega_B}\frac{\mathrm{d}\,[\varphi_{sx}]_M}{\mathrm{d}t} &= [u_{sx}]_M - [R_s]_M \otimes [i_{sx}]_M + \omega_1 \left[\varphi_{sy}\right]_M \\
\frac{1}{\omega_B}\frac{\mathrm{d}\left[\varphi_{sy}\right]_M}{\mathrm{d}t} &= \left[u_{sy}\right]_M - [R_s]_M \otimes \left[i_{sy}\right]_M - \omega_1 [\varphi_{sx}]_M \\
\frac{1}{\omega_B}\frac{\mathrm{d}\,[\varphi_{rx}]_M}{\mathrm{d}t} &= [u_{rx}]_M - [R_r]_M \otimes [i_{rx}]_M + [s]_M \otimes \left[\varphi_{ry}\right]_M \\
\frac{1}{\omega_B}\frac{\mathrm{d}\left[\varphi_{ry}\right]_M}{\mathrm{d}t} &= \left[u_{ry}\right]_M - [R_r]_M \otimes \left[i_{ry}\right]_M - [s]_M \otimes [\varphi_{rx}]_M
\end{aligned}\right. \tag{2.4}$$

where the $[\cdot]_M$ means the M dimension vector variables. The $\otimes$ represents the element-wise product. Other notations in (2.4) have the same meaning as Eq. (2.2).

The residual operations of the DFIG can be similarly articulated akin to Eq. (2.4). Through this approach, the various parameters dispersed across distinct wind turbines within the wind farm can be collectively encapsulated by a unified vector variable. This strategy retains adaptability, allowing each turbine to maintain its specific parameters. The diagram illustrating the DFIG submodule within a wind farm, utilizing the vector variable, is shown in Fig. 2.5. The thick line signifies that the variable is represented as a vector.

2.2 Model of Collector System Considering Dynamics

To build a collector network model that considers dynamic behaviors and all topologies, the initial step is to define the functions of the kth line along with the nth bus. Typically, the cables are viewed as π sections. The function for the kth line can be represented in per-unit outcomes as below:

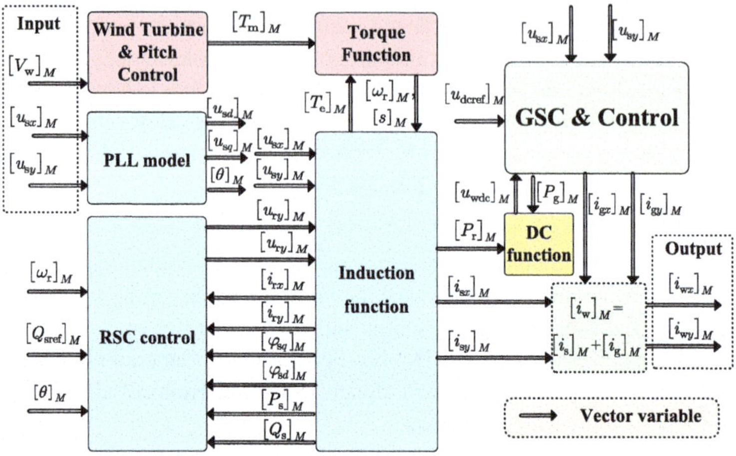

Fig. 2.5 Single DFIG submodule based on M dimensions vector variable, to constitute the whole M DFIGs in the wind farm

$$\begin{cases} \frac{L_{bk}}{\omega_B}\frac{di_{bkx}}{dt} = u_{fx} - u_{tx} + \omega_1 L_{bk} i_{bky} - R_{bk} i_{bkx} \\ \frac{L_{bk}}{\omega_B}\frac{di_{bky}}{dt} = u_{fy} - u_{ty} - \omega_1 L_{bk} i_{bkx} - R_{bk} i_{bky} \end{cases} \tag{2.5}$$

where the subscripts f, t, x, and y mean the sending bus, receiving bus, x-axis and y-axis, respectively. u means the voltage of bus. i means the current of the line. L_{bk}, R_{bk} are the inductance and resistance of the kth line, respectively.

At the nth node of the electrical grid, a capacitance C_n and a conductance G_n are utilized to represent the compensating capacitor and the load, respectively, serving as compensatory elements connected to each wind turbine. In parallel, the capacitance values of the transmission lines connected to this nth node are combined. Consequently, the characteristics of the node can be expressed in per-unit terms as follows:

$$\begin{cases} \frac{1}{\omega_B}\left(C_n + \sum_{k\in\mathbb{M}} C_{bk}\right)\frac{du_{nx}}{dt} = i_{nx} - \sum_{k\in\mathbb{M}} i_{bkx} - \left(C_n + \sum_{k\in\mathbb{M}} C_{bk}\right)\omega_1 u_{ny} - G_n u_{nx} \\ \frac{1}{\omega_B}\left(C_n + \sum_{k\in\mathbb{M}} C_{bk}\right)\frac{du_{ny}}{dt} = i_{ny} - \sum_{k\in\mathbb{M}} i_{bky} + \left(C_n + \sum_{k\in\mathbb{M}} C_{bk}\right)\omega_1 u_{nx} - G_n u_{ny} \end{cases} \tag{2.6}$$

where the term i_{nxy} denotes the electrical current supplied by the connected devices at the device-specific nodes. If there are no such devices, these currents default to zero for the nth coupling points within the grid. The symbol u_{nxy} stands for the voltage at the bus. The notation $\mathbb{M}$ refers to the collection of branches that are linked to the nth bus. Lastly, C_{bk} designates the capacitance associated with the kth transmission line.

Utilizing the characteristics of the lines and buses previously outlined, a dynamic model of the collector system can be constructed manually. Nonetheless, the independence of the line and bus functions necessitates that researchers repeatedly assemble the components in accordance with the network's structure, a process that is not particularly efficient. To address the variable topology of the wind farm's collector system, the incidence matrix T is employed. A network comprising n nodes and b connections can be represented by an incidence matrix $T \in \mathbb{R}^{n \times b}$. Within this matrix, the entry T_{ik} equals 1 if connection k departs from node i, -1 if it arrives at node i, and 0 if there is no direct link between connection k and node i. As depicted in Fig. 2.6, a segment of the network is showcased to provide a clear visual representation of T. The related section of the incidence matrix is articulated in Eq. (2.7).

$$T = \begin{bmatrix} 1 & 0 & 0 & \cdots \\ -1 & 1 & 1 & \cdots \\ 0 & -1 & 0 & \cdots \\ 0 & 0 & -1 & \cdots \\ \vdots & \vdots & \vdots & \ddots \end{bmatrix} \in \mathbb{R}^{n \times \mathrm{b}} \tag{2.7}$$

Additionally, by employing the incidence matrix T, the individual functions for a single line (2.5) and a single bus (2.6) can be integrated into vectorized functions. This allows for the modeling of the entire collector system, regardless of its specific layout or topology (2.8).

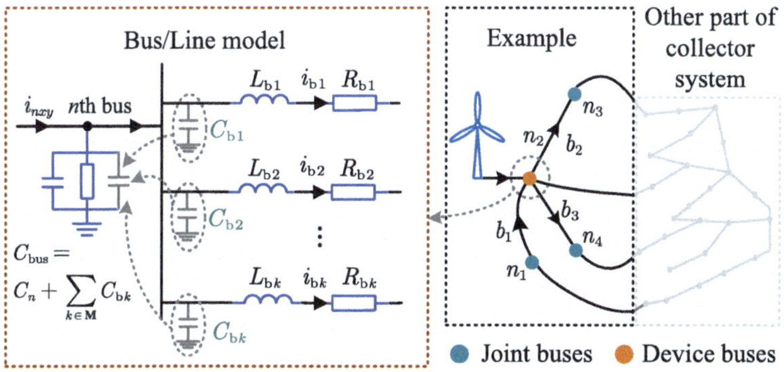

Fig. 2.6 Structure of wind farm collector network model, including the model of bus and line

$$
\begin{cases}
\dfrac{\operatorname{diag}(|T|C_{\mathrm{b}}+C_n)}{\omega_{\mathrm{B}}}\dfrac{\mathrm{d}u_x}{\mathrm{d}t}=i_x-Ti_{\mathrm{b}x}+\operatorname{diag}\left(|T|\,C_{\mathrm{b}}+C_n\right)\omega_1 u_y-\operatorname{diag}(G_n)u_x\\
\dfrac{\operatorname{diag}(|T|C_b+C_n)}{\omega_{\mathrm{B}}}\dfrac{\mathrm{d}u_y}{\mathrm{d}t}=i_y-Ti_{\mathrm{b}y}-\operatorname{diag}(|T|C_{\mathrm{b}}+C_n)\omega_1 u_x-\operatorname{diag}(G_n)u_y\\
\dfrac{\operatorname{diag}\left(L_{\mathrm{b}}\right)}{\omega_B}\dfrac{\mathrm{d}i_{\mathrm{b}x}}{\mathrm{d}t}=T^T u_x+\omega_1\operatorname{diag}\left(L_{\mathrm{b}}\right)i_{\mathrm{b}y}-\operatorname{diag}\left(R_{\mathrm{b}}\right)i_{\mathrm{b}x}\\
\dfrac{\operatorname{diag}\left(L_{\mathrm{b}}\right)}{\omega_B}\dfrac{\mathrm{d}i_{\mathrm{b}y}}{\mathrm{d}t}=T^T u_y-\omega_1\operatorname{diag}\left(L_{\mathrm{b}}\right)i_{\mathrm{b}x}-\operatorname{diag}\left(R_{\mathrm{b}}\right)i_{\mathrm{b}y}
\end{cases}
\tag{2.8}
$$

where the vectors $C_b \in \mathbb{R}^b$, $R_b \in \mathbb{R}^b$, and $L_b \in \mathbb{R}^b$ represent the branch's capacitance, resistance, and inductance, respectively. The bus voltage is denoted by $u_x \in \mathbb{R}^n$ and $u_y \in \mathbb{R}^n$. The currents being injected into the bus are represented by $i_x \in \mathbb{R}^n$ and $i_y \in \mathbb{R}^n$. Additionally, $C_n \in \mathbb{R}^n$ and $G_n \in \mathbb{R}^n$ correspond to the capacitance and conductance at the bus, which are associated with the compensating capacitor and the load, respectively.

Employing the DFIG model and the previously discussed collector system model, a comprehensive model of a wind farm has been assembled. This model facilitates the examination of the dynamics and stability issues across the entire wind farm. Moreover, the VSC-HVDC technology is regarded as a crucial method for transmitting power from wind farms, particularly those located far offshore. In order to scrutinize the dynamics of a wind farm system integrated with VSC-HVDC, this chapter provides a concise overview of the VSC model in this section, despite it not being an essential component of a wind farm. The VSC model is delineated as [2]:

$$
\begin{cases}
\dfrac{L_{\mathrm{v}}}{\omega_{\mathrm{B}}}\dfrac{\mathrm{d}i_{\mathrm{v}x}}{\mathrm{d}t}=u_{\mathrm{p}x}-u_{\mathrm{c}x}-R_{\mathrm{v}}i_{\mathrm{v}x}+\omega_1 L_{\mathrm{v}}i_{\mathrm{v}y}\\
\dfrac{L_{\mathrm{v}}}{\omega_{\mathrm{B}}}\dfrac{\mathrm{d}i_{\mathrm{v}y}}{\mathrm{d}t}=u_{\mathrm{p}y}-u_{\mathrm{c}y}-R_{\mathrm{v}}i_{\mathrm{v}y}-\omega_1 L_{\mathrm{v}}i_{\mathrm{v}x}
\end{cases}
\tag{2.9}
$$

In this context, L_{v} and R_{v} denote the inductance and resistance of the phase reactor, respectively. The terms $u_{\mathrm{p}xy}$ and $u_{\mathrm{c}xy}$ represent the voltage levels on the system side and the converter side, respectively. Additionally, i_{vxy} signifies the current flowing through the VSC.

The VSC that interfaces with the wind farm, often referred to as the rectifier VSC, employs an AC voltage control strategy to regulate the wind farm's AC voltage. The schematic of the VSC, along with its control system, is depicted in Fig. 2.7.

Typically, the inverter VSC adopts a fixed DC control method to regulate the DC voltage in the VSC-HVDC system. As this research focuses on the modeling of the wind farm, the inverter VSC is simplified and represented as a DC voltage source. The DC voltage control dynamics of the inverter VSC have minimal impact on the AC-side dynamics of the rectifier VSC, particularly when the inverter VSC is connected to a strong AC grid. Thus, this simplification is considered reasonable [3].

Historically, in some widely used simulation tools like PSCAD, PSASP, and others, variables and parameters were confined to a single dimension, which was inadequate for implementing the discussed model. However, Simulink's vector simulation capabilities permit the expression and calculation of variables and parameters in multiple dimensions, such as arrays or matrices, making it suitable for deploying the previously described model.

The core components of the detailed wind farm model have been developed as described earlier. Moreover, an example of a wind farm model, consisting of M DFIGs and an N-bus collector system, is illustrated in Fig. 2.8. By using the models shown in Figs. 2.5 and 2.6, the multi-dimensional wind generator module and the vector-based collector system module can be combined to construct the full wind farm model. Since the entire wind farm model is encapsulated within just two modules, the graphical interface's memory usage is reduced, thereby maximizing the allocation of computer memory for simulation data storage.

Next, the VSC module presented in Fig. 2.7 is combined with the collector system module. The connections within the model are depicted in Fig. 2.9. The DFIG and VSC modules take the voltage at the device buses as input, and their output is the

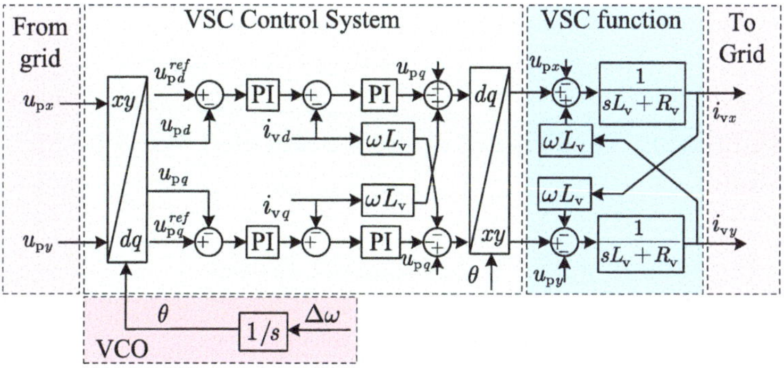

Fig. 2.7 The control system and AC side function of VSC station

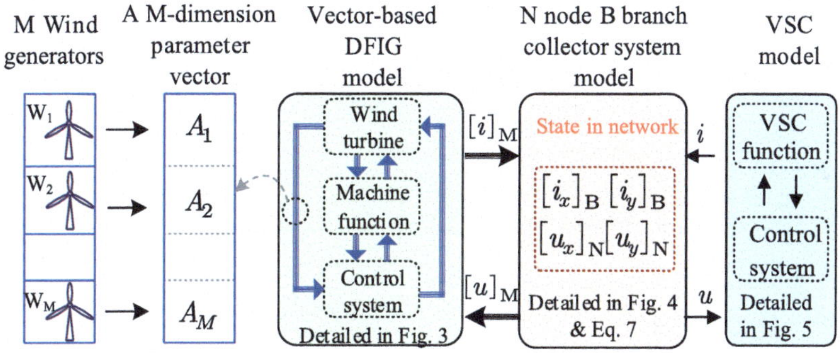

Fig. 2.8 Two typical modules used to model the wind farm, DFIG module and collector system module

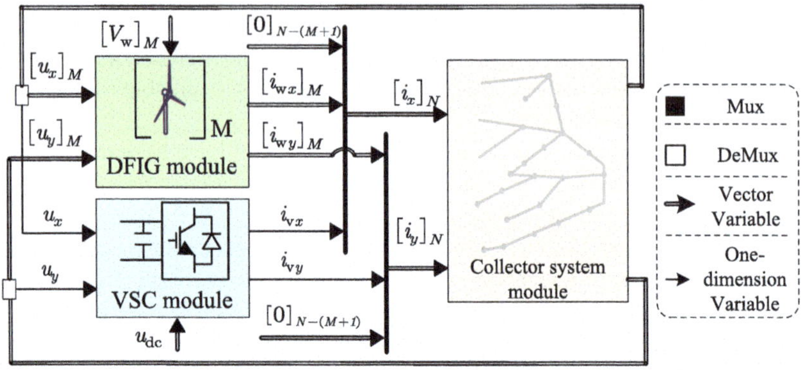

Fig. 2.9 The structure of the connections between the wind farm main components and VSC station

current injected into these device buses. Since the current at the joint bus is zero, a null vector of dimensions $N - (M + 1)$ is incorporated into the current vector to represent the zero current injection at the joint bus.

2.3 Simulation Verification of the Established Wind Farm Model

2.3.1 Verification with Electromagnetic Model

To affirm the efficacy and accuracy of the modeling technique introduced in this Chapter, a Simscape electromagnetic model featuring four DFIGs and a ring-shaped collector system is established, as portrayed in Fig. 2.10.

This scenario is designed to test the system's response to variations in wind velocity. At the one-second mark, $t = 1$s, a shift in wind speed occurs across the wind

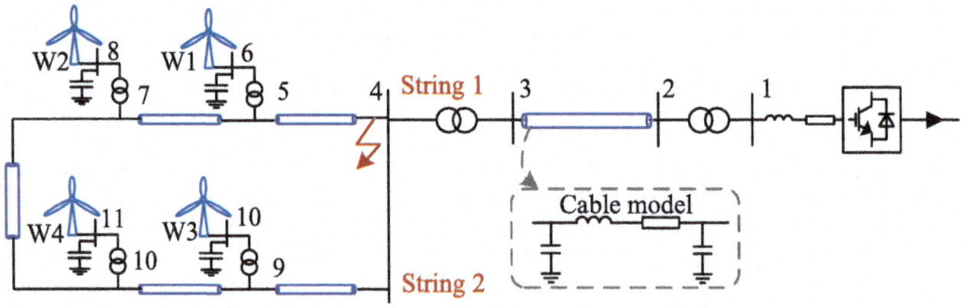

Fig. 2.10 System for electromagnetic validation: a wind farm consisting of four DFIGs and a ringed design collector system

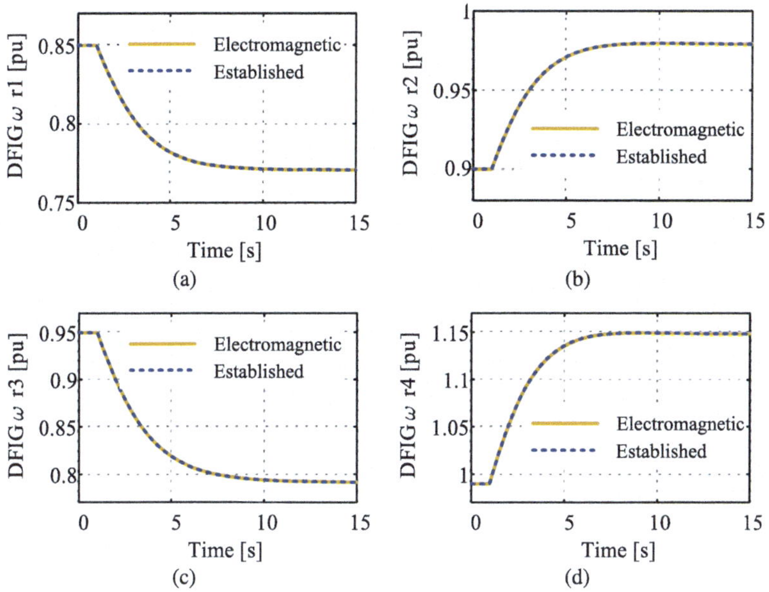

Fig. 2.11 Dynamic results comparison under wind speed change: **a** Rotor speed of DFIG 1; **b** Rotor speed of DFIG 2; **c** Rotor speed of DFIG 3; **d** Rotor speed of DFIG 4

farm. Specifically, DFIGs 1 and 3 experience a reduction in wind speed by 1 m/s and 2 m/s, respectively, while DFIGs 2 and 4 see an increase of 1 m/s and 2 m/s, respectively. The outcomes are displayed in Fig. 2.11. Upon examination, it is evident that the developed model yields results that closely match those of the electromagnetic model.

To confirm the dynamic behavior of the developed model during a fault condition, a 3.5-ohm three-phase-to-ground fault is simulated at bus 4, with a fault duration of 200 ms. The rotor speed of DFIG 1 and the VSC control signal are depicted in Fig. 2.12. Despite the presence of high-frequency oscillations in the electromagnetic model, the outcomes of our model are strikingly similar to those of the electromagnetic model.

The methodology introduced can be broadened to encompass unbalanced conditions, as the functions for negative and zero sequences can be articulated through vector variables as well. In addressing unbalanced issues, our model can be enhanced by incorporating the functions of these sequences. It is intended that future research will delve into the unbalanced phenomena and the associated wind farm models more thoroughly and appropriately.

In this instance, a comparison of the sub-synchronous dynamics within the developed model is conducted. The damping ratio for the sub-synchronous oscillations is diminished by fine-tuning certain parameters. The scenario begins with varying wind speeds across the DFIGs and distinct initial power outputs from the wind turbines. At the one-second mark, $t = 1s$, a perturbation is introduced to the reactive power

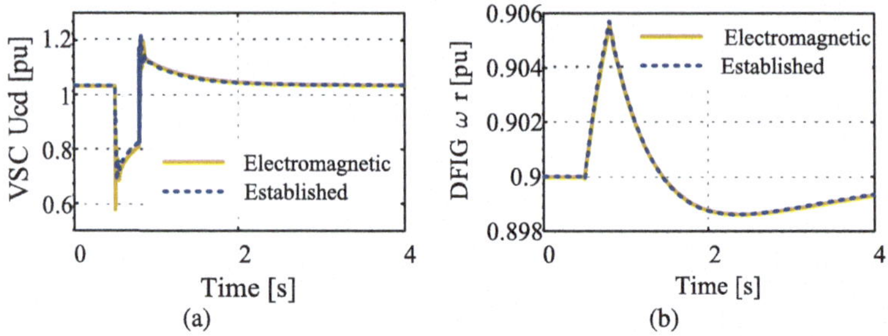

Fig. 2.12 Dynamic results comparison under the three-phase-to-ground fault: **a** The control signal of VSC; **b** The rotor speed of DFIG 1

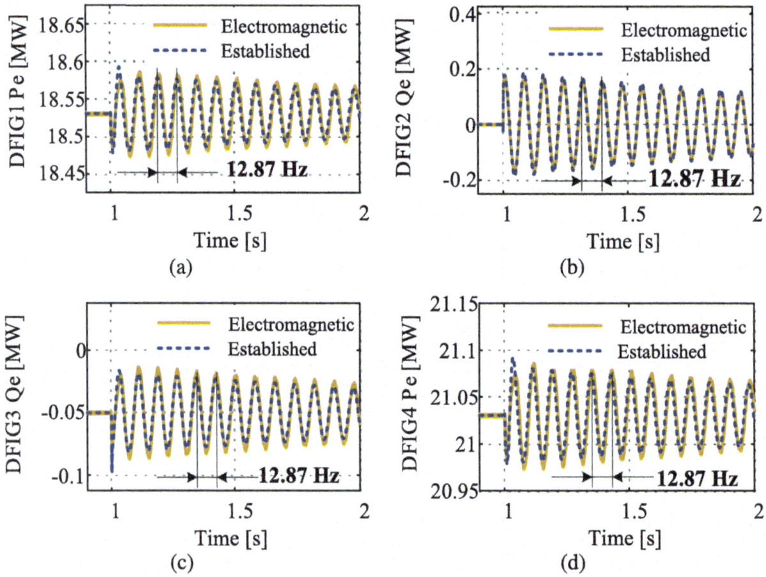

Fig. 2.13 Dynamics comparison of sub-synchronous phenomenon: **a** The active power of DFIG 1; **b** The reactive power of DFIG 2; **c** The reactive power of DFIG 3; **d** The active power of DFIG 4

setpoint of DFIG 2. The time-domain responses are illustrated in Fig. 2.13. Upon analysis, it is evident that the outcomes from our model closely mirror those of the electromagnetic model.

Observations indicate that the developed model is capable of illustrating the dynamics for each individual DFIG. The outcomes from our model align closely with those from the electromagnetic model, thereby substantiating the credibility of our model. Additionally, to quantify the discrepancies between the two models, the mean relative error (MRE) metric, as outlined in the manuscript and expressed in Eq. (2.10), is utilized.

Table 2.1 The mean relative error of DFIG active power between established model and the electromagnetic model

Scenarios	DFIG 1 (%)	DFIG 2 (%)	DFIG 3 (%)	DFIG 4 (%)
Wind speed change	0.053	0.051	0.054	0.053
Three-phase-to-ground fault	0.22	0.25	0.24	0.21
Sub-synchronous oscillation	0.086	0.085	0.083	0.087

$$E_r = \frac{1}{N_s} \sum_{i=1}^{N_s} \left| \frac{A_{\text{est}}(i) - A_{\text{emt}}(i)}{A_{\text{emt}}(i)} \right| \tag{2.10}$$

where A_{emt} corresponds to the measurement obtained from the electromagnetic model, while A_{est} signifies the measurement from our developed model. The term N_s denotes the total count of sampled data points.

In this section, a comparison of the power output of the DFIGs as simulated by the two models is presented. The mean relative error (MRE) for the comparison is detailed in Table 2.1. The data indicates that the average discrepancy in the power output between the models is below 0.25% across various conditions. It is noteworthy that the MRE is slightly higher during the fault scenario, reaching approximately 0.23%, compared to less than 0.1% in the other scenarios. This increase can be attributed to the presence of high-frequency oscillations in the electromagnetic model's results. However, the MRE remains within acceptable limits.

2.3.2 Simulation Efficiency of Established Model

Owing to the vector variable-based approach, the model boasts enhanced simulation efficiency. It's worth highlighting that for a 15-second simulation, the actual duration for the electromagnetic model, which includes four wind turbines, is roughly 223 s. In contrast, the developed model, of similar scale, requires only about 5 s to simulate under identical solver settings. Consequently, the proposed model offers substantial advantages in terms of simulation speed.

To further assess the simulation duration for models of varying sizes, additional wind farm models were constructed and tested with a 15-second simulation time. All tests were conducted using Matlab 2021B on a system equipped with an Intel Core i7 10700 CPU and running Windows 10. The outcomes are depicted in Fig. 2.14. It is observable that the model with 175 wind turbines exhibits a simulation time comparable to that of the electromagnetic model with just four turbines, indicating that the simulation time for the developed model is deemed satisfactory. Moreover, the computational time escalates in a roughly linear fashion with the model's size, rather than growing exponentially. The proposed method leverages Matlab's robust vectorized computation capabilities, which are inherently optimized in Matlab's logic, thereby significantly enhancing the simulation speed.

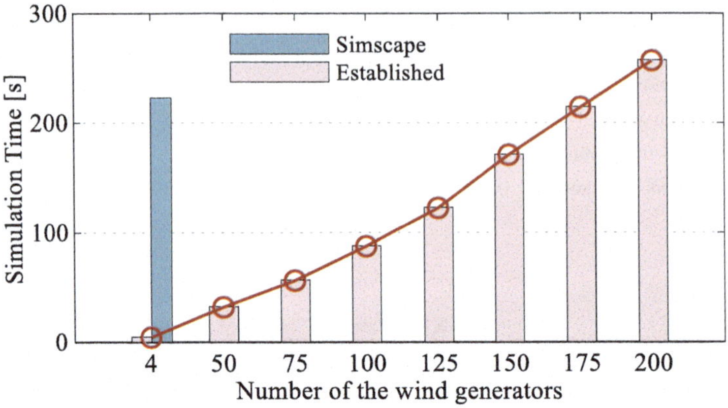

Fig. 2.14 Duration of simulation in the established model under varying wind farm scale and comparison with the electromagnetic model

Table 2.2 The simulation time between the phasor model and the proposed model

Wind farm scale	Phasor model (s)	Proposed model (s)	Reduced time (s)
50	87.08	37.88	49.12
75	163.03	56.59	106.44

To demonstrate the superior simulation speed of the proposed methodology, wind farm models comprising 50 and 75 DFIGs were constructed using both the conventional phasor model and the new approach. A comparison of the simulation times for these two models is presented in Table 2.2. As the scale of the wind farm (i.e., the number of turbines) grows from 50 to 75, the time saved by employing the proposed model ranges from 50 s to approximately 105 s. This indicates that the proposed method enhances simulation efficiency as the size of the simulation expands.

2.4 Summary

This chapter introduces a vector-variable-based approach for dynamic modeling of large-scale wind farms. The approach consolidates parameters dispersed among various wind turbines into a single vector variable by leveraging the fact that each turbine within a wind farm shares similar structures and functionalities. Consequently, a wind farm model encompassing any desired number of wind turbines and an arbitrary topology collector system that accounts for dynamics can be reduced to two primary modules: the wind turbine module and the collector system module. This model facilitates the analysis of the wind farm's dynamics and small-signal stability. The accuracy and simulation efficiency of the developed model are validated through a small-scale electromagnetic model. A detailed wind farm comprising 25 wind turbines is constructed using the proposed approach. Comparisons with the aggregation model demonstrate that as the variance in wind speed distribution increases, the relative error between the established model and the aggregation model also increases. Additionally, the established model is compared with a wind farm model that uses a static collector model. The findings indicate that disregarding the collector network dynamics can introduce errors when analyzing sub-synchronous modes, potentially leading to incorrect assessments of instability. The results confirm that the developed model accurately captures the characteristics of the wind farm. Based on this model, further studies will be conducted, such as examining the impact of topology on stability and exploring coordinated control algorithms within the wind farm. Although this method significantly enhances simulation efficiency, the simulation time may still be lengthy for analyzing a very large-scale wind farm (comprising thousands of turbines). However, it is unnecessary to model all turbines individually, as some can be aggregated without greatly compromising accuracy. Therefore, the number of turbines can be reduced (e.g., from thousands to hundreds) by applying an aggregation method. The proposed approach can then be used to model these aggregated turbines. As a result, large-scale wind farms can be modeled with high accuracy and simulation efficiency. To achieve this, future research should explore the methodology and impact of the aggregation model.

References

1. Fan L, Miao Z (2021) Analytical model building for type-3 wind farm subsynchronous oscillation analysis. Electr Pow Syst Res 201:107566. https://doi.org/10.1016/j.epsr.2021.107566
2. Cole S, Beerten J, Belmans R (2010) Generalized dynamic VSC MTDC model for power system stability studies. IEEE Trans Power Syst 25(3):1655–1662. https://doi.org/10.1109/TPWRS.2010.2040846
3. Sun K, Yao W, Fang J, Ai X, Wen J, Cheng S (2020) Impedance modeling and stability analysis of grid-connected DFIG-based wind farm with a VSC-HVDC. IEEE J Emerg Sel Top Power Electron 8(2):1375–1390. https://doi.org/10.1109/JESTPE.2019.2901747

Part II
Fault Analysis and Active Suppression in Large-Scale Offshore Wind Farms Connected Via VSC-HVDC

Offshore wind power offers several advantages, including proximity to load centers, no land resource usage, and suitability for large-scale development [1, 2]. High-voltage direct current (HVDC) transmission using voltage source converters (VSC), particularly modular multilevel converters (MMC), is recognized as an ideal technology for deep offshore wind projects [3]. Notable examples include Borwin 1-4, Dolwin 1-6, Helwin 1-2 in Germany, and Dogger Bank in the UK [4, 5]. Offshore wind projects employing MMC-HVDC are often situated in challenging environments and are vulnerable to various faults, including grid-side, valve-side, DC-side, and wind farm-side faults. Ensuring the safety and stability of MMC-HVDC systems in these projects is crucial for maintaining reliable grid operations.

Grid-side faults: The MMC at the sending end is unable to regulate the power injected into the HVDC transmission system from renewable energy sources. During a short-circuit fault in the receiving-end AC system, the interruption of the power output channel results in surplus power on the HVDC line [1]. Energy-consuming devices like crowbars in wind turbines are designed primarily for low-voltage ride-through and are inadequate for dissipating wind power during severe AC faults [6]. Consequently, safety and stability control devices are typically required to disconnect some renewable energy units, which generally takes about 150–160 ms [7]. However, before these offshore wind turbines are disconnected, the surplus power can cause the DC grid voltage to rapidly rise to the overvoltage protection threshold within mere tens of milliseconds [8]. If the converter station is blocked due to overvoltage and the AC circuit breaker trips, it leads to a substantial disconnection of renewable energy units, thereby reducing the availability of the renewable energy system [9].

To address the power surplus issue in DC grids, existing methods are categorized into four types: wind farm active power reduction [10], wind-farm-side MMC (WFMMC) active voltage reduction [1], WFMMC active frequency increase [11], and the installation of dynamic braking resistors (DBR) [12]. Active power reduction in wind farms faces challenges in completing rapid load shedding within 100 ms due to time delays and turbine inertia. Active voltage reduction of WFMMC leverages wind farms' low-voltage ride-through capabilities [13], but demands high performance from energy-consuming devices like crowbars [14]. Moreover, voltage drops at WFMMC cause inconsistent voltage drops at different turbine locations due to transformer and line differences, complicating accurate active power control [15]. Increasing WFMMC frequency can force active power to land [16], yet this method is limited since new energy units lack primary frequency modulation capability [11]. DBR installation is currently the mainstream method [17, 18], but as DC grid voltage levels and transmission capacities rise, the demand for energy-consuming equipment grows, making it costly [19]. Given the limitations of these methods, further solutions for surplus power in MMC-HVDC systems are needed. Fully utilizing MMC internal energy storage devices, such as sub-module (SM) capacitors, can provide active energy recovery, allowing time for energy-consuming device input or wind farm power reduction, thus reducing system costs.

The SM capacitors within an MMC serves as the primary energy storage element for converter stations. Currently, half-bridge MMCs (HBMMC) utilize the DC voltage control strategy of VSC, which closely couples the MMC capacitor voltage with the output DC voltage. This coupling poses a challenge in maintaining the DC voltage near the rated value while also performing energy control. To address this, it is essential to decouple the MMC capacitor voltage from the DC side voltage for effective energy recovery. Initial investigations into decoupling schemes for MMC converter stations used in overhead lines are presented in [20]. Although voltage fluctuations are inevitable, it is feasible to manage MMC energy by actively adjusting the capacitor voltage within the system's safety margin [21].

Addressing the power surplus issue in offshore transmission systems during grid-side AC faults is crucial for providing time for existing load shedding schemes and reducing the capacity cost of DBR and other equipment. Therefore, Chap. 3 introduces an Active Energy Control (AEC) scheme for MMC, leveraging the decoupling of SM capacitor voltage and DC voltage. Initially, it presents a steady-state control strategy for MMC-HVDC based on the principle of decoupling these voltages. Subsequently, it details the AEC schemes for both grid-side MMC (GSMMC) and WFMMC. The chapter illustrates the application of AEC in cooperative ACFRT with wind farms, and simulation results confirm the effectiveness of AEC in enhancing ACFRT.

Chapter 3 highlights the significant energy storage capabilities of the numerous SMs in MMCs. Leveraging this advantage, a control method was developed to decouple DC voltage from sub-module energy, enabling the absorption of excess power. This approach enhances the fault ride-through (FRT) capability of MMC-HVDC systems without incurring additional hardware costs. However, during FRT, the onshore converter station's reactive power output command remains zero, failing to support the onshore voltage. The active power output of the MMC-HVDC system depends on both the PCC voltage and the active current. Consequently, maximizing the active power output of MMC is challenging without providing voltage support. Moreover, in cases where the onshore fault is not severe, MMCs retain a considerable energy margin post-FRT. Utilizing part of this residual energy to boost onshore voltage can create a more favorable operating environment for all onshore equipment and reduce the frequency of low-voltage ride-through events. This strategy not only maximizes active power output but also ensures better voltage stability for the entire system.

To overcome the limitations discussed in Chap. 3, Chap. 4 presents a multi-mode matching control (MMMC) strategy. This method evaluates the fault level and offshore wind farm (OWF) conditions by monitoring energy fluctuations and DC-side power in the onshore GSMMC. Based on these evaluations, it selects the most suitable MMC-HVDC control mode. Mode I: AEC—When the fault level is high, and the PCC voltage is low with limited reactive power support, the transient control remains aligned with the methods in Chap. 3. Mode II: Dynamic Reactive Current Optimization (DRO)—As the fault level diminishes, DRO is activated. DRO dynamically fine-tunes the active and reactive currents of the GSMMC to optimize active power output. Mode III: Energy Coordinated Synchronous (ECS)—With a further decrease in fault level, ECS is employed. ECS coordinates the GSMMC with the offshore wind farm-side MMC (WFMMC) to collectively manage the output power of OWF. This coordination reduces the active current output from the GSMMC while increasing reactive current, thereby enhancing voltage support for the onshore grid. The MMMC strategy optimizes the energy utilization of the submodules (SMs), markedly enhancing the fault ride-through and voltage support capabilities of the OWF-MMC-HVDC system.

valve-side faults: Valve-side faults, which occur within the onshore converter station between the converter and the converter transformer, possess characteristics that differ from grid-side faults due to the influence of the converter transformer [22]. Among these, single-phase-to-ground (SPG) faults are especially prevalent [23]. Typically, the converter transformer is located outside the MMC enclosure, with valve-side winding bushings passing through the hall wall to connect with the converter AC bus [24]. Insulation failures and flashovers in these through-wall bushings are key contributors to valve-side SPG faults [25]. Furthermore, the high-humidity environment in the onshore converter station can cause water ingress and corrosion in cable joints, which significantly contribute to the occurrence of valve-side SPG faults.

Valve-side SPG faults in converter stations, unlike grid-side AC faults, remain an under-researched area. Previous studies, such as [26], analyzed SPG faults in line-commutated converters (LCCs) and proposed fault suppression schemes that are not applicable to MMCs due to differences in power electronics. Similarly, [27] examined SPG faults in two-level VSCs, but their topological differences also make these findings unsuitable for MMCs. Studies like [28] and [29] addressed MMC valve-side AC faults but did not offer theoretical analysis on the large oscillations and overvoltages in the DC pole-to-ground voltage. Further, [30] and [31] provided theoretical analysis of MMC valve-side SPG faults, focusing on system structures commonly used for overhead transmission lines with asymmetric monopole or symmetrical bipolar topologies [32, 33]. The proposed fault suppression methods in these studies depend on the use of full-bridge sub-modules (FB-MMC). While these methods are effective, they significantly increase the number of IGBTs and the spatial requirements for converter stations, which in turn escalates project costs. Given the high cost of submarine cables, MMC-HVDC integrated offshore wind farms typically use symmetric monopole topology [34], and due to offshore environmental constraints, grounding poles are not feasible for offshore converter stations [17]. These topological differences result in fault characteristics for offshore MMC-HVDC systems that are markedly different from those of onshore systems using overhead lines. To enhance the security of MMC-HVDC integrated offshore wind farms, it is crucial to explore the valve-side SPG fault characteristics and design control schemes to suppress potential fault overvoltage or overcurrent. However, a systematic investigation into these areas has not yet been conducted.

Chapter 5 delves into the characteristics of valve-side SPG faults in onshore converter stations and presents an overvoltage suppression scheme for MMC-HVDC integrated offshore wind farms. Initially, it examines the impact of onshore SPG faults on the MMC-HVDC system by developing an equivalent circuit model. This analysis reveals that onshore valve-side SPG faults can cause overvoltage issues in both the offshore transmission line and the onshore converter station. To address this, a fault overvoltage suppression strategy that incorporates zero-sequence voltage modulation based on the control characteristics of MMC is proposed.

DC-side faults: Chaps. 3–5 discuss symmetrical monopole MMC-HVDC systems. Another important configuration for connecting offshore wind farms is the symmetrical bipolar MMC-HVDC. In symmetrical monopole MMC-HVDC systems, the positive and negative transmission lines form the same current circuit, which means that a fault in either cable will result in the complete shutdown of the system [8]. However, in a symmetrical bipolar structure, the positive and negative poles have independent control characteristics, allowing the system to maintain some transmission capacity on the non-fault pole even when one pole experiences a breakage fault. When a symmetrical bipolar MMC-HVDC system encounters a single-pole breakage fault, the offshore fault pole converter is deactivated due to protective actions. Offshore wind turbines typically have a cut-off time of over 100 ms [7]. During this period, the non-fault pole must handle the entire power output of the offshore wind farm, leading to a potential DC overcurrent on the non-fault line. If the overcurrent exceeds the relay protection thresholds in terms of peak or duration, the non-fault pole converter will also be blocked, causing a complete system shutdown [9]. Therefore, addressing the overcurrent issue in symmetrical bipolar MMC-HVDC systems is crucial for ensuring continuous fault-tolerant operation following a single-pole submarine cable breakage.

To suppress overcurrent on the non-fault pole, it's crucial to manage the excess power being delivered by wind farms. Current solutions to this overpower issue fall into two primary categories: installing AC energy-consuming devices to dissipate the wind farms' output, or controlling wind farms to reduce their output power. The first approach, installing AC energy-consuming devices, faces significant challenges, primarily the need to expand the offshore platform [17]. This expansion is complicated by the difficulties of deep offshore construction and the requirement to purchase and install large equipment, which substantially increases project costs. The second approach involves controlling wind farms to reduce their output power. The most common method is to adjust turbine parameters such as rotational speed and pitch angle for load shedding. However, this method is constrained by various delays and turbine inertia, typically requiring over 100 ms [10]. An alternative method proposed in [1] involves indirectly controlling wind turbines to achieve rapid power reduction by leveraging low-voltage ride-through (LVRT) characteristics through voltage control at the offshore converter station. However, this method demands high standards for LVRT equipment within the turbines, such as crowbars [14]. Additionally, inconsistent voltage drops detected by wind turbines at different locations complicate accurate short-term active power control [15]. Another suggestion, as seen in [16], is to reduce wind farm output power by increasing the AC-side frequency of the offshore converter station and utilizing the frequency response capability of the wind farm. Unfortunately, the response speed and power reduction achieved by this method are insufficient to effectively suppress transient overcurrent.

All the previously mentioned control methods do not effectively leverage the power or energy control capabilities of the MMC. Presently, all WFMMCs are HBMMCs. In grid-forming mode, WFMMCs use AC voltage control to provide voltage amplitude and frequency for wind farms [35]. This configuration prevents WFMMCs from controlling the power input to the DC line from the wind farm [2]. Therefore, designing a control scheme that fully utilizes the energy storage elements within the MMC (e.g., sub-module capacitors) to enable energy control capabilities in the MMC can address this issue. By doing so, the power from the wind farm entering the DC line can be regulated, allowing sufficient time for wind farms to reduce their power output and achieve transient overcurrent suppression [21].

Therefore, Chap. 6 presents a multi-level coordinated control (MLCC) strategy designed for symmetrical bipolar MMC-HVDC systems integrated with offshore wind farms. The chapter starts by examining the fault characteristics associated with single-pole breakage in a symmetric bipolar MMC-HVDC system. Based on this analysis, a steady-state control scheme for the offshore converter station is developed, incorporating MMC sub-module capacitor energy control. The MLCC scheme is then elaborated across four levels: the offshore fault pole converter, the onshore converter station, the offshore non-fault pole converter, and the wind farms. Simulation results are provided to illustrate the effectiveness and wide applicability of the MLCC strategy.

The unique geographical location and complex operational dynamics of OWFs mean that offshore faults can include both DC and AC faults. Of these, AC asymmetric faults are the most frequent and account for a significant portion of offshore faults. Addressing the challenges of AC asymmetric FRT for OWFs and WFMMC is therefore critical. The fault characteristics of offshore AC asymmetric faults differ significantly from those of onshore grid faults [36]. Offshore faults typically occur between the OWFs and the WFMMC, with the OWFs lacking the support of the onshore grid, leading to complex, two-sided power electronic characteristics during faults [36, 37]. Additionally, the low overcurrent and overvoltage characteristics of power electronics exacerbate the multi-stage voltage and current FRT issues between WFMMC and OWFs under offshore AC faults [38]. Ensuring the stability and FRT capability of grid-connected OWFs is essential for their reliable and economical operation [37].

To address these challenges, current engineering solutions often focus on negative sequence (NS) current elimination control [39]. However, this approach overlooks the positive sequence (PS) current of the WFMMC coupled with the GSC output current, which can lead to an excessive NS voltage in the grid, potentially triggering NS protection during a fault. There is also a risk of overvoltage during the fault and subsequent recovery. Reference [40] primarily focuses on WFMMC, neglecting the fault current contribution from OWFs and only considering the fault current characteristics injected by WFMMC. Reference [41] introduces negative sequence current control (NSCC) for both OWFs and WFMMC, with defined limits for NS current injection during faults. However, this approach does not fully consider the sequence network coupling constraint for asymmetric faults or the current source characteristics of the offshore converter station. When WFMMC exhibits current source characteristics, it may block due to overcurrent and lose voltage support to the offshore wind farms, potentially leading to large-scale turbine disconnections and converter damage under severe conditions. Enhanced dynamic voltage control for OWFs has been proposed [42] to prevent excessive AC voltage rise in the healthy phase during AC asymmetric faults. Another approach involves the use of parallel supercapacitor energy storage with a sequential component coordination control strategy [43], incorporating a DC-DC converter and supercapacitor energy storage. However, this increases both engineering and offshore platform construction costs, and the reliability of supercapacitor energy storage requires further evaluation. Reference [44] employs a linearized analysis method to study the steady-state characteristics of faults during symmetrical and asymmetrical conditions, ultimately deriving an analytical formula for the fault current of the converter under various fault scenarios. However, this analytical formula depends on a fixed control combination, limiting its generalizability. To address the issue of power electronics overcurrent during faults, some scholars have proposed a low-voltage current-limiting control strategy on the WFMMC side. However, given the low-voltage FRT requirements, the active voltage drop may cause turbines to disconnect due to low voltage [45, 46]. In conclusion, research on offshore AC asymmetric short-circuit faults is still incomplete but is vital for the stability of grid-connected OWF systems. There is an urgent need to further investigate fault mechanisms and develop effective FRT schemes.

Chapter 7 proposes the MSNEC for FRT, based on the topology of OWFs connected to the grid via MMC-HVDC. The chapter begins with an analysis of the characteristics of offshore AC asymmetric faults, revealing their underlying mechanisms. The principles of overcurrent and overvoltage during faults are derived, and the energy control potential of WFMMC is explored from the perspective of energy transformation and transfer. By combining sequence network control with energy control, AC asymmetric fault ride-through is achieved through four key strategies: power supply equivalence, negative sequence suppression, energy control, and voltage stabilization.

References

1. Li W, Zhu M, Chao P, Liang X, Xu D (2020) Enhanced FRT and postfault recovery control for MMC-HVDC connected offshore wind farms. IEEE Trans Power Syst 35(2):1606–1617. 10.1109/TPWRS.2019.2944940
2. Xiong Y, Yao W, Wen J, Lin S, Ai X, Fang J, Wen J, Cheng S (2021) Two-level combined control scheme of VSC-MTDC integrated offshore wind farms for onshore system frequency support. IEEE Transactions on Power Systems 36(1):781–792. 10.1109/TPWRS.2020.2998579
3. Sun K, Yao W, Fang J, Ai X, Wen J, Cheng S (2020) Impedance modeling and stability analysis of grid-connected DFIG-based wind farm with a VSC-HVDC. IEEE J Emerg Sel Topics Power Electron. 8(2):1375–1390. 10.1109/JESTPE.2019.2901747
4. Hennig T, von Bonin M, Rohrig K, Stock S, Hofmann L (2016) Investigation on reliability-driven network expansions of offshore transmission systems. In: 2016 IEEE Innovative Smart Grid Technologies—Asia (ISGT-Asia), pp 653–658. 10.1109/ISGT-Asia.2016.7796462
5. Yang B, Liu B, Zhou H, Wang J, Yao W, Wu S, Shu H, Ren Y (2022) A critical survey of technologies of large offshore wind farm integration: Summarization, advances, and perspectives. Protection and Control of Modern Power Systems. In press
6. Guo G, Wang H, Song Q, Zhang J, Wang T, Ren B, Wang Z (2020) HB and FB MMC based onshore converter in series-connected offshore wind farm. IEEE Trans Power Electron 35(3):2646–2658. 10.1109/TPEL.2019.2929689
7. Shi L, Adam GP, Li R, Xu L (2020) Control of offshore MMC during asymmetric offshore ac faults for wind power transmission. IEEE J. Emerg. Sel. Topics Power Electron 8(2):1074–1083. 10.1109/JESTPE.2019.2930399
8. Cai L, Karaagac U, Mahseredjian J (2017) Simulation of startup sequence of an offshore wind farm with MMC-HVDC grid connection. IEEE Trans. Power Del 32(2):638–646. 10.1109/TPWRD.2016.2517184
9. Meng Y, Zou Y, Wang H, Kong Y, Du Z, Wang X (2021) Novel submodule topology with large current operation and dc-fault blocking capability for MMC-HVDC. IEEE Trans Power Del 36(3):1542–1551. 10.1109/TPWRD.2020.3011228
10. Deng F, Chen Z (2013) Operation and control of a DC-Grid offshore wind farm under DC transmission system faults. IEEE Trans. Power Del 28(3):1356–1363. 10.1109/TPWRD.2013.2261561
11. Silva B, Moreira CL, Leite H, Pe?as Lopes JA (2014) Control strategies for AC fault ride through in multiterminal HVDC grids. IEEE Trans Power Del 29(1): 395–405. 10.1109/TPWRD.2013.2281331
12. Egea-¨¤lvarez A, Arag¨1¨¦s-Pe?alba M, Prieto-Araujo E, Gomis-Bellmunt O (2017) Power reduction coordinated scheme for wind power plants connected

with VSC-HVDC. Renewable Energy 107:1–13. https://doi.org/10.1016/j.renene.2017.01.027
13. Karaagac U, Mahseredjian J, Cai L, Saad H (2017) Offshore wind farm modeling accuracy and efficiency in MMC-based multiterminal hvdc connection. IEEE Trans. Power Del 32(2):617–627. 10.1109/TPWRD.2016.2522562
14. Jing Y, Li R, Xu L, Wang Y (2018) Enhanced AC voltage and frequency control of offshore MMC station for wind farm connection. IET Renew Power Gener 12(15):1771–1777. https://doi.org/10.1049/iet-rpg.2018.5391
15. Arag¨1¨¦s Pe?alba M, Gomis-Bellmunt O, Martins M (2014) Coordinated control for an offshore wind power plant to provide fault ride through capability. IEEE Trans Sustain Energy 5(4):1253–1261. 10.1109/TSTE.2014.2344172
16. Hu X, Liang J, Rogers DJ, Li Y (2013) Power flow and power reduction control using variable frequency of offshore AC grids. IEEE Trans Power Syst 28(4):3897–3905. 10.1109/TPWRS.2013.2257884
17. Moawwad A, El Moursi MS, Xiao W (2016) Advanced fault ride-through management scheme for VSC-HVDC connecting offshore wind farms. IEEE Trans Power Syst 31(6):4923–4934. 10.1109/TPWRS.2016.2535389
18. Moawwad A, El Moursi M, Xiao W (2015) A novel transient control strategy for VSC-HVDC connecting offshore wind power plant. In: 2015 IEEE Power Energy Society General Meeting, pp 1–1. 10.1109/PESGM.2015.7285804
19. Davidson C, Mukhedkar R (2014) A comparison of different types of dynamic braking system for HVDC systems for offshore wind power. In: Proc. of Cigr¨¦ Brussels, pp 1–8
20. Saad H, Guillaud X, Mahseredjian J, Dennetiere S, Nguefeu S (2015) MMC capacitor voltage decoupling and balancing controls. IEEE Trans. Power Del. 30(2):704–712. 10.1109/TPWRD.2014.2338861
21. Shinoda K, Benchaib A, Dai J, Guillaud X (2018) Virtual capacitor control: Mitigation of dc voltage fluctuations in MMC-based hvdc systems. IEEE Trans. Power Del. 33(1):455–465. 10.1109/TPWRD.2017.2723939
22. Liu W, Li G, Liang J, UgaldeLoo CE, Li C, Guillaud X (2020) Protection of single-phase fault at the transformer valve side of FB-MMC-based bipolar HVdc systems. IEEE Trans. Ind. Electron. 67(10):8416–8427. 10.1109/TIE.2019.2947853
23. Li G, Liu W, Joseph T, Liang J, Song Z (2020) Double-thyristor-based protection for valve-side single-phase-to-ground faults in HB-MMC-based bipolar HVDC systems. IEEE Trans. Ind. Electron. 67(7):5810–5815. 10.1109/TIE.2019.2931502
24. Xue S, Gu C, Liu B, Fan B (2020) Analysis and protection scheme of station internal AC grounding faults in a bipolar MMC-HVDC system. IEEE Access 8, 26:536–548. 10.1109/ACCESS.2020.2971516
25. Li G, Liang J, Ma F, Ugalde-Loo CE, Liang H, Li H (2017) Analysis of single-phase-to-ground faults at the valve-side of HB-MMCs in bipolar HVDC systems. In: Proc. 2017 IEEE Energy Conver. Cong. Exposi. (ECCE), pp 2659–2665. 10.1109/ECCE.2017.8096501

26. Li X, Liu Y, Cai Z, Han K (2011) Research on single phase grounding fault at HVDC converter transformer valve side. In: Proc. 2011 4th Int. Conf. Elect. Utility Deregulation Restruct. Power Technol., pp 471–476. 10.1109/DRPT.2011.5993937
27. Tang L, Ooi B (2002) Managing zero sequence in voltage source converter. In: Proc. Conf. Rec. IEEE Ind. Appl. Conf. 37th IAS Annu. Meeting, vol 2, pp 795–802 vol 2. 10.1109/IAS.2002.1042650
28. Shi X, Wang Z, Liu B, Liu Y, Tolbert LM, Wang F (2015) Characteristic investigation and control of a modular multilevel converter-based HVDC system under single-line-to-ground fault conditions. IEEE Trans. Power Electron 30(1):408–421. 10.1109/TPEL.2014.2323360
29. Chen X, Zhao C, Cao C (2011) Research on the fault characteristics of HVDC based on modular multilevel converter. In: Proc. 2011 IEEE Elect. Power Energy Conf., pp 91–96. 10.1109/EPEC.2011.6070260
30. Li G, Liang J, Ma F, UgaldeLoo CE, Liang H (2019) Analysis of single-phase-to-ground faults at the valve-side of HB-MMCs in HVDC systems. IEEE Trans. Ind. Electron. 66(3):2444–2453. 10.1109/TIE.2018.2829666
31. Li G, Liang J, UgaldeLoo CE, Ma F, Liang H, Song Z (2020) Protection for submodule overvoltage caused by converter valve-side single-phase-to-ground faults in FB-MMC based bipolar HVDC systems. IEEE Trans. Power Del. 35(6):2641–2650. 10.1109/TPWRD.2020.2974413
33. Xiang W, Yang S, Adam GP, Zhang H, Zuo W, Wen J (2021) DC fault protection algorithms of MMC-HVDC grids: Fault analysis, methodologies, experimental validations, and future trends. IEEE Trans. Power Electron. 36(10):11,245–11,264. 10.1109/TPEL.2021.3071184
34. Lu Z, Ye Y, Qiao Y (2019) An adaptive frequency regulation method with grid-friendly restoration for VSC-HVDC integrated offshore wind farms. IEEE Trans. Power Syst. 34(5):3582–3593. 10.1109/TPWRS.2019.2901986
35. Wang W, Yin X, Jiang L, Cao Y, Li Y (2022) Perturbation observer-based nonlinear control of VSC-MTDC systems. Int. J. Electr. Power Energy Syst. 134. 10.1016/j.ijepes.2021.107387
36. Li R, Yu L, Xu L (2019) Offshore ac fault protection of diode rectifier unit-based hvdc system for wind energy transmission. IEEE Trans. Ind. Electron. 66(7): 5289–5299. 10.1109/TIE.2018.2869357
37. Sun K, Yao W, Wen J, Jiang L (2022) A two-stage simultaneous control scheme for the transient angle stability of vsg considering current limitation and voltage support. IEEE Trans. Power Syst. 37(3):2137–2150. 10.1109/TPWRS.2021.3121474
38. Xiao H, Sun K, Pan J, Li Y, Liu Y (2021) Review of hybrid hvdc systems combining line communicated converter and voltage source converter. Int. J. Electr. Power Energy Syst. 129:106,713. 10.1016/j.ijepes.2020.106713
39. Seifi K, Moallem M (2019) An adaptive pr controller for synchronizing grid-connected inverters. IEEE Trans. Ind. Electron. 66(3):2034–2043. 10.1109/TIE.2018.2838098

40. Li W, Zhu M, Chao P, Liang X, Xu D (2020) Enhanced frt and postfault recovery control for mmc-hvdc connected offshore wind farms. IEEE Trans. Power Syst. 35(2):1606–1617. 10.1109/TPWRS.2019.2944940
41. Xiang W, Zhang H, Yang S, Zhou M, Lin W, Wen J (2021) A differential pilot protection scheme for mmc-based dc grid resilient to communication failure. IEEE J. Emerg. Sel. Top. Power Electron. 9(5):5631–5645. 10.1109/JESTPE.2020.3035672
42. Cazzaro D, Trivella A, Corman F, Pisinger D (2022) Multi-scale optimization of the design of offshore wind farms. Appl. Energy 314, 118, 830. 10.1016/j.apenergy.2022.118830
43. Neumann T, Wijnhoven T, Deconinck G, Erlich I (2015) Enhanced dynamic voltage control of type 4 wind turbines during unbalanced grid faults. IEEE Trans. Energy Convers. 30(4):1650–1659. 10.1109/TEC.2015.2470126
44. Ramirez D, Zarei ME, Gupta M, Serrano J (2020) Fast model-based predictive control (fmpc) for grid connected modular multilevel converters (mmc). Int. J. Electr. Power Energy Syst. 119, 105, 951. 10.1016/j.ijepes.2020.105951
45. Peng X, Yao W, Yan C, Wen J, Cheng S (2020) Two-stage variable proportion coefficient based frequency support of grid-connected dfig-wts. IEEE Trans. Power Syst. 35(2):962–974. 10.1109/TPWRS.2019.2943520
46. Zhou H, Yao W, Ai X, Zhang J, Wen J, Li C (2022) Coordinated power control of electrochemical energy storage for mitigating subsequent commutation failures of hvdc. Int. J. Electr. Power Energy Syst. 134:107, 455. 10.1016/j.ijepes.2021.107455

Chapter 3
Active Energy Control of Converter Stations to Cope with Onshore Grid-Side Faults

Abstract Modular Multilevel Converter-based High Voltage Direct Current (MMC-HVDC) systems connected to offshore wind farms (WFs) face a critical challenge during grid-side AC faults: the management of surplus power. This chapter introduces an Active Energy Control (AEC) scheme that leverages MMC sub-module (SM) capacitors to actively absorb excess power, thereby improving AC fault ride-through (ACFRT) capability. The chapter begins by analyzing the energy decoupling principle in MMCs and proposing a steady-state control strategy for MMC-HVDC systems. Based on this analysis, the AEC scheme is developed for both onshore and offshore converter stations, functioning across four stages: passive energy recovery, active energy recovery, active energy maintenance, and active energy release. Additionally, coordinated control strategies are designed to integrate AEC with the active power reduction of offshore WFs, showing that the AEC scheme significantly enhances fault ride-through capability. The effectiveness of these control strategies is validated through simulations using the PSCAD/EMTDC platform.

3.1 MMC-HVDC Steady-State Control Scheme with Energy Decoupling Capability

The configuration and control framework for MMC-HVDC systems connected to offshore WFs (OWFs) are illustrated in Fig. 3.1. The WFs use direct drive permanent magnet synchronous generators (PMSG), with their output aligned with the rated transmission power of the WFMMC. Both the WFMMC and GSMMC employ half-bridge MMC (HBMMC) technology.

3.1.1 Principle of Energy Decoupling Control

Based on the HBMMC topology depicted in Fig. 3.2, the differential mode voltage $u_{\text{diff}j}$, common mode voltage $u_{\text{com}j}$, AC current $i_{\text{v}j}$, and interphase circulating current $i_{\text{cir}j}$ for phase j (j= a, b, c) are defined as:

W. Yao et al., *Active Control of Large-Scale Offshore Wind Farms Connected Via VSC-HVDC*, Power Systems, https://doi.org/10.1007/978-981-97-9346-4_3

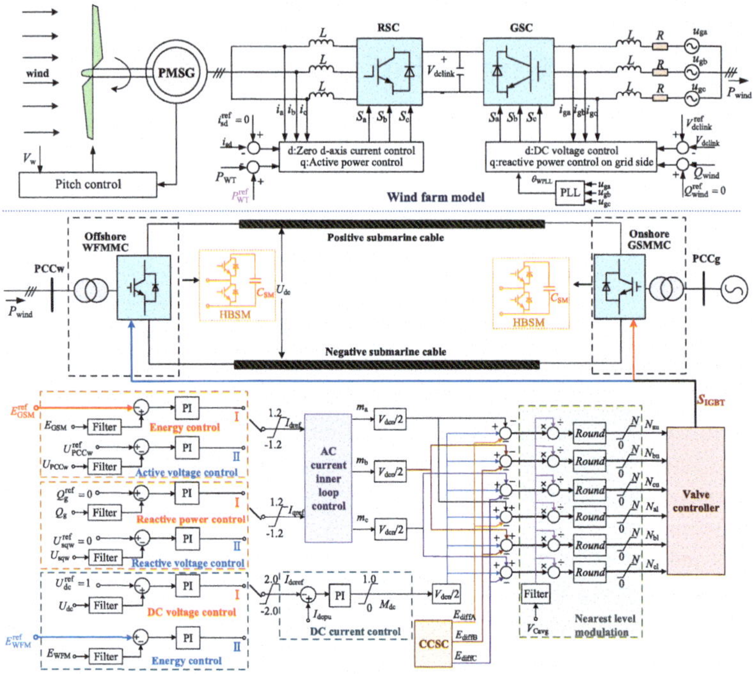

Fig. 3.1 Topology and control structure of MMC-HVDC connected with OWFs

$$
\begin{cases}
u_{\text{diff}\,j} = \frac{1}{2}(u_{\text{n}j} - u_{\text{p}j}) \\
u_{\text{com}j} = \frac{1}{2}(u_{\text{n}j} + u_{\text{p}j}) \\
i_{\text{v}j} = i_{\text{p}_j} - i_{\text{n}_j} \\
i_{\text{cir}j} = \frac{i_{\text{p}_j} + i_{\text{n}_j}}{2}
\end{cases}
\tag{3.1}
$$

where, $u_{\text{p}j}$, $u_{\text{n}j}$, $i_{\text{p}j}$, and $i_{\text{n}j}$ denote the voltages and currents of the upper and lower bridge arms for phase j. The dynamic equations for phase j can be expressed as:

$$
\begin{cases}
(L_{\text{ac}} + \frac{L_0}{2})\frac{di_{\text{v}j}}{dt} + \frac{R_0}{2} i_{\text{v}j} = -u_{\text{s}j} + u_{\text{diff}\,j} \\
L_0 \frac{di_{\text{cir}j}}{dt} + R_0 i_{\text{cir}j} = \frac{U_{\text{dc}}}{2} - u_{\text{com}j}
\end{cases}
\tag{3.2}
$$

where, L_{ac} represents the equivalent inductance from the converter outlet to the AC system's equivalent potential, L_0 denotes the bridge arm reactor, R_0 represents the equivalent resistance representing the bridge arm loss, and U_{ac} signifies the DC voltage.

Define the SM-input ratio of the upper and lower arm of the phase j

$$
\alpha_{\text{p,n}_j} = \frac{N_{\text{p,n}_j}}{N}
\tag{3.3}
$$

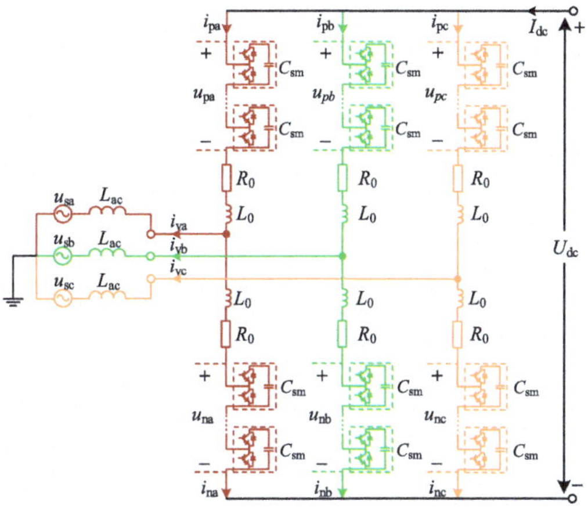

Fig. 3.2 Three-phase HBMMC topology

where, $N_{\mathrm{p,n}j}$ represents the number of SMs in the active state within a bridge arm, and N is the total number of SMs in a bridge arm. Consequently, the relationship between the voltages of the upper and lower arms $u_{\mathrm{p,n}j}$ of phase j and the sum of the capacitive voltages of the arm's SMs ($u_{\mathrm{Cs}_{\mathrm{p,n}j}}$) is as follows:

$$u_{\mathrm{p,n}_j} = \alpha_{\mathrm{p,n}_j} u_{\mathrm{Cs}_{\mathrm{p,n}_j}} \tag{3.4}$$

According to (3.1) and (3.4), $u_{\mathrm{diff}\,j}$ and $u_{\mathrm{com}j}$ are further expressed as:

$$\begin{cases} u_{\mathrm{diff}\,j} = \frac{1}{2}(\alpha_{\mathrm{n}j} u_{\mathrm{Cs}_{\mathrm{n}j}} - \alpha_{\mathrm{p}j} u_{\mathrm{Cs}_{\mathrm{p}j}}) \\ u_{\mathrm{com}j} = \frac{1}{2}(\alpha_{\mathrm{n}j} u_{\mathrm{Cs}_{\mathrm{n}_j}} + \alpha_{\mathrm{p}j} u_{\mathrm{Cs}_{\mathrm{p}_j}}) \end{cases} \tag{3.5}$$

It has been observed that when the power imbalance between the upper and lower bridge arms of the MMC is minimal, $u_{\mathrm{Cs}_{\mathrm{p}j}}$ and $u_{\mathrm{Cs}_{\mathrm{n}j}}$ can be approximately considered equal and are close to the average value of the two ($\bar{u}_{\mathrm{Cs}j}$) [1]:

$$u_{\mathrm{Cs}_{\mathrm{n}j}} \approx u_{\mathrm{Cs}_{\mathrm{p}j}} = \overline{u}_{\mathrm{Cs}j} \tag{3.6}$$

Define the AC/DC side input ratio α_{ac_j} and α_{dc_j} of HBMMC

$$\begin{cases} \alpha_{\mathrm{ac}_j} = \frac{1}{2}(\alpha_{\mathrm{n}_j} - \alpha_{\mathrm{p}_j}) \\ \alpha_{\mathrm{dc_j}} = \frac{1}{2}(\alpha_{n_j} + \alpha_{\mathrm{p}_j}) \end{cases} \tag{3.7}$$

thus, (3.5) can be further expressed as:

$$\begin{cases} u_{\mathrm{diff}\,j} = \alpha_{\mathrm{ac}_j} \overline{u}_{\mathrm{Cs}j} \\ u_{\mathrm{com}j} = \alpha_{\mathrm{dc}_j} \overline{u}_{\mathrm{Cs}j} \end{cases} \tag{3.8}$$

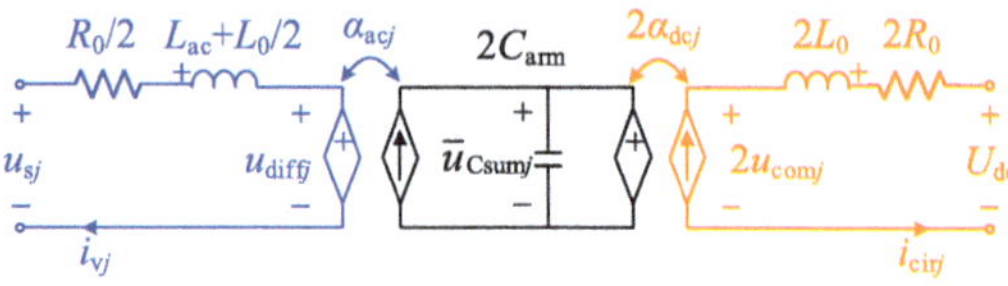

Fig. 3.3 Equivalent dynamic circuit of phase j

The relationship between the upper and lower bridge arm current of phase j (i_{p,n_j}) and the average value of the capacitive current flowing through all SMs of the bridge arm (i_{Cp,n_j}) is $i_{Cp,n_j} = \alpha_{p,n_j} i_{p,n_j}$. The equivalent capacitance of the bridge arm is $C_{arm} = C_{sm}/N$, where C_{sm} is the capacitance of a single SM. Therefore, the dynamic equivalent equation of the bridge arm can be expressed as follows:

$$C_{arm}\frac{du_{Cs_{p,n_j}}}{dt} = i_{C_{p,n_j}} \tag{3.9}$$

The combinations of (3.6) with (3.9), and the summation of upper and lower arm equations gives

$$2C_{arm}\frac{d\bar{u}_{Csj}}{dt} = \alpha_{p_j} i_{p_j} + \alpha_{n_j} i_{n_j} \tag{3.10}$$

The previous (3.10) is further combined with (3.1) and (3.7), to write

$$2C_{arm}\frac{d\bar{u}_{Csj}}{dt} = \alpha_{ac_j} i_{v_j} + 2\alpha_{dc_j} i_{cir_j} \tag{3.11}$$

Combining (3.2), (3.8) and (3.11), the equivalent circuit of phase j is obtained, as shown in Fig. 3.3.

Figure 3.3 illustrates the control method for decoupling the DC voltage of HBMMC from the capacitor voltage of the SMs by adjusting the input ratios α_{ac_j} and α_{dc_j} of each phase bridge arm.

3.1.2 Control Strategy of GSMMC

The control strategy for GSMMC employs Control Mode I, illustrated in Fig. 3.1. In this mode, the AC control loop regulates the MMC's internal energy, while the DC control loop manages the DC voltage of the MMC-HVDC system. The principle of Control Mode I is as follows:

The relationship between its internal energy and the power exchanged at the AC and DC sides can be described as:

$$\frac{d}{dt}E_{MMC} = C_{arm}\frac{d}{dt}\sum_{j=1}^{3}(u_{Cs_{nj}}^2 + u_{Cs_{pj}}^2) = U_{dc}I_{dc} - P_{ac} \tag{3.12}$$

where, P_{ac} is the three-phase AC output power of MMC. For the MMC operating stably in equilibrium, the P_{ac} can be expressed as:

$$P_{ac} = U_{gd} I_{gd} \tag{3.13}$$

where U_{gd} and I_{gd} are the d-axis components of voltage and current at the AC side of the MMC respectively. Therefore, based on (3.12) and (3.13), the AEC of GSMMC can generate the reference value of AC current (I_{dref}):

$$\begin{gathered} I_{dref} = \left(-C_{PI}\left(E_{GSM}^{ref} - E_{GSM}\right) + U_{dc} I_{dc}\right) \frac{1}{U_{gd}} \\ \Leftrightarrow I_{dref} = -C_{PIGS}\left(E_{GSM}^{ref} - E_{GSM}\right) \end{gathered} \tag{3.14}$$

where, C_{PI} is PI control function.

DC voltage control can be used to generate DC current reference value I_{dcref}:

$$I_{dcref} = -C_{PI}\left(U_{dc}^{ref} - U_{dc}\right) \tag{3.15}$$

The Control Mode I illustrated in Fig. 3.1 can be derived from the energy control equation (3.14) and the DC voltage control equation (3.15).

3.1.3 Control Strategy of WFMMC

The WFMMC operates in Control Mode II, detailed in Fig. 3.1. This mode integrates AC and DC control loops: the AC loop regulates the offshore AC grid voltage, while the DC loop manages the MMC's internal energy. The operational concept of Control Mode II is outlined as follows.

The AC voltage control mechanism is akin to that of a traditional MMC feeding power to a passive grid [2]. It stabilizes the offshore AC voltage amplitude by adjusting the current command I_{dref} on the WFMMC's power grid side, ensuring $U_{sqw} = 0$ by controlling the current command I_{qref}.

The internal energy control of WFMMC is devised from the dynamic equation governing HBMMC's internal energy and the power exchange on the AC/DC sides. Based on (3.12), energy control of MMC generates the reference DC current value I_{dcref}.

$$\begin{gathered} I_{dcref} = \left(C_{PI}\left(E_{WFM}^{ref} - E_{WFM}\right) + P_{ac}\right) \frac{1}{U_{dc}} \\ \Leftrightarrow I_{dcref} = C_{PIWF}\left(E_{WFM}^{ref} - E_{WFM}\right) \end{gathered} \tag{3.16}$$

For both GSMMC and WFMMC, the capacitor voltage of SMs will vary during MMC energy control. To prevent this from impacting the calculation of the input ratio of upper and lower bridge arm SMs (α_{p,n_j}) and the decoupling control of DC voltage and SM capacitor voltage, the nearest level modulation in the control strategy uses variable SM capacitor voltage modulation [3]. Due to the capacitor voltage equalization strategy, it is assumed that all SMs are identical. The average value of the SM capacitors voltage (V_{Cavg}) is detected in real-time and used as the base value for SM capacitors voltage modulated by the nearest level, ensuring the correct number of SMs is input. The CCSC in Fig. 3.1 indicates the circulating current suppression control.

3.2 AEC for Enhancing AC Fault Ride-Through Capability

3.2.1 Maximum Recovery Time of Surplus Power

The SM capacitor is the primary energy storage component of the MMC converter station. If only the SM capacitors are considered, the rated energy storage capacities of WFMMC and GSMMC are:

$$E_{\mathrm{MMC0}} = \frac{1}{2} \times C_{\mathrm{SM}} \times U_{\mathrm{SM}}^2 \times 6N \tag{3.17}$$

where, U_{SM} is the rated value of capacitor voltage of single SM.

Table 3.1 presents the voltage protection settings and actions for the MMC submodules (SMs) used in an MMC-HVDC project in China. It shows that while preventing the activation of SM overvoltage protection, there are safety margins of 0.7 pu and 0.6 pu for the single SM capacitor voltage and the average SM capacitor voltage, respectively. By temporarily reducing these voltage safety margins, the energy storage capacity of the SM capacitors can be utilized, allowing the MMC to actively absorb and recover surplus power.

To prevent voltage overshoot from activating SM protection during Active Energy Control (AEC), an additional voltage margin is reserved, setting the upper voltage control limit at 1.5 pu. This approach allows a single HBMMC converter station to store energy exceeding its rated capacity:

$$\Delta E_{\mathrm{MMC}} = \frac{1}{2} \times C_{\mathrm{SM}} \times (1.5^2 - 1) \times U_{\mathrm{SM}}^2 \times 6N = 1.25 E_{\mathrm{MMC0}} \tag{3.18}$$

If the offshore surplus power is defined as P_{s}, the maximum time (Δt) for which the MMC-HVDC system shown in Fig. 3.1 can absorb residual power is expressed as:

Table 3.1 Voltage protection threshold and action of the SM

Protection type	Overvoltage protection of single module capacitor	Average value protection of capacitor voltage of all SMs
Threshold	1.7 pu	1.6 pu
Action	Bypass + Blocking	Bypass + Blocking

$$\Delta E = \Delta E_{\mathrm{GSM}} + \Delta E_{\mathrm{WFM}} = \int_{t_1}^{t_1+\Delta t} P_{\mathrm{s}} dt \tag{3.19}$$

where, t_1 is the fault start time, ΔE_{WSM} and ΔE_{GFM} are the additional energy stored in the capacitors of WFMMC and GSMMC respectively, P_{s} can be given by the expression:

$$P_{\mathrm{s}} = P_{\mathrm{wind}} - U_{\mathrm{PCCgr}} I_{\mathrm{m}} \tag{3.20}$$

where, U_{PCCgr} represents the residual voltage at the point of common coupling (PCC) on the grid side post-fault. I_{m} is the current amplitude on the AC side of the GSMMC during the fault, generally constrained to 1.2 pu. P_{wind} denotes the power generated by the wind farms (WFs).

3.2.2 Active Energy Recovery Rate Calculation of WFMMC

If the energy absorption rate of WFMMC matches the generation rate of unbalanced power, there will be no surplus power issue in the offshore transmission system during WFMMC's active energy recovery.

In the case of three-phase metal fault, the maximum active energy recovery time of WFMMC (t_{WFmax}) is

$$\Delta t_{\mathrm{WFmax}} = \frac{\Delta E_{\mathrm{WFM}}}{P_{\mathrm{wind}}} \tag{3.21}$$

within the maximum energy margin of 1.25 pu, the energy recovery rate is:

$$v_{\mathrm{WFM0}} = \frac{1.25}{\Delta t_{\mathrm{WFmax}}} = 1.25 \frac{P_{\mathrm{wind}}}{\Delta E_{\mathrm{WFM}}} \tag{3.22}$$

Further considering the surplus power value caused by different fault degrees, the energy recovery rate of WFMMC (v_{WFM}) is:

$$v_{\mathrm{WFM}} = 1.25 \frac{P_{\mathrm{wind}} P_{\mathrm{s}}}{\Delta E_{\mathrm{WFM}}}. \tag{3.23}$$

3.2.3 AEC Scheme Based on Preset Curve

The AEC scheme for offshore wind power using MMC-HVDC grid connection is illustrated in Fig. 3.4, with the algorithm flow chart shown in Fig. 3.5. This control scheme encompasses three main components: fault detection, AEC for GSMMC, and AEC for WFMMC. In Fig. 3.4, $\dot{U}_{\mathrm{GSMa/b/c}}$ indicates the three-phase AC voltage at the grid side of GSMMC. The fault signal is denoted as S_{f}, while $S_{\mathrm{GSMs/m/r}}$ and

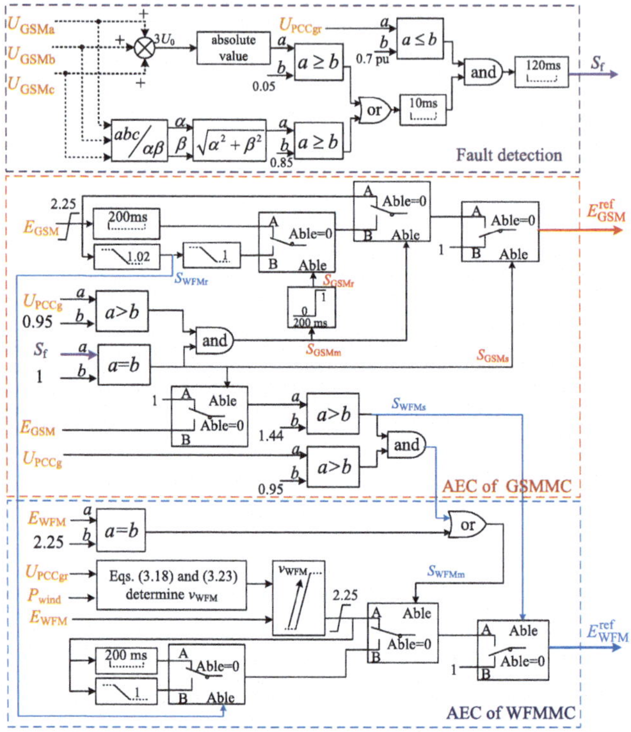

Fig. 3.4 AEC scheme of MMC-HVDC connected with OWFs

$S_{\mathrm{WFMs/m/r}}$ represent the energy recovery, maintenance, and release signals for the GSMMC and WFMMC AEC, respectively.

To clarify the AEC scheme's principle, Fig. 3.6 presents a schematic of the preset AEC command value curves for GSMMC and WFMMC across three different AC fault durations. In this figure, i in t_i and t_{i-j} refers to common action stages across the scenarios, and j indicates the varying moments these stages occur in each scenario. Figure 3.6a details the AEC timing for Scenario 1, illustrating the sequence for GSMMC and WFMMC.

Stage I ($t < t_1$): The MMC-HVDC is in steady state operation and the internal energies of the MMCs are all stabilized near their 1 pu ratings.

Stage II ($t_1 \leq t < t_2$): At t_1, a fault occurs, charging the GSMMC SM capacitors with unbalanced power. GSMMC begins passive energy recovery upon detecting the fault signal (S_{f}=1), adjusting its energy command value $E_{\mathrm{GSMMC}}^{\mathrm{ref}}$ accordingly. The AC fault detection method, borrowed from commutation failure prevention (CFPREV) in Line-Commutated-Converter HVDC [4, 5], uses the bus voltage at GSMMC's AC side for calculation, resulting in a small signal transmission delay. The energy control

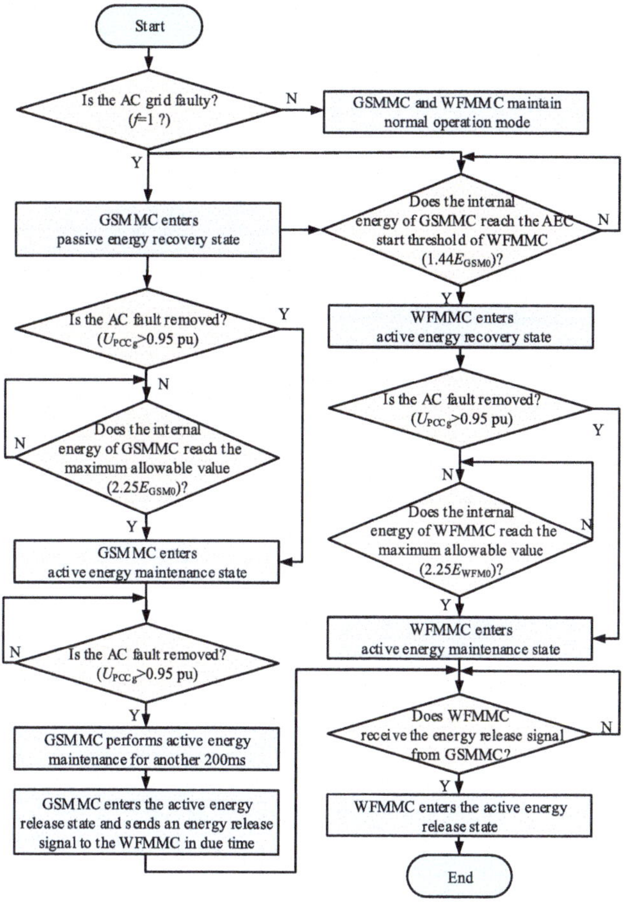

Fig. 3.5 AEC algorithm of GSMMC and WFMMC

command of GSMMC responds approximately 5 ms after the fault, considering fault detection time.

Stage III ($t_2 \leq t < t_3$): At t_2, GSMMC's internal energy exceeds 1.44 pu, triggering the transmission of an active energy recovery start command and residual voltage (U_{PCCgr}) information to WFMMC. WFMMC enters an active energy recovery state. Since seabed communication uses optical fiber, the delay is negligible, with the main consideration being data conversion delay ($\Delta t_{\text{L}_\text{W}}$ = 10 ms). The energy recovery rate of WFMMC is calculated to ensure complete absorption of unbalanced power. At this point, GSMMC theoretically has no power injection, and its energy remains constant. If the residual voltage is high, GSMMC can independently complete the active energy recovery. Therefore, if the collected residual voltage exceeds 0.7 pu, GSMMC will cease transmitting the start command to WFMMC.

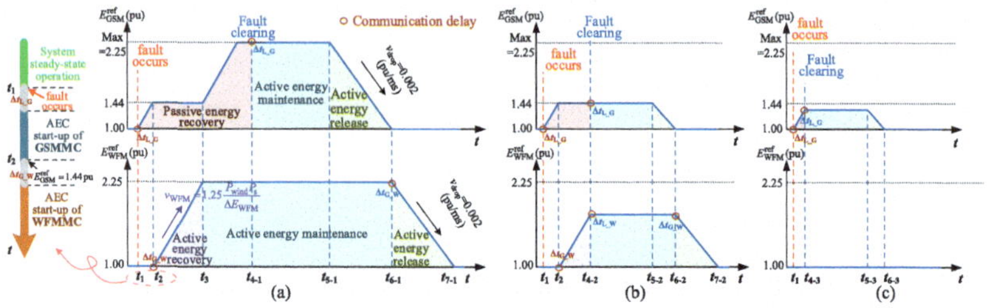

Fig. 3.6 AEC preset curves for MMCs. **a** Scenario 1: Fault clears after WFMMC completes active energy recovery. **b** Scenario 2: Fault clears during WFMMC active energy recovery. **c** Scenario 3: Fault clears before AEC starts of WFMMC

Stage IV ($t_3 \leq t < t_{4-1}$): When the energy of WFMMC reaches the limit of 2.25 pu, WFMMC enters the active energy maintenance state to maintain a constant energy level. Meanwhile, GSMMC continues to recover energy. As the energy of GSMMC approaches the limit of 2.25 pu, its energy command value $E_{\mathrm{GSM}}^{\mathrm{ref}}$ remains unchanged.

Stage V ($t_{4-1} \leq t < t_{7-1}$): The fault is cleared at moment t_{4-1}. To minimize the impact of converter stations on the recovery of the onshore system, GSMMC enters the active energy maintenance stage when the grid-side PCC voltage (U_{PCCg}) is detected to have recovered to above 0.95 pu, and maintains this state for 200 ms ($t_{4-1} - t_{5-1}$). Subsequently, it actively releases energy ($t_{5-1} - t_{6-1}$), with the release rate set to 0.002 pu/ms to avoid significant impact on the onshore system. When the energy of GSMMC is released to 1.02 pu (slightly after t_{6-1}), GSMMC signals WFMMC to begin actively releasing energy ($t_{6-1} - t_{7-1}$).

In Scenario 2, depicted in Fig. 3.6b, the fault clears during WFMMC's active energy recovery state. Once the AC bus voltage exceeds 0.95 pu, indicating grid recovery, the onshore power grid directly sends an energy maintenance signal to WFMMC (considering a 10 ms communication delay, $\Delta t_{\mathrm{L_W}}$). WFMMC then enters the energy maintenance stage ($t_{4-2} - t_{6-2}$). After fault removal, the control timing of active energy maintenance and recovery for GSMMC and WFMMC is the same as in Scenario 1.

In Scenario 3, illustrated in Fig. 3.6c, the fault clears before GSMMC's energy reaches 1.44 pu. GSMMC stops sending start signals to WFMMC and relies solely on GSMMC to complete the ACFRT.

To enhance the understanding of the AEC algorithm, Fig. 3.5 is further explained in conjunction with the aforementioned scenarios and stages.

- When no AC fault is detected, both GSMMC and WFMMC operate in Stage I, maintaining steady-state operation.
- Upon detecting an AC fault, GSMMC initiates passive energy recovery. If GSMMC's internal energy exceeds 1.44 pu, it sends an activation signal to WFMMC, allowing WFMMC to enter the active energy recovery state. If

GSMMC's internal energy remains below 1.44 pu, GSMMC independently completes the ACFRT (Scenario 3).

- If the AC fault persists, GSMMC and WFMMC continuously monitor their internal energy levels to determine if they reach the maximum limit of 2.25 pu. Upon reaching this limit, both converters switch to the active energy maintenance state. If the fault is cleared, they immediately transition to the active energy maintenance state (Scenario 2). If the GSMMC detects AC fault removal, both converters enter the active energy maintenance state (Scenario 2).
- After GSMMC has maintained active energy for 200 ms, it enters the active energy release state. As this state concludes, an active energy release signal is transmitted to WFMMC to prompt WFMMC to release energy. Eventually, the energy levels of both WFMMC and GSMMC return to their pre-fault steady state (Stage V).

By applying the AEC algorithm described above and combining Eqs. (3.19) and (3.20), we can derive the maximum energy recovery time for the offshore wind power system converter station as:

$$\Delta t_{\text{AEC}} = \frac{\Delta E_{\text{GSM}} + \Delta E_{\text{WFM}}}{P_{\text{wind}} - U_{\text{PCCgr}} I_{\text{m}}}. \tag{3.24}$$

3.2.4 Coordinated Scheme Between AEC and WFs

The AEC scheme for MMCs described above aims to optimize investment time or reduce costs associated with existing ACFRT strategies, such as DBR installation and WF power reduction. This subsection uses the collaboration between WF active power reduction and AEC as a case study to demonstrate the beneficial impact of AEC on ACFRT.

Wind turbines (WTs) have limited output power reduction rates due to their inertia, typically taking from tens to hundreds of milliseconds to decrease from rated power to zero [6]. Therefore, AEC can be employed to delay WF power reduction. The coordinated control command preset curve between WTs and MMCs is depicted in Fig. 3.7, where t_{III} and t_{V} represent the times when the WT reaches the 0.5 pu limit and completes power recovery, respectively. The corresponding WT control strategy is illustrated in Fig. 3.8, where β, V_{w}, and ω denote the pitch angle, wind speed, and turbine rotational speed, respectively. P_{Mref} is the WT output power reference value determined by the maximum power point tracking (MPPT) curve. F_{L1} indicates the start signal for WTs active power reduction, F_{L2} represents the power recovery signal for WTs, and i_{Gq}^{*} is the reference value of the q-axis current of GSVSC.

In Scenario 1, the sequence of actions for WTs can still be divided into the following stages.

Stage I ($t_1 \leq t < t_2$): The fault arises at t_1, triggering passive energy recovery in the GSMMC. The WF's power reduction control remains inactive as its internal energy fails to exceed the 1.44 pu threshold.

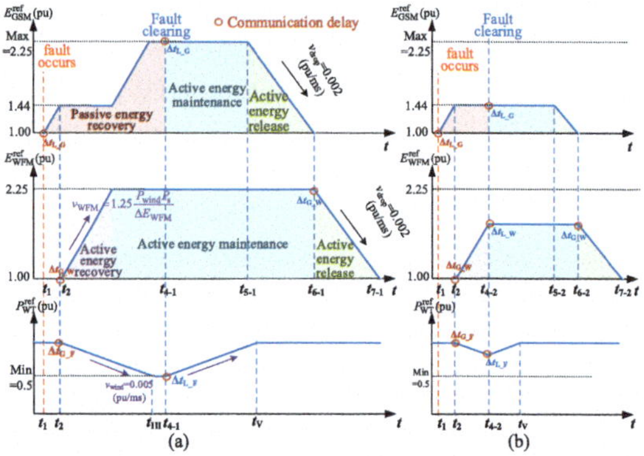

Fig. 3.7 Control command preset curve between WTs and AEC: **a** Scenario 1: Fault clears after completion of active energy recovery by WFMMC. **b** Scenario 2: Fault clears during active energy recovery by WFMMC

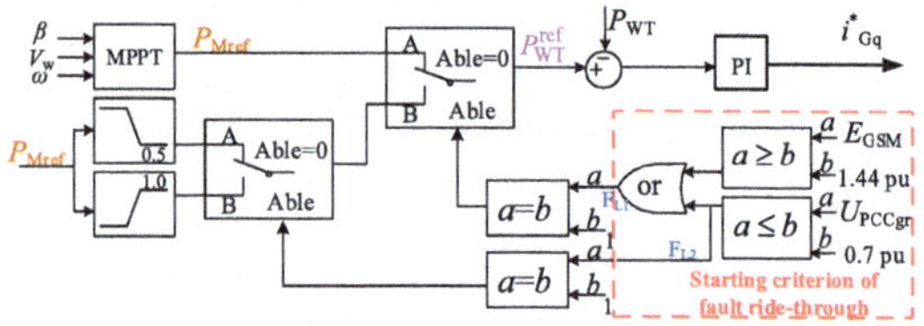

Fig. 3.8 ACFRT control strategy of WTs

Stage II ($t_2 \le t < t_{\text{III}}$): If the GSMMC's energy E_{GSM} exceeds 1.44 pu or U_{PCCgr} drops below 0.7 pu, a control signal F_{L1} is sent to the WFs through a time delay $\Delta t_{\text{L\_F}}$ (assumed to be 15 ms) to initiate power reduction in the WTs.

Stage III ($t_{\text{III}} \le t < t_{4-1}$): To prevent adverse effects from excessive reduction of WT output power, such as prolonged power recovery time, the output power of the WT remains unchanged once it decreases to 0.5 pu.

Stage IV ($t_{4-1} \le t < t_{\text{V}}$): Once the fault is cleared, if the U_{PCCg} rises to more than 0.7 pu, the WT power returns to near the rated value.

Scenario 2 involves the WT immediately resuming its output power upon receiving a fault clearance signal.

It is important to note that the coordinated control of AEC and WFs described above is just one example of AEC application, not the only one. For instance, AEC could also work in conjunction with DBR or with WFMMC for voltage drop control [2].

In DC grids, as the number of converter stations neighboring a single converter station increases, there are more energy control margins available to achieve ACFRT. Thus, applying AEC in the DC grid can further enhance the success of ACFRT compared to dual-terminal MMC-HVDC, provided that a suitable power allocation strategy is devised. This could lead to a further reduction in the required capacity of DBR or a decrease in the impact of WF power reduction.

3.3 Simulation Verification

Based on the parameters of the Shantou-Zhongpeng II OWF project under construction in China, the main parameters of the test system shown in Fig. 3.1 are set as detailed in Table 3.2. In case of a fault, the output power of WFs is set to their rated power (P_{windN}) to simulate the most severe power surplus problem caused by an AC fault.

3.3.1 Steady-State Control Scheme Validation

To validate the MMC-HVDC steady-state control scheme, simulations were conducted with the following setup: during steady-state operation, the WFMMC energy was increased to 1.5 pu at 3 s and then restored to 1 pu after 0.5 s. Similarly, the GSMMC energy was raised to 1.5 pu at 4 s and then returned to 1 pu after 0.5 s. The dynamic system behavior is depicted in Fig. 3.9.

As illustrated in Fig. 3.9, the DC voltage remains essentially constant under the voltage control of the GSMMC, even when the internal energy of the MMCs varies. The DC power adjusts according to the energy output of the WFMMC. This demonstrates that the proposed control scheme effectively decouples DC voltage control from submodule energy management in steady-state mode, while also providing robust DC power regulation capabilities.

3.3.2 Role of AEC in AC Fault Ride-Through

To demonstrate the impact of AEC on ACFRT, simulations are conducted comparing a system without AEC to one with AEC.

Fault scenario: A three-phase metallic grounding fault (TPMF) is introduced at the onshore AC bus at 6 s, and the fault lasts for 100 ms. Figure 3.10 illustrates the energy variation of MMCs, power on both sides of GSMMC, and DC current.

Table 3.2 Main parameters of test system

Parameter	Value
DC voltage U_{dc}	±320 kV
SM capacitance C_{SM}	12500 μF
Number of SMs per arm N	320
Rated voltage of SM U_{SM}	2 kV
Rated power of WFs P_{windN}	1200 MW
Rated voltage of onshore PCC bus U_{PCCgN}	230 kV
AC current amplitude limit of GSMMC I_{m}	1.2 pu

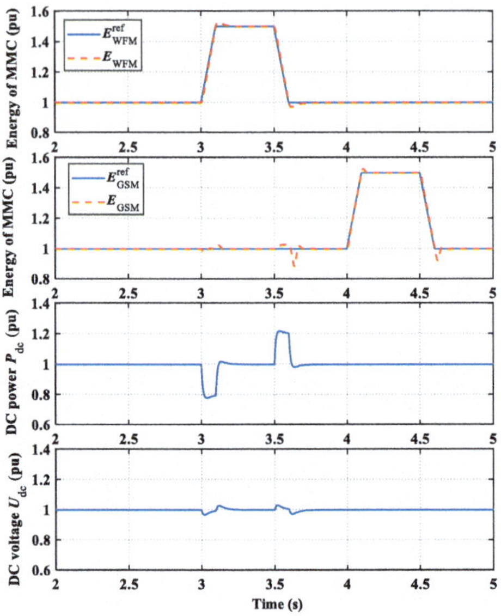

Fig. 3.9 Characterization of MMC-HVDC under the proposed steady-state control scheme

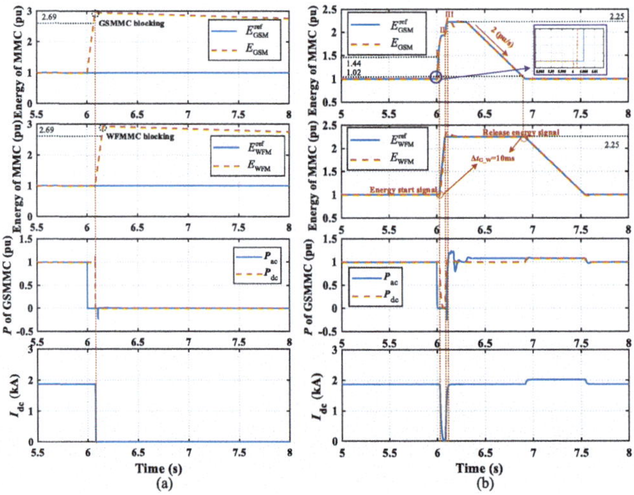

Fig. 3.10 Influence of AEC on ACFRT: **a** Three-phase metallic grounding fault without AEC. **b** Three-phase metallic grounding fault with AEC enabled

From Fig. 3.10a:

- Prior to the fault, the system parameters were at their rated operating levels, confirming the effectiveness of the proposed steady-state control strategy.
- Without AEC, the energy of GSMMC surges after an AC fault, leading to a rapid increase in SM voltage (or energy) until it exceeds the protection threshold, causing the converter station to trip. Subsequently, DC transmission power drops to zero. The WFMMC then passively absorbs energy until it is also tripped.

From Fig. 3.10b:

- With the AEC configured, GSMMC can detect the fault signal approximately 5 ms after the fault occurrence and then initiate AEC. The energy command value $E_{\mathrm{GSM}}^{\mathrm{ref}}$ changes in real-time with the actual energy level.
- When the GSMMC energy surpasses 1.44 pu, WFMMC receives a start signal from GSMMC to begin active energy recovery (I–II). The rate of active energy recovery for WFMMC, calculated using (3.23), is set at 0.025 pu/ms. Consequently, the DC power or DC current will decrease to 0, indicating that the energy recovery rate of WFMMC aligns with the energy generation rate of the WFs.
- Once the WFMMC reaches 2.25 pu, the WFMMC changes to an active energy maintenance state. The GSMMC continually accelerates energy recovery until it reaches the 2.25 pu threshold (II–III) and moves to an active energy maintenance state. Both the GSMMC and the WFMMC reaches the 2.25 pu energy recovery limit during the 100 ms fault duration.
- Upon receiving the fault clearance signal (U_{PCCg} >0.95 pu), GSMMC continues to maintain energy for 200 ms. Once the energy of GSMMC is reduced to 1.02 pu, an energy release signal is sent to the WFMMC.

Based on the previous analysis, AEC can achieve a response time of around 100 ms for other ACFRT measures under the most severe fault conditions. Additionally, during AEC operation, GSMMC effectively maintains power balance between its AC and DC sides, efficiently suppressing surplus power.

3.3.3 AEC Performance Under Different Fault Durations

Fault scenario: A TPMF is simulated, occurring at 6 s on the onshore AC bus. And the fault lasts for 10 ms, 60 ms, 120 ms respectively. Figure 3.11 illustrates the energy variations of MMCs and the power levels on both sides of GSMMC. The control scheme demonstrates stability under different fault durations.

From Fig. 3.11a: For very short fault durations, the energy of GSMMC remains below 1.44 pu, so no start command is transmitted to WFMMC. Around 6.015 s, GSMMC receives fault removal information and initiates energy maintenance. After maintaining energy for 200 ms, active energy release occurs, leading to an increase in power at the AC side of GSMMC. The amplitude of this power rise can be adjusted

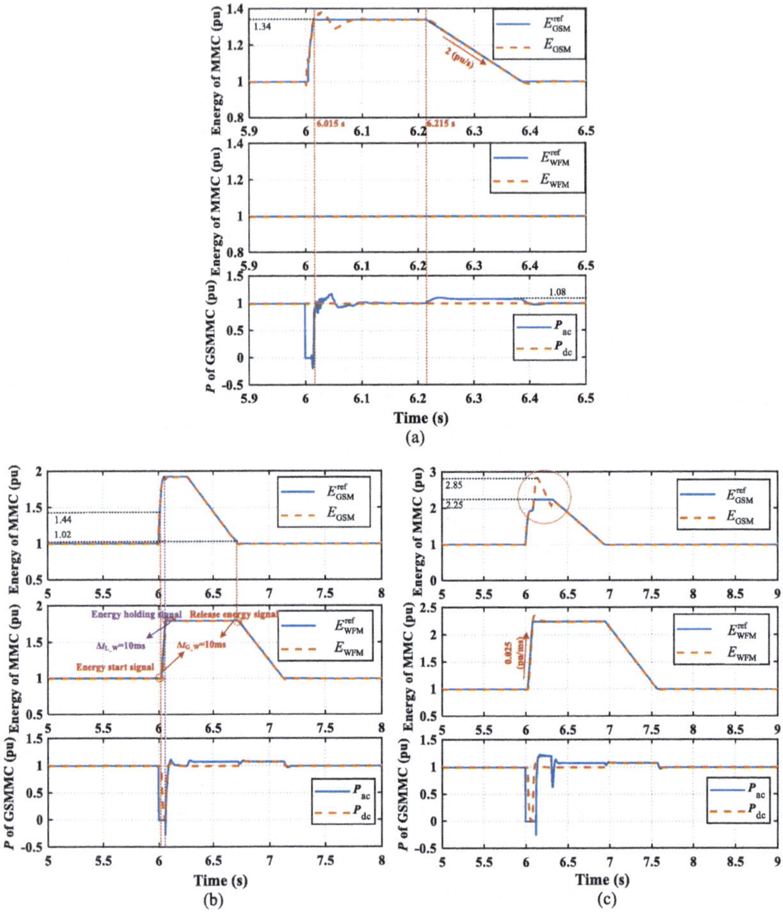

Fig. 3.11 AEC performance for a three-phase metallic ground fault with various fault durations: **a** 10 ms, **b** 60 ms, **c** 120 ms

by modifying the active energy release rate. Hence, if the fault is quickly removed, GSMMC alone can complete the ACFRT.

From Fig. 3.11b: When the fault persists for a longer duration and is eventually cleared onshore, WFMMC delays receiving the energy holding signal, thus entering the active energy maintenance state.

From Fig. 3.11c: When the energy control command value of GSMMC reaches 2.25 pu and the fault remains uncleared, the energy control within GSMMC may become unstable, potentially triggering the relay protection of the converter station (relay protection actions are not simulated here). In this scenario, completing ACFRT solely relying on AEC becomes challenging.

3.3.4 AEC Performance Under Different Fault Severities

According to Eq. (3.24), the relationship between the onshore residual voltage and the maximum energy recovery time is depicted in Fig. 3.12. As AEC initiates when the residual voltage falls below 0.7 pu, Fig. 3.12 displays this relationship for residual voltages ranging from 0 to 0.7 pu.

For systems with an onshore AC bus rated voltage of 230 kV, the State Grid Corporation of China stipulates that AC faults should be cleared within 120 ms. Hence, the residual voltage should not drop below 0.139 pu.

Fault scenario: a three-phase ground fault occurs at 6 s on the onshore AC bus, resulting in residual voltages of 75%, 50%, and 25% of the rated voltage, respectively. The fault is cleared after 500 ms, 250 ms, and 142 ms, respectively. Figure 3.13 illustrates the energy variations of the MMCs, as well as the power and voltage characteristics on both sides of the GSMMC. The control scheme demonstrates stability under varying fault severity levels.

From Fig. 3.13a:

- Given that the grid-side AC voltage of GSMMC remains above 0.7 pu after the fault, the AEC is not initiated.
- GSMMC undergoes two stages: passive energy recovery and active energy release. Due to the mild fault severity, ACFRT is achieved through these two stages of GSMMC.
- Throughout the fault duration, the system's DC voltage remains near the rated voltage and does not fluctuate with the internal energy of the MMC. This observation further confirms the decoupling characteristics of the capacitor voltage and DC voltage of the MMC SM as analyzed previously.

From Fig. 3.13b:

- Following the fault (I–II), the energy of GSMMC rises rapidly, and at 10 ms after the residual voltage drops below 0.7 pu, GSMMC samples the value to determine the fault severity.
- When the energy of GSMMC reaches 1.44 pu (II–III), the fault severity information is sent to WFMMC along with a start signal. WFMMC calculates the active energy recovery rate based on the fault severity to balance the power on both sides of GSMMC.

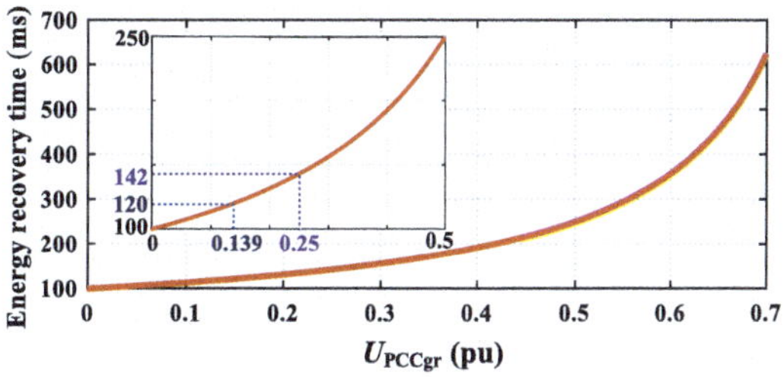

Fig. 3.12 The relationship between the onshore residual voltage and the maximum energy recovery time

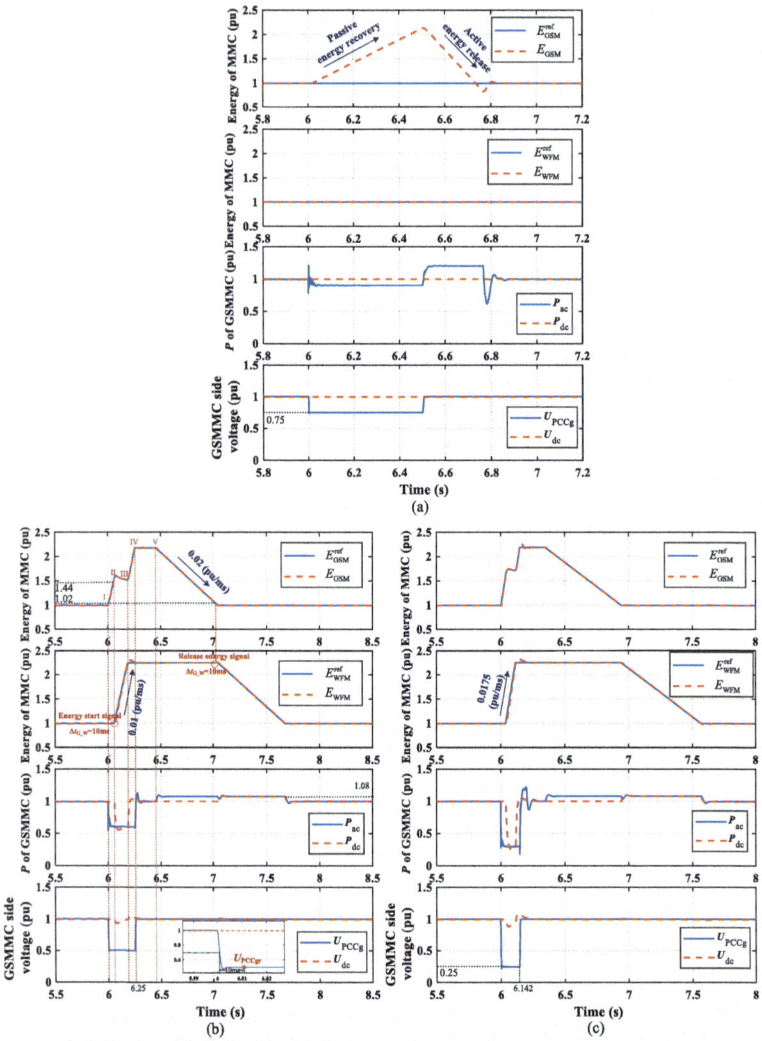

Fig. 3.13 Performance of active energy control (AEC) under various fault degrees. **a** AC voltage drops to 75%. **b** AC voltage drops to 50%. **c** AC voltage drops to 25%

- Upon reaching the energy limit (III–IV), WFMMC enters the active energy maintenance state, while GSMMC continues energy recovery. Subsequently, GSMMC transitions into the energy maintenance state (IV–V). Finally, the energy of GSMMC and WFMMC is released sequentially, concluding the ACFRT process.
- When the fault lasts 250 ms, the energy of GSMMC and WFMMC just reaches 2.25 pu, confirming the accuracy of the theoretical energy recovery time.

From the figure, as the fault severity increases, WFMMC's active energy recovery rate also increases, effectively mitigating higher levels of unbalanced power on the AC/DC side of GSMMC. Consequently, the active energy recovery time is further shortened. This indicates that AEC can effectively respond to varying fault severities, actively facilitating ACFRT completion.

3.3.5 AEC Performance Under Different Fault Types

To assess the effectiveness of AEC under various fault conditions, two types of asymmetric faults are examined.

Fault scenarios: a single line-to-ground fault and a line-to-line-to-ground fault on the onshore AC bus occur at 6 s, each lasting 120 ms. The energy profiles of MMCs and the power distribution on both sides of GSMMC are depicted in Fig. 3.14.

For various fault types lasting 120 ms, AEC effectively manages surplus power in the HVDC system, ensuring that the energy in each converter station does not exceed the limit. Thus, AEC can successfully achieve ACFRT within the fault clearance time for asymmetric faults.

3.3.6 AEC Coordination with WFs

When a severe fault occurs with an extended fault clearance time, it becomes necessary to combine other strategies to achieve ACFRT. The impact of WF power reduction alone is compared to the coordinated approach between WFs and AEC, as described in subsection III-D, on ACFRT.

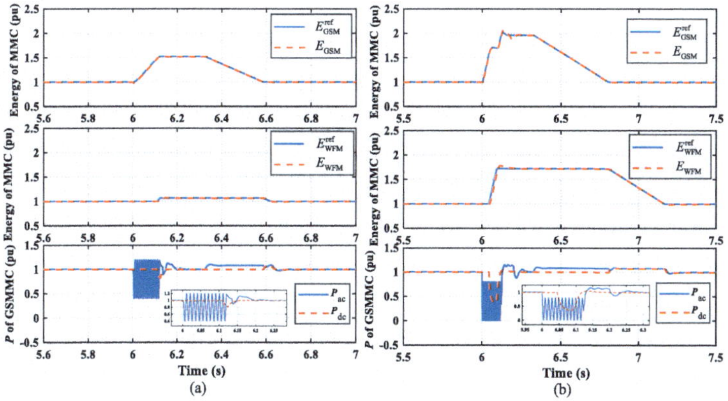

Fig. 3.14 AEC performance under different fault types. **a** Single line-to-ground fault. **b** Line-to-line-to-ground fault

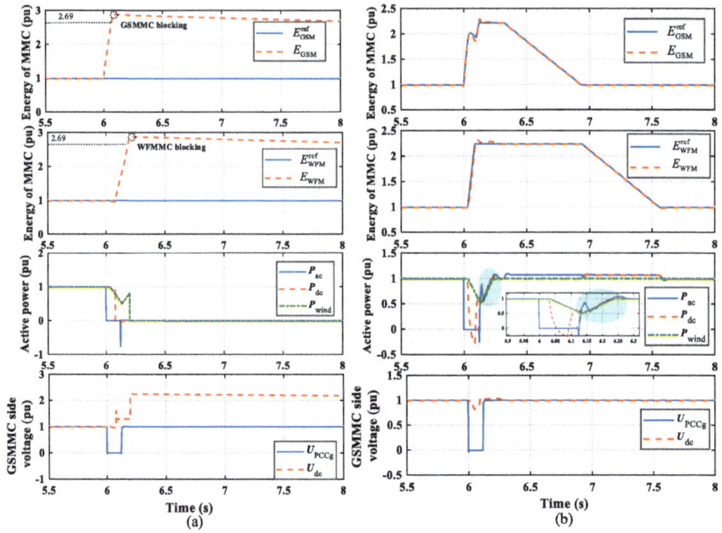

Fig. 3.15 Coordinated ACFRT capability between WFs and AEC. **a** WFs power reduction only. **b** AEC cooperation with WFs

Fault Scenario: A three-phase metallic grounding fault occurs on the shore AC bus at 6 s. The offshore wind power system undergoes ACFRT using two approaches: WF power reduction only and coordinated WF power reduction with AEC. The fault duration is 120 ms. Figure 3.15 shows the energy of MMCs, the active power in the system, and the voltage on both sides of GSMMC.

From Fig. 3.15a:

- Due to the limited rate at which WFs can reduce their power output, the energy stored in GSMMC increases rapidly after a fault occurs. When the energy of the submodule exceeds the protection threshold, GSMMC is blocked, causing WFMMC to also be blocked. As a result, the OWF loses voltage support, and its output power rapidly drops to 0.
- After the converter station is blocked, the surplus power charges the capacitance of the DC submarine cable, leading to a rapid increase in DC voltage. This increase in DC voltage poses a threat to the safety of power grid equipment. Therefore, achieving ACFRT solely through WF power reduction is challenging.

From Fig. 3.15b:

- The cooperation between WF power reduction and MMCs' AEC further reduces the unbalanced power. The energy of MMCs does not exceed the limit within 120 ms.
- During WFMMC's active energy recovery, GSMMC passively releases energy to maintain the residual power on the AC side because WFs are reducing their output energy.

The preceding analysis provides an illustration of the collaborative potential between WFs and AEC, demonstrating that AEC, when combined with other ACFRT measures, can significantly bolster the ACFRT capacity of offshore wind power systems.

3.4 Summary

This chapter introduces an Active Energy Control (AEC) scheme designed for MMC-HVDC systems connected to offshore wind farms (OWFs). In the event of an AC fault in the onshore power grid, AEC coordinates the operations of both offshore and onshore converter stations to address the resulting power imbalance. The AEC process is structured into four primary stages: passive energy recovery, active energy recovery, active energy maintenance, and active energy release. A timing control algorithm is also proposed to optimize the absorption of offshore residual power, thereby enhancing the AC fault ride-through (ACFRT) capability of MMC-HVDC system connected to OWFs.

To validate the effectiveness of the AEC scheme, a simulation model based on the Shantou-Zhongpeng II OWF project in China was developed. The results demonstrate that AEC successfully achieves ACFRT for a three-phase metallic grounding fault with a duration of approximately 100 ms. The scheme effectively manages unbalanced power across various fault severities and types, confirming its reliability.

Moreover, by illustrating the collaboration between wind farms and AEC, it is shown that AEC can be integrated with other existing ACFRT strategies to further enhance the ACFRT capability of MMC-HVDC systems connected to OWFs. Unlike the installation of Dynamic Braking Resistors (DBR), the proposed AEC scheme utilizes the existing capacitors within the MMC submodules for ACFRT, avoiding the need for additional equipment. Consequently, implementing AEC for ACFRT offers a more cost-effective solution compared to DBR installation.

References

1. Saad H, Guillaud X, Mahseredjian J, Dennetire S, Nguefeu S (2015) MMC capacitor voltage decoupling and balancing controls. IEEE Trans Power Del 30(2):704–712. https://doi.org/10.1109/TPWRD.2014.2338861
2. Li W, Zhu M, Chao P, Liang X, Xu D (2020) Enhanced FRT and postfault recovery control for MMC-HVDC connected offshore wind farms. IEEE Trans Power Syst 35(2):1606–1617. https://doi.org/10.1109/TPWRS.2019.2944940
3. Saad H, Ould-Bachir T, Mahseredjian J, Dufour C, Dennetire S, Nguefeu S (2015) Real-time simulation of MMCs using CPU and FPGA. IEEE Trans Power Electron 30(1):259–267. https://doi.org/10.1109/TPEL.2013.2282600
4. Yao W, Liu C, Fang J, Ai X, Wen J, Cheng S (2020) Probabilistic analysis of commutation failure in LCC-HVDC system considering the CFPREV and the initial fault voltage angle. IEEE Trans Power Del 35(2):715–724. https://doi.org/10.1109/TPWRD.2019.2925399

5. Zhou H, Yao W, Ai X, Li D, Wen J, Li C (2022) Comprehensive review of commutation failure in HVDC transmission systems. Electr Power Syst Res 205:107768. https://doi.org/10.1016/j.epsr.2021.107768
6. Deng F, Chen Z (2013) Operation and control of a DC-Grid offshore wind farm under DC transmission system faults. IEEE Trans Power Del 28(3):1356–1363. https://doi.org/10.1109/TPWRD.2013.2261561

Chapter 4
Enhancement Method for Grid-Side Fault Ride-Through Capability Based on Multi-mode Matching

Abstract To address onshore AC faults in MMC-HVDC integrated offshore wind farms (OWF-MMC-HVDC), this chapter introduces a Multi-mode Matching Control (MMMC) strategy, which operates across three distinct modes. In Mode I, Active Energy Control is deployed during severe faults when the MMC's ability to output reactive power is constrained. This mode utilizes the energy stored in the MMC submodules (SMs) to absorb surplus offshore power, thereby preventing DC overvoltage. In Mode II, Dynamic Reactive Current Optimization (DRO) Control is applied during moderate faults. This mode dynamically adjusts the MMC's active and reactive current commands within acceptable limits, providing voltage support to the onshore grid and maximizing offshore active power output, which in turn reduces surplus offshore power. In Mode III, Energy Coordinated Synchronous (ECS) Control is implemented for milder faults. This mode simultaneously manages the energy absorption rates of both the WFMMC and the GSMMC, effectively using the energy margins of the MMCs to support the onshore grid while maintaining fault ride-through capability. The accuracy and effectiveness of the MMMC strategy are validated through simulations conducted on the PSCAD/EMTDC platform.

4.1 System Structure and Control Framework of OWF-MMC-HVDC

4.1.1 System Structure and Control Basics

In line with Chapter III, the typical OWF-MMC-HVDC structure is depicted in Fig. 4.1 [1]. Both WFMMC and GSMMC utilize half-bridge SMs (HBSM). For the onshore GSMMC, the AC control loop regulates internal energy E_G and reactive power Q_G, while the DC control loop manages DC voltage U_G. For the offshore WFMMC, the AC control loop supplies grid-connected voltage U_W to the OWFs, and the DC control loop oversees internal energy E_W. These controlled quantities are maintained at the commanded values via PI control.

W. Yao et al., *Active Control of Large-Scale Offshore Wind Farms Connected Via VSC-HVDC*, Power Systems, https://doi.org/10.1007/978-981-97-9346-4_4

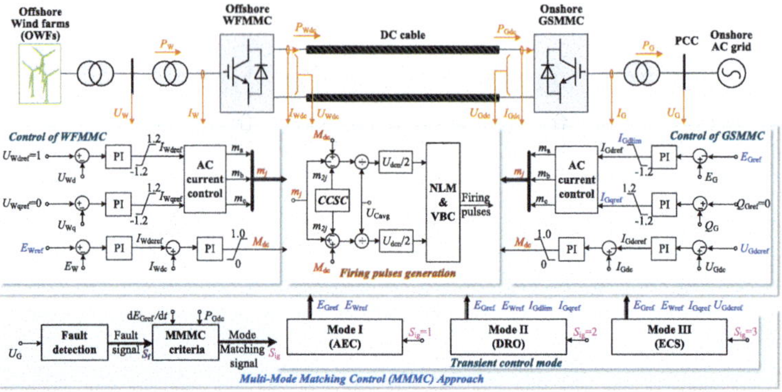

Fig. 4.1 System structure and control framework with multi-mode matching control (MMMC) of OWF-MMC-HVDC

$$\begin{cases} I_{\text{Gdref}} = k_{\text{p}}(E_{\text{G}} - E_{\text{Gref}}) + k_{\text{i}} \int (E_{\text{G}} - E_{\text{Gref}})dt \\ I_{\text{Gdcref}} = k_{\text{p}}(U_{\text{G}} - U_{\text{Gref}}) + k_{\text{i}} \int (U_{\text{G}} - U_{\text{Gref}})dt \\ I_{\text{Wdcref}} = k_{\text{p}}(E_{\text{Wref}} - E_{\text{W}}) + k_{\text{i}} \int (E_{\text{Wref}} - E_{\text{W}})dt \end{cases} \tag{4.1}$$

The number of SMs engaged by the upper and lower arms ($N_{\text{p}j}$ and $N_{\text{n}j}$ respectively) of each MMC phase in the firing pulse generation link is

$$\begin{cases} N_{\text{p}j} = Round\left(\frac{U_{\text{dcn}}}{2U_{\text{Cavg}}}(M_{\text{dc}} - m_j - m_{2j})\right) \\ N_{\text{n}j} = Round\left(\frac{U_{\text{dcn}}}{2U_{\text{Cavg}}}(M_{\text{dc}} + m_j - m_{2j})\right) \end{cases} \tag{4.2}$$

where, U_{dcn} is the rated DC voltage, U_{Cavg} is the average SM capacitor voltage, M_{dc} is the DC current modulation ratio, m_j is the AC current modulation ratio, m_{2j} is the circulating current suppression control modulation ratio and $Round(x)$ is the rounding function.

Further since $m_{2j} \ll m_j$, the small AC component m_j can be ignored. And considering that the three phases are balanced [2], according to (4.2) we can obtain:

$$U_{\text{dc}} = (N_{\text{p}j} + N_{\text{n}j})U_{\text{Cavg}} = M_{\text{dc}}U_{\text{dcn}} \tag{4.3}$$

It is evident that the MMC DC-side voltage U_{dc} can be regulated by adjusting M_{dc}, with U_{dc} being decoupled from U_{Cavg}. This means that, unlike traditional MMC control, the three-dimensional control enables independent regulation of the DC voltage and the SM capacitor voltage (or energy).

The total energy stored in the MMC's SM capacitors can be determined using (4.4) [3], where E_{MMC} denotes the voltage across the SM capacitors, C_{SM} represents the capacitance of each SM capacitor, and N signifies the number of SMs per arm.

$$E_{\text{MMC}} = \frac{1}{2}C_{\text{SM}}U_{\text{SM}}^2 \times 6N \tag{4.4}$$

Thus, setting the energy command value (E_{Gref} and E_{Wref} both set to 1 pu) during normal operation ensures that the OWFs output closely matches the active power output from the onshore GSMMC to the grid. During faults, leveraging the over-voltage capability of the SMs allows for controlled adjustment of E_{Gref} and E_{Wref}, facilitating the absorption of excess offshore power by the WFMMC and GSMMC.

4.1.2 Multi-mode Matching Montrol (MMMC) Framework

The MMMC strategy enhances the fault ride-through capability and voltage support of the OWF-MMC-HVDC system during transients. Figure 4.1 illustrates its three key components: fault detection, MMMC criteria, and transient control modes.

- Fault Detection Module: This module continuously monitors the PCC voltage to detect any deviations that indicate a fault. When a fault is detected, the module sends a signal to the MMMC criteria module.
- MMMC Criteria Module: Upon receiving the fault signal, this module determines the appropriate transient control mode by evaluating factors such as the rate of change in internal energy and the DC power received by the GSMMC.
- Transient Control Mode Module: Mode I: Activated during severe faults, this mode uses the SM capacitor voltage margin to absorb excess offshore power, thereby improving fault ride-through capability. Mode II: Applied during moderate faults, this mode dynamically adjusts the d-axis and q-axis current commands of the GSMMC, enhancing both fault ride-through capability and voltage support for the onshore grid. Mode III: Used for milder faults, this mode facilitates easier fault ride-through by converting the MMC energy margin, designated for partial fault ride-through, into additional voltage support for the onshore grid, thereby enhancing overall voltage support while maintaining fault ride-through capability.

4.2 Multi-mode Matching Control (MMMC) Approach

4.2.1 Fault Detection

This study leverages the fault detection approach from the commutation failure prevention (CFPREV) mechanism in line-commutated converter (LCC) based high voltage direct current (HVDC) systems [4]. The proposed fault detection module integrates two key components: a zero sequence detector designed to identify asymmetric faults (i.e., (4.5)), and a three-phase fault detector utilizing the $abc - \alpha\beta$ transformation method (i.e., (4.6)). A fault condition is determined when the computed values of (4.5) or (4.6) exceed a predefined threshold.

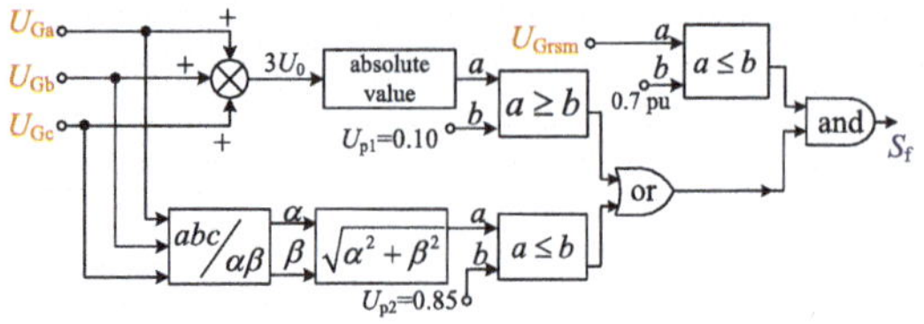

Fig. 4.2 Fault detection module

$$3U_0 = U_{Ga} + U_{Gb} + U_{Gc} \tag{4.5}$$

$$\begin{cases} |U_{\alpha\beta}| = \sqrt{U_\alpha^2 + U_\beta^2} \\ U_\alpha = \frac{2}{3}U_{Ga} - \frac{1}{3}(U_{Gb} + U_{Gc}) \\ U_\beta = \frac{\sqrt{3}}{3}(U_{Gb} - U_{Gc}) \end{cases} \tag{4.6}$$

Additionally, according to Ref. [5], for very minor faults (where the PCC voltage root mean square (RMS) U_{Grms} exceeds 0.7 pu), transient control actions are unnecessary, and fault ride-through can be achieved solely by utilizing the internal energy of the GSMMC. Hence, the fault detection module, depicted in Fig. 4.2, sets U_{p1} and U_{p2} to 0.1 and 0.85 respectively, enabling fault detection typically within 5 ms [6].

4.2.2 MMMC Criteria

the maximum short-term withstand voltage of the SM (U_{SMmax}) is specified as 1.5 times U_{SM0}. This setting ensures that SM capacitor overvoltage does not trigger hardware protection mechanisms. Based on (4.4), the available energy margin for recovering surplus power for each MMC is calculated as follows.

$$\Delta E_{MMC} = \frac{E_{max} - E_{MMC0}}{E_{MMC0}} = (U_{SMmax}^2 - 1) = 1.25\text{pu} \tag{4.7}$$

where, E_{max} is the MMC short-time maximum allowable energy and E_{MMC0} is the MMC rated energy.

According to design principles for MMC capacitance, the energy storage time constant T_{MMC} in engineering converter stations typically ranges from 30 to 60 ms [7]. This means that the rated energy capacitance E_{MMC0} of a single MMC, discharged at rated power P_{MMC0}, can sustain continuous discharge for $T_{MMC} = 30 \sim 60$ ms, as described by (4.8).

$$T_{MMC} = \frac{E_{MMC0}}{P_{MMC0}} = 30 \sim 60\,\text{ms} \tag{4.8}$$

Typically, the rated power of OWFs equals the rated power of the MMC (i.e., $P_{WF0} = P_{MMC0}$). Therefore, under the most severe fault scenario (e.g., a three-phase ground fault at the GSMMC outlet PCC bus) and with OWF output at rated power, the minimum duration available for the energy absorption process in the dual-terminal MMC-HVDC system can be expressed as

$$\Delta t_{\min} = 2 \times 1.25 \times T_{\mathrm{MMC}} = 75 \sim 150\,\mathrm{ms} \tag{4.9}$$

thus, approximate 100 ms energy absorption can be achieved using only the energy margin of the MMC.

If the fault level and the OWF state remain constant during the fault, the rate of energy increase in the GSMMC can be calculated.

$$\frac{\mathrm{d}E_{\mathrm{G}}}{\mathrm{d}t} = \frac{\Delta E_{\mathrm{MMC}}}{\Delta t} = \frac{1.25}{1.25E_{\mathrm{MMC0}}/(P_{\mathrm{Gdc}} - P_{\mathrm{G}})} \tag{4.10}$$

where, P_{Gdc} and P_{G} represent the DC power and AC power of the GSMMC, respectively. The severity of the fault can be assessed based on the rate at which GSMMC absorbs energy. Under the most severe operating conditions, $P_{\mathrm{G}} = 0$ and $P_{\mathrm{Gdc}} \approx P_{\mathrm{WF0}}$. Therefore, according to (10), the maximum energy absorption rate of the GSMMC can be expressed as:

$$\left.\frac{\mathrm{d}E_{\mathrm{G}}}{\mathrm{d}t}\right|_{\max} = \frac{P_{\mathrm{Gdc}}}{E_{\mathrm{MMC0}}} \tag{4.11}$$

Upon fault detection and before completing the control mode alignment, the GSMMC energy command value (E_{Gref}) is set equal to the measured energy value (E_{G}). This facilitates fault detection based on the rate of change of the energy command value $\frac{\mathrm{d}E_{\mathrm{Gref}}}{\mathrm{d}t}$.

During severe fault conditions, where the PCC voltage is significantly low and reactive support is challenging, all available current margins should be utilized for active power output. Hence, Control Mode I should be activated when the following condition is met:

$$\left.\frac{\mathrm{d}E_{\mathrm{Gref}}}{\mathrm{d}t}\right|_{t=t_{\mathrm{f}}+5ms} \geq 90\% \frac{P_{\mathrm{Gdc}}}{E_{\mathrm{MMC0}}} \tag{4.12}$$

At the moment of receiving the fault signal S_{f}, denoted as t_{f}, a delay of 5 ms is implemented to mitigate unstable transient energy fluctuations at the onset of the fault.

Typically, AC faults can be cleared within 100 to 120 ms. With a conservative approach, when the energy absorption duration exceeds 200 ms, a significant energy margin remains within the MMC after completing fault ride-through. Mode III is designed to utilize a portion of this energy margin during low fault levels to further support onshore AC voltage. This control mode adjusts the MMCs' energy absorption rate such that, after 200 ms, both GSMMC and WFMMC reach their maximum energy $E_{\max}$. By increasing the total energy absorption rate of GSMMC and WFMMC, the active power output of GSMMC decreases while reactive power output increases. The activation condition for Mode III is:

$$\left.\frac{\mathrm{d}E_{\mathrm{Gref}}}{\mathrm{d}t}\right|_{t=t_{\mathrm{f}}+5ms} \leq \frac{2\Delta E_{\mathrm{MMC}}}{200 \times 10^{-3}} \tag{4.13}$$

When the combined energy absorption time of GSMMC and WFMMC is less than 200 ms, maximizing fault ride-through time becomes crucial, emphasizing the output of active power from GSMMC. Control Mode II achieves this by optimizing the relationship between active (d-axis) and reactive (q-axis) currents of GSMMC. Therefore, the activation criteria for Control Mode II are as follows.

$$\frac{2\Delta E_{\mathrm{MMC}}}{200 \times 10^{-3}} < \left.\frac{\mathrm{d}E_{\mathrm{Gref}}}{dt}\right|_{t=t_{\mathrm{f}}+5ms} < 90\% \frac{P_{\mathrm{Gdc}}}{E_{\mathrm{MMC0}}} \tag{4.14}$$

Therefore, (4.12)–(4.14) collectively define the MMMC criteria. The range of control mode selections is depicted in Fig. 4.3, while the MMMC criteria module is illustrated in Fig. 4.4. During normal operation, the mode selection signal S_{ig} remains at 0, but during transients, S_{ig} switches to 1, 2, or 3 to activate Modes I, II, or III, respectively. Each transient involves the activation of only one control mode, and the three modes operate independently without interaction during a single transient event.

Notably, the DC power received by GSMMC determines the selection of the transient control mode, while the state of OWFs influences the DC power transmission. Therefore, the transient control modes of the system may vary depending on the operational state of OWFs, reflecting differences in offshore system capabilities to support onshore voltage.

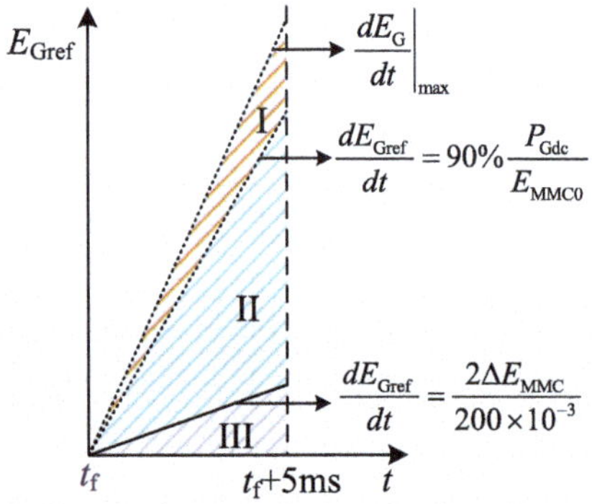

Fig. 4.3 Matching range of three transient control modes

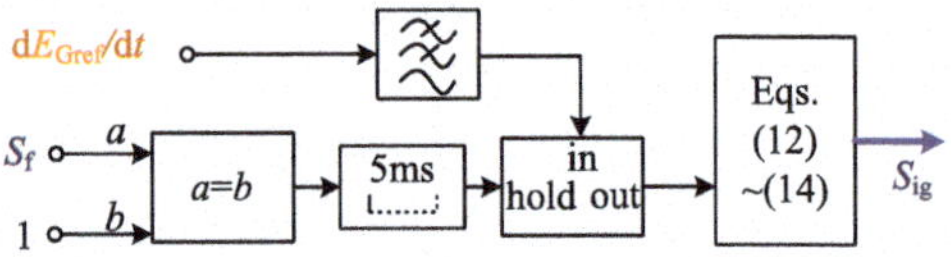

Fig. 4.4 MMMC criteria module

4.2.3 Control Mode I (AEC)

Control Mode I (AEC) remains consistent with the control strategy outlined in Chap. 3. Under normal operating conditions, the maximum AC output power of the MMC typically does not exceed 1.2 pu. This implies that the output power of GSMMC should be kept below 1.2 pu. Additionally, during energy discharge from WFMMC, precautions are necessary to prevent excessive rise in GSMMC energy due to high power on the DC side. Therefore, the maximum allowable DC output power of WFMMC is also limited to 1.2 pu. Consequently, both GSMMC and WFMMC are constrained to a maximum output power of 1.2 pu. Therefore, under rated operating conditions of the wind farm, the energy discharge rates of GSMMC and WFMMC can be expressed as:

$$\begin{aligned} k_{\mathrm{re}} &= \frac{\mathrm{d}E_{\mathrm{MMC}}}{\mathrm{d}t} = \frac{S_{\mathrm{N}}}{E_{\mathrm{MMC0}}}(P_{\mathrm{in}} - P_{\mathrm{out}}) \\ &\geq \frac{S_{\mathrm{N}}}{E_{\mathrm{MMC0}}}(1 - 1.2) = -0.2\frac{S_{\mathrm{N}}}{E_{\mathrm{MMC0}}} \end{aligned} \tag{4.15}$$

where, S_{N} is the rated capacity of the MMC, and P_{in} and P_{out} are the input and output power of the MMC, respectively.

4.2.4 Control Mode II (DRO)

The AEC control mode does not consider the impact of reactive currents throughout the fault ride-through process. The active power and reactive power output by GSMMC to the AC grid during the transient period can be expressed as (4.16).

$$\begin{cases} P_{\mathrm{G}} = \frac{3}{2}U_{\mathrm{G}}I_{\mathrm{Gd}} + p_{\mathrm{c2}}\cos(2\omega t) + p_{s2}\sin(2\omega t) \\ Q_{\mathrm{G}} = -\frac{3}{2}U_{\mathrm{G}}I_{\mathrm{Gq}} + q_{\mathrm{c2}}\cos(2\omega t) + q_{\mathrm{s2}}\sin(2\omega t) \end{cases} \tag{4.16}$$

where, ω represents the AC system angular frequency. p_{s2}, p_{c2}, q_{s2}, and q_{c2} denote coefficients of the sine and cosine components for active and reactive power, respectively. These coefficients are all zero under symmetric system operation.

It is evident that P_{G} is influenced by both active current and PCC voltage. Conversely, PCC voltage is influenced by reactive current. Therefore, neglecting the impact of reactive current can hinder achieving maximum power output from GSMMC.

The maximum total current limit (I_{lim}) of MMC typically ranges from 1.2 to 1.5 pu, set to 1.2 pu in this study. Hence, the reactive current command value (I_{Gdref}) during the transient period can be determined using (4.17).

$$I_{\mathrm{Gqref}} = \sqrt{I_{\mathrm{lim}}^2 - I_{\mathrm{Gdref}}^2} \tag{4.17}$$

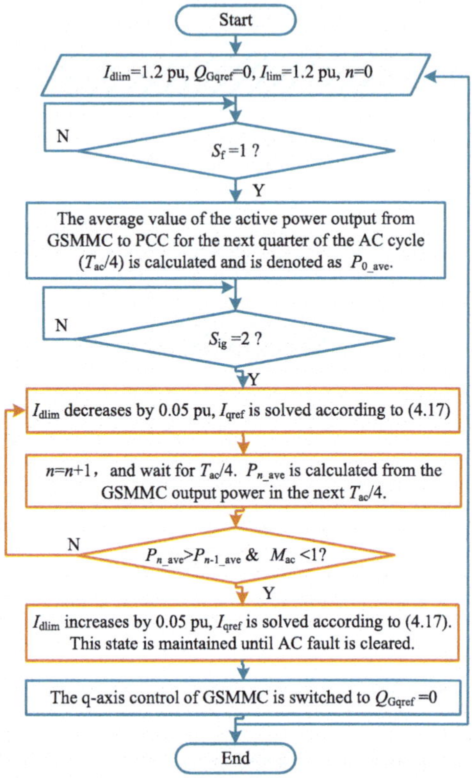

Fig. 4.5 DRO algorithm flow chart

The DRO still employs the control framework of the AEC. Reactive current output is optimized solely during the energy absorption stage of GSMMC. The flowchart for DRO is depicted in Fig. 4.5.

The GSMMC operates normally until a fault signal is received. The d-axis current upper limit of GSMMC is $I_{dlim} = 1.2$ pu, and the number of DRO iterations is set to $n = 0$. Upon receiving the fault signal, the active power output of GSMMC is collected for a quarter of the AC cycle (i.e., $T_{ac}/4$), and its average value, denoted as P_{0_ave}, is calculated. The choice of $T_{ac}/4$ is based on (4.16), where the active power output from MMC under an asymmetric fault includes a second harmonic component. $T_{ac}/4$ precisely corresponds to half of the second harmonic cycle, effectively reflecting the average power output within one second harmonic cycle. Upon receiving the DRO start signal ($S_{ig} = 2$), dynamic optimization of reactive current begins. The optimization process unfolds in the following four steps:

- Step 1: Decrease the active current upper limit value I_{dlim} by 0.05 pu. Solve for the reactive current command value I_{qref} using (4.17).
- Step 2: Increment n by 1 and wait for $T_{ac}/4$. Calculate P_{n_ave} from the GSMMC output power in the subsequent $T_{ac}/4$ period.

- Step 3: Compare the active power output and the AC modulation ratio M_{ac} of GSMMC in the current optimization cycle with those in the previous cycle. If $P_{n\_\mathrm{ave}} > P_{n-1\_\mathrm{ave}}$, it indicates that the current I_{qref} allows GSMMC to produce more active power. Return to step 1 to continue optimizing I_{qref}. If $P_{n\_\mathrm{ave}} \leq P_{n-1\_\mathrm{ave}}$, retain the I_{qref} from the previous optimization cycle and proceed to step 4. Additionally, ensure $M_{\mathrm{ac}} < 1$ to prevent over modulation.
- Step 4: Increase I_{dlim} by 0.05 pu. Solve for I_{qref} using (4.17). Maintain this state until the AC fault is cleared.

After the fault is cleared, GSMMC resumes normal operation with $Q_{\mathrm{Gref}} = 0$. All other processes in Control Mode II (such as active energy maintenance, active energy release, etc.) remain fully inhibited as in Control Mode I.

4.2.5 Control Mode III (ECS)

When $S_{\mathrm{ig}} = 3$ is selected, it indicates that the GSMMC and WFMMC have enough energy margin to absorb surplus power for more than 200 ms under the fault condition. The principle of Energy Support Control (ESC) is to adjust the energy absorption rate of the GSMMC and WFMMC precisely to absorb exactly 200 ms of surplus power. This approach has dual benefits: firstly, it increases the overall power absorption capability of the MMCs; secondly, it allows GSMMC to reduce its active current output, thereby increasing its reactive current output to support AC voltage. Moreover, ensuring a fault ride-through time of 200 ms is sufficient to clear AC faults completely.

In (4.13), when the rate of change of energy takes its maximum value, i.e.

$$\frac{\mathrm{d}E_{\mathrm{Gref}}}{\mathrm{d}t} = \frac{\Delta E_{\mathrm{MMC}}}{100 \times 10^{-3}} \tag{4.18}$$

This implies that within an OWF-MMC-HVDC system, GSMMC can absorb precisely 100 ms of surplus power at this energy change rate. Activating WFMMC simultaneously to absorb surplus power alongside GSMMC can achieve a system-wide absorption duration of 200 ms if the energy change rate of both GSMMC and WFMMC satisfies (4.19).

$$\begin{cases} K_{\mathrm{e}} = \frac{\mathrm{d}E_{\mathrm{Gref}}}{dt} = \frac{\mathrm{d}E_{\mathrm{Wref}}}{dt} = \frac{\Delta E_{\mathrm{MMC}}}{200 \times 10^{-3}} (P_{\mathrm{Gdc}} > P_{\mathrm{Gdc\_t}}) \\ K_{\mathrm{e}} = \frac{\mathrm{d}E_{\mathrm{Gref}}}{dt} = \frac{\mathrm{d}E_{\mathrm{Wref}}}{dt} = \frac{P_{\mathrm{Gdc}}}{P_{\mathrm{Gdc\_t}}} \frac{\Delta E_{\mathrm{MMC}}}{200 \times 10^{-3}} (P_{\mathrm{Gdc}} \leq P_{\mathrm{Gdc\_t}}) \\ P_{\mathrm{Gdc\_t}} = \frac{2 \times 1.25 E_{\mathrm{MMC0}}}{200 \times 10^{-3}} \end{cases} \tag{4.19}$$

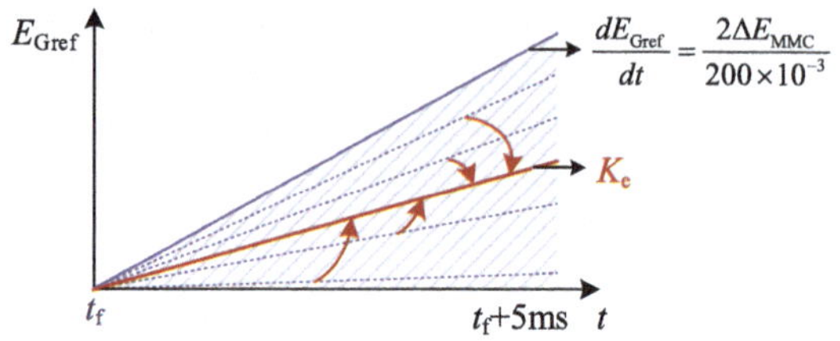

Fig. 4.6 Energy change rate of GSMMC and WFMMC in ECS

where, P_{Gdc_c} is the critical DC power of the GSMMC. If the OWF output power is small, the value of K_e needs to be adjusted. The P_{Gdc_c} prevents the power absorption rate of GSMMC and WFMMC from being greater than the output power of the OWFs.

Therefore, when control mode III is applied, the energy change rate of both GSMMC and WFMMC should be adjusted to meet the criteria specified in (4.19). The schematic representation is illustrated in Fig. 4.6.

Since only the onshore GSMMC can detect the fault severity, communication-free transmission of energy control commands can be facilitated using the offshore DC voltage [8]. When the ECS is activated, the DC voltage command value of the GSMMC is set to

$$U_{Gdcref} = (1 + k_{GW})U_{dc0} \tag{4.20}$$

where, k_{GW} is the voltage control factor and U_{dc0} is the nominal DC link voltage. Considering that the DC link voltage generally does not exceed 1.1 pu, and to avoid the value of k_{GW} being too small to trigger control errors, the value of k_{GW} in this chapter is 0.08 pu.

The parameter k_{GW} serves as the voltage control factor, while U_{dc0} represents the nominal DC link voltage. Typically, the DC link voltage does not surpass 1.1 pu. To prevent k_{GW} from becoming excessively small, which could lead to control errors, it is set to 0.08 pu in this study.

For the WFMMC, estimating the DC voltage of the GSMMC involves applying Kirchhoff's voltage law, which determines the change in DC voltage.

$$\begin{cases} U_{Gdc\_E} = U_{Wdc} + I_{Gdc}Z_{cab} \\ \Delta U_{dc\_E} = U_{Gdc\_E} - U_{dc0} \end{cases} \tag{4.21}$$

where, U_{Gdc_E} represents the estimated DC voltage of the GSMMC by the WFMMC, U_{Wdc} denotes the DC voltage measured by the WFMMC, I_{Wdc} is the DC current measured by the WFMMC, Z_{cab} stands for the impedance of the submarine cable, and ΔU_{dc_E} signifies the estimated DC voltage variation. When the estimated ΔU_{dc_E} satisfies the condition specified in Eq. (4.14) consistently over a sustained period of 2 ms, the WFMMC initiates active energy absorption control. This control mode allows it to absorb surplus power in accordance with the energy change rate described

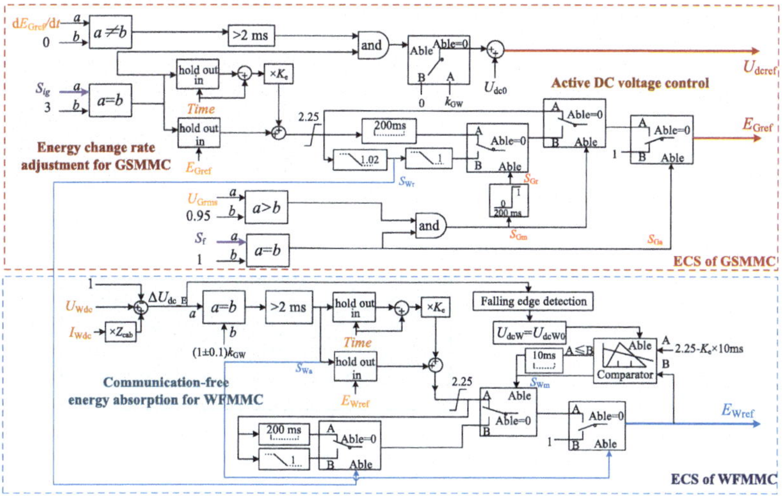

Fig. 4.7 Control structure of control mode III (ECS)

in (4.19). This approach enables coordinated synchronous control of the WFMMC and GSMMC without the need for communication.

$$\Delta U_{\mathrm{dc\_E}} = (1 \pm 0.1)k_{\mathrm{GW}} \tag{4.22}$$

The overall control block diagram for the ECS is depicted in Fig. 4.7. In this diagram, $S_{\mathrm{Gs,m,r}}$ and $S_{\mathrm{Ws,m,r}}$ denote the start signals for active energy recovery, maintenance, and release for GSMMC and WFMMC, respectively. ECS activation occurs upon receipt of $S_{\mathrm{ig}} = 3$. The GSMMC absorbs energy based on the energy change rate in (4.2), simultaneously triggering the WFMMC to initiate energy absorption via voltage control.

Upon fault clearance or when the energy command value of the GSMMC reaches 2.25 pu, the energy command value change rate of the GSMMC becomes 0. As the WFMMC begins energy absorption approximately 10 ms after the GSMMC, residual energy remains in the WFMMC when the GSMMC energy reaches 2.25 pu. After the WFMMC detects the DC voltage returning to its rated state, it compares E_{Wref} with $2.25 - k_{\mathrm{e}} \times 10\,\mathrm{ms}$. If $2.25 - k_{\mathrm{e}} \times 10\,\mathrm{ms} \leq \mathrm{E}_{\mathrm{Wref}}$, it suggests the GSMMC has likely reached its energy limit without fault clearance. In this scenario, the WFMMC continues absorbing energy for an additional 10 ms to maximize its internal energy margin. Conversely, if the GSMMC stops energy recovery due to fault clearance, the WFMMC immediately enters the energy maintenance state. Subsequent stages of active energy maintenance and release align with control modes I and II.

During the transient recovery stage after AC fault clearance, the AC grid undergoes a brief period of recovery where voltage and power levels have not yet stabilized. To prevent detrimental effects on grid recovery, a 200 ms delay is set before releasing power back to the grid. This delay ensures the AC grid returns to a relatively stable state. The delay blocks in Fig. 4.7 process single-byte signals, eliminating the need for high-performance processors and sensors due to their low bandwidth requirements.

4.3 Simulation Verification

The OWF-MMC-HVDC model depicted in Fig. 4.1 is implemented using PSCAD/EMTDC v4.5. The AC system parameters on the inverter side follow those defined in the CIGRE-HVDC model [9, 10]. Key parameters are detailed in Tables 4.1 and 4.2, which includes the PI parameters for both the AC and DC loops of GSMMC and WFMMC. In the AC loop, the d-axis parameters are identical to the q-axis parameters.

Given the system parameters and Eqs. (4.8)–(4.11), the MMC energy storage time constant is $T_{\text{MMC}} = 40$ ms. The minimum time for the MMC-HVDC system to

Table 4.1 Main parameters of OWF-MMC-HVDC

Parameters	Value
Rated DC voltage U_{dc0}	±320 kV
Rated DC transmission power P_{dc0}	1200 MW
Rated capacity of MMC S_{N}	1200 MW
Number of SMs per arm N	320
Rated capacitance of SM C_{sm}	12500 μF
Rated voltage of SM U_{sm}	2 kV
Rated power of OWFs P_{W0}	1200 MW
Rated voltage of onshore PCC bus P_{G0}	230 kV
Modulation ratio of GSMMC M_{ac}	0.85
Rated AC frequency f_{ac}	50 Hz

Table 4.2 PI control parameters for the AC and DC loops of GSMMC and WFMMC

Control parameters	Value
AC outer loop for GSMMC $k_{\text{GAop}}, k_{\text{GAoi}}$	4, 400
AC inner loop for GSMMC $k_{\text{GAip}}, k_{\text{GAii}}$	1, 100
DC outer loop for GSMMC $k_{\text{GDip}}, k_{\text{GDii}}$	4, 100
DC inner loop for GSMMC $k_{\text{GDip}}, k_{\text{GDii}}$	2, 100
AC outer loop for WFMMC $k_{\text{WAop}}, k_{\text{WAoi}}$	2, 100
AC inner loop for WFMMC $k_{\text{WAip}}, k_{\text{WAii}}$	0.5, 50
DC outer loop for GSMMC $k_{\text{GDip}}, k_{\text{GDii}}$	15, 400
DC inner loop for GSMMC $k_{\text{GDip}}, k_{\text{GDii}}$	2, 100

Table 4.3 Control mode matching range

Control mode	dE_{Gref}/dt
Mode I (AEC)	[0.0225, 0.025) pu/ms
Mode II (DRO)	(0.0125, 0.0225) pu/ms
Mode III (ECS)	[0, 0.0125] pu/ms

absorb surplus power is $\Delta t_{\min} = 100$ ms, and the maximum energy recovery rate is $dE_G/dt = 0.025$ pu/ms. Based on (4.15), the release rate for the three control modes must meet $k_{\mathrm{re}} \le -0.005$ pu/ms. Therefore, k_{re} is set to -0.002 pu/ms. According to Eqs. (4.12)–(4.14), the activation ranges for the three control modes are outlined in Table 4.3.

4.3.1 Characteristics Comparison of the Three Control Modes

To compare the characteristics of different control modes, this section employs control modes I, II, and III (i.e., AEC as proposed in [5], DRO, and ECS) for transient control under fault conditions, rather than the proposed MMMC. Simulation setup: A three-phase-to-ground fault occurs in the onshore AC grid at 4 s, causing the PCC bus voltage to drop to 0.5 pu. The fault is cleared at 4.1 s. The grid characteristics for the three control modes are illustrated in Fig. 4.8.

From Fig. 4.8*:*

- Based on the MMMC criterion, the energy change rate at the sampling point is determined to be 0.0099 pu/ms, making control mode III (ECS) the appropriate choice.
- In all three control modes, the energy levels in the GSMMC and WFMMC remain below the 2.25 pu threshold, ensuring effective fault ride-through for the OWF-MMC-HVDC system. ECS optimizes the energy margins of both GSMMC and WFMMC while maintaining fault ride-through capability.
- Compared to the AEC control method discussed in [5], both DRO and ECS controls improve PCC voltage. ECS stands out by using both GSMMC and WFMMC to absorb offshore power simultaneously, achieving the highest total absorption rate among the three modes. This minimizes the active current output from GSMMC to the PCC, allowing for greater reactive power output and, therefore, the highest level of voltage support.
- DRO focuses on optimizing active and reactive currents, maximizing the GSMMC's active power output. Consequently, after fault ride-through, the remaining energy margin between GSMMC and WFMMC is largest under DRO control. This makes DRO more effective in supporting onshore AC voltage and reducing offshore surplus power compared to AEC.

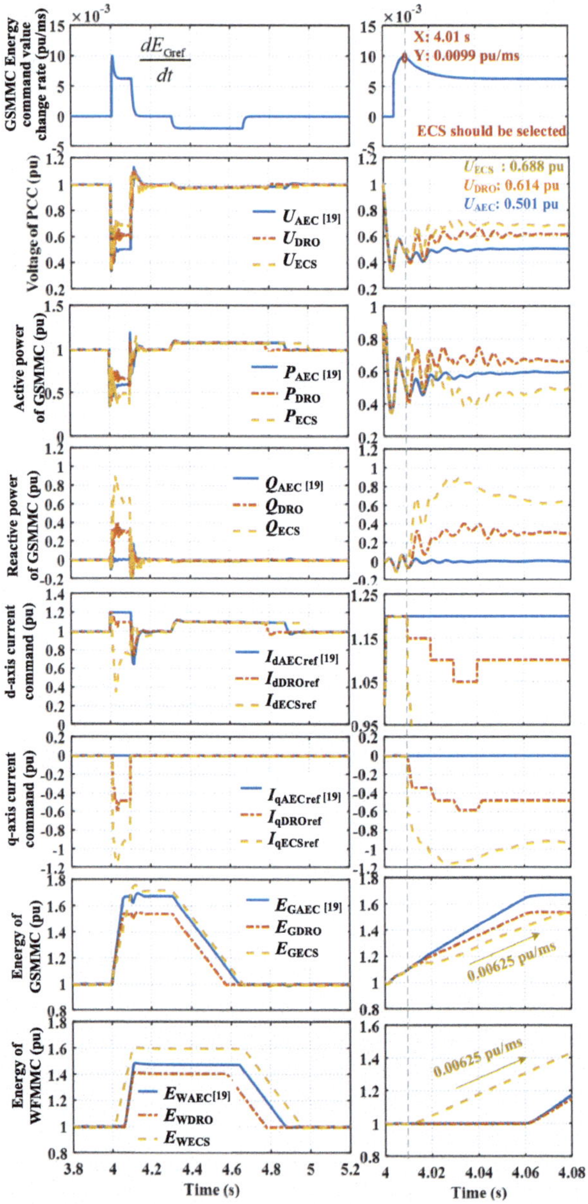

Fig. 4.8 Characteristics comparison of the three control modes

- DRO successfully optimizes active and reactive currents, while ECS effectively controls the energy rise rate of WFMMC and GSMMC to 0.00625 pu, in line with (4.19). This confirms the accuracy and effectiveness of both DRO and ECS.
- During the energy maintenance phase, the GSMMC's active power output to the AC grid matches the incoming DC power. If the fault is not cleared, the GSMMC continues to supply reactive power to support PCC voltage under DRO or ECS. In contrast, under AEC, the offshore system does not influence onshore voltage since the GSMMC does not output reactive power. Once the fault is cleared, the GSMMC ceases reactive power output, resulting in minimal impact on PCC voltage.

4.3.2 Control Characteristics of MMMC

The critical voltage for mode matching in the MMMC can be determined using Eqs. (4.12) through (4.16) when the OWF is operating at rated power. Additionally, the minimum energy recovery time for the MMMC across different modes can be plotted and compared with the minimum energy absorption time of the AEC, as discussed in [5], as shown in Fig. 4.9. The figure shows that DRO control can enhance PCC voltages, while ECS reduces the margin typically allocated for fault ride-through in AEC, thereby providing improved voltage support onshore.

To further verify the MMMC's mode-matching capability, a simulation was conducted where a three-phase-to-ground fault occurs in the onshore AC grid at 4 s, causing the PCC voltage to drop to 0 pu and 0.2 pu, respectively. The fault is cleared at 4.1 s, and the grid behavior after applying MMMC is shown in Fig. 4.10.

When the voltage drops to 0 pu, the MMMC detects an energy change rate of 0.0245 pu/ms, leading it to select Mode I. In this mode, the reactive current output command is set to 0, while the active current reaches its limit of 1.2 pu. When the voltage drops to 0.2 pu, the MMMC shifts to Mode II, optimizing both active and reactive currents. This adjustment increases the PCC voltage to 0.24 pu and boosts the active power output from the GSMMC during the transient period.

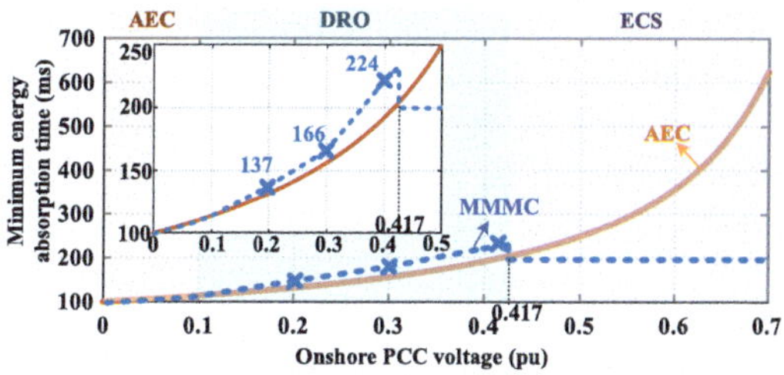

Fig. 4.9 Minimum energy absorption time of AEC and MMMC at different PCC voltages

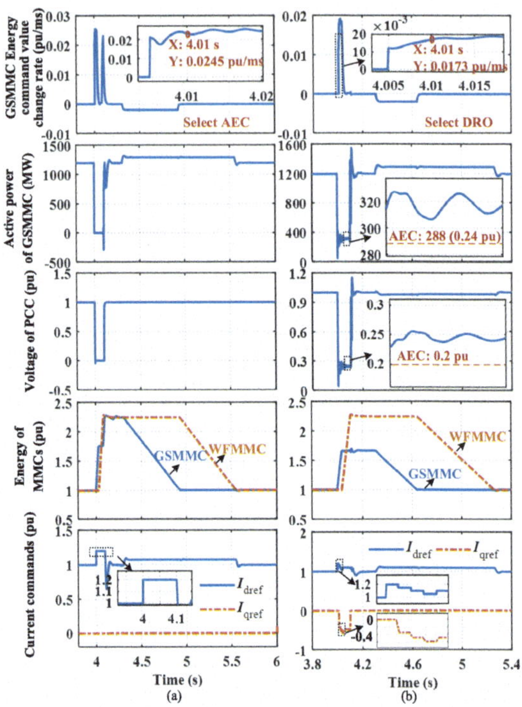

Fig. 4.10 MMMC response characteristics at different levels of voltage drops. **a** Voltage drops to 0. **b** Voltage drops to 0.2 pu

4.3.3 Multi-case Validation of MMMC

First, the performance of the MMMC under varying fault durations is assessed. In the simulation, a three-phase-to-ground fault occurs in the onshore AC grid at 4 s, causing the PCC bus voltage to drop to 0.6 pu. Two fault durations are tested: 200 ms and 50 ms. The simulation results are shown in Fig. 4.11.

For the 200 ms fault duration, the MMMC operates in Mode III, with both the WFMMC and GSMMC precisely managing their internal energies at 2.25 pu, effectively supporting the PCC voltage. During the 50 ms fault duration, the MMMC also responds effectively, maintaining a substantial energy margin between the GSMMC and WFMMC. This adaptability of the MMMC highlights its capability to adjust and stabilize the system under different fault durations.

Next, the MMMC's performance under varying OWF output conditions is evaluated. In this scenario, a three-phase-to-ground fault occurs in the onshore AC grid at 4 s, reducing the PCC bus voltage to 0.2 pu, with a fault duration of 100 ms. The OWFs are set to output 600 MW and 400 MW of active power, respectively, during steady state. The simulation results are depicted in Fig. 4.12.

Under this fault condition, the MMMC operates in Mode III. When the OWFs output lower power, the WFMMC and GSMMC adjust their absorption rates to ensure that the MMC absorbs power without exceeding the OWF output. As shown in Figs. 4.10b and 4.12, when the OWFs generate 1200 MW, the system switches to

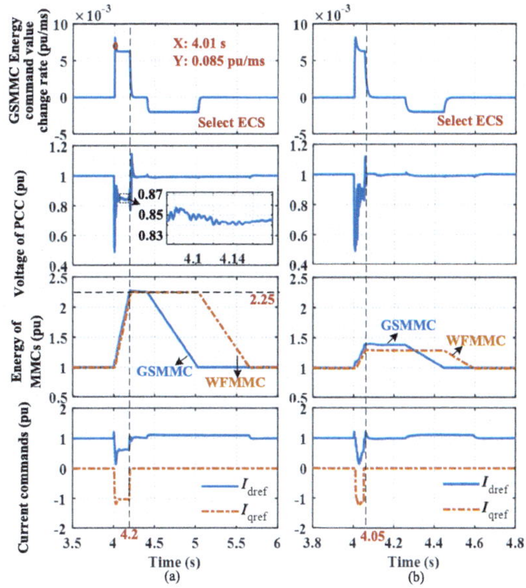

Fig. 4.11 MMMC response characteristics for different fault durations. **a** The fault lasts 200 ms. **b** The fault lasts 50 ms

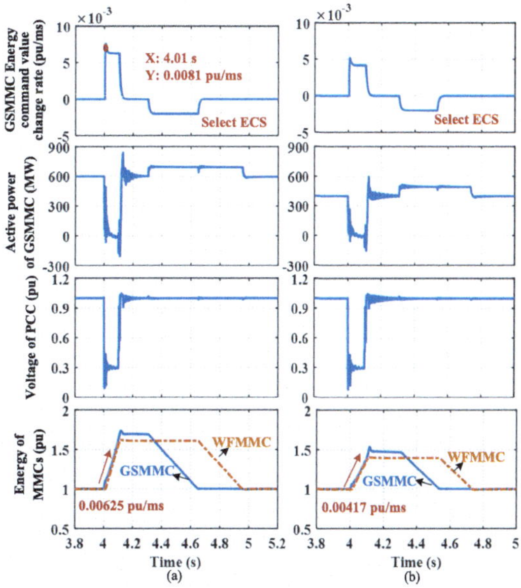

Fig. 4.12 MMMC response characteristics for different OWFs output states. **a** The OWFs outputs 600 MW active power. **b** The OWFs output 400 MW active power

DRO control mode, maintaining the PCC voltage around 0.24 pu. When the OWFs generate 600 MW and 400 MW, the system operates in ECS control mode, stabilizing the PCC voltage around 0.3 pu. This demonstrates the MMMC's ability to provide adaptive voltage support under different OWF output conditions.

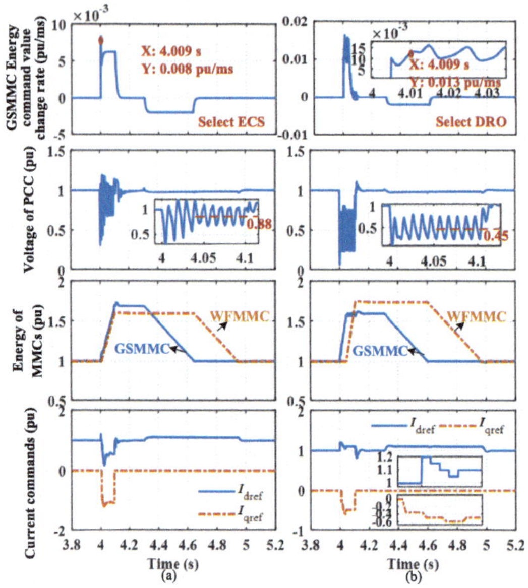

Fig. 4.13 MMMC response characteristics for different fault types. **a** Single-phase-to-ground fault. **b** phase-to-phase-to-ground fault

Lastly, the MMMC's performance under different fault types is tested. Specifically, a single-phase-to-ground fault and a phase-to-phase-to-ground fault are simulated on the PCC bus at 4 s, each lasting 100 ms. The results, shown in Fig. 4.13, confirm the MMMC's effectiveness in adapting to these faults.

After an asymmetric fault in the onshore AC grid, the MMMC selects the appropriate control mode. During a single-phase-to-ground fault, Mode III is activated, raising the average PCC voltage to about 0.88 pu. For a phase-to-phase-to-ground fault, Mode II is selected, boosting the average PCC voltage to approximately 0.45 pu.

The MMMC demonstrates strong performance across a variety of complex operating conditions, significantly enhancing the fault ride-through and voltage support capabilities of the OWF-MMC-HVDC system.

4.4 Summary

This chapter introduces a MMMC strategy specifically designed for the OWF-MMC-HVDC system. This strategy adapts to onshore AC faults by selecting one of three control modes based on the severity of the fault:

Control Mode I: AEC is used during severe faults. It taps into the energy storage capacity of MMC sub-module capacitors to absorb excess energy from the offshore system, ensuring fault ride-through for approximately 100 ms.

Control Mode II: DRO is activated in moderate fault conditions. DRO dynamically optimizes the active and reactive currents of the GSMMC to maximize power delivery to the grid, extending fault tolerance and enhancing AC voltage support beyond what AEC provides.

Control Mode III: ECS is applied during minor faults. In this mode, the OWF-MMC-HVDC system shows strong fault tolerance, with ECS using part of this capability to improve onshore AC voltage support while maintaining fault ride-through.

Simulation results validate that the MMMC strategy effectively handles a wide range of operating conditions, greatly improving the fault ride-through capability and voltage support of the OWF-MMC-HVDC system.

References

1. Pan E, Yue B, Li X, Zhao Z, Zhu Q (2020) Integration technology and practice for long-distance offshore wind power in China. Energy Convers Econ 1(1):4–19. https://doi.org/10.1049/enc2.12003
2. Zhang H, Xiang W, Wen J (2023) Dual grid-forming control with energy regulation capability of MMC-HVDC system integrating offshore wind farms and weak grids. IEEE Trans Power Syst 1–11. Early Accesshttps://doi.org/10.1109/TPWRS.2023.3244807
3. Zhou H, Yao W, Sun K, Zhao Y, Ai X, Wen J (2023) A multi-level coordinated DC overcurrent suppression scheme for symmetrical bipolar MMC-HVDC integrated offshore wind farms. Int J Electr Power Energy Syst 147:108880. https://doi.org/10.1016/j.ijepes.2022.108880
4. Zhou H, Yao W, Ai X, Li D, Wen J, Li C (2022) Comprehensive review of commutation failure in HVDC transmission systems. Electr Power Syst Res 205:107768. https://doi.org/10.1016/j.epsr.2021.107768
5. Zhou H, Yao W, Zhou M, Ai X, Wen J, Cheng S (2023) Active energy control for enhancing AC fault ride-through capability of MMC-HVDC connected with offshore wind farms. IEEE Trans Power Syst 38(3):2705–2718. https://doi.org/10.1109/TPWRS.2022.3179443
6. Tian D, Xiong X (2022) Corrections of original cfprev control in LCC-HVDC links and analysis of its inherent plateau effect. CSEE J Power Energy Syst 8(1):10–16. https://doi.org/10.17775/CSEEJPES.2020.05700
7. Xu Z (2016) Flexible DC transmission, 2nd edn. China Machine Press, Beijing
8. Zhang H, Xiang W, He Y, Wen J (2023) Optimal energy utilization of MMC-HVDC system integrating offshore wind farms for onshore weak grid inertia support. IEEE Trans Power Syst 1–14. Early Access https://doi.org/10.1109/TPWRS.2023.3239166
9. Zhou H, Yao W, Ai X, Zhang J, Wen J, Li C (2022) Coordinated power control of electrochemical energy storage for mitigating subsequent commutation failures of HVDC. Int J Electr Power Energy Syst 134: 107455. https://doi.org/10.1016/j.ijepes.2021.107455
10. Szechtman M, Wess T, Thio C (1991) A benchmark model for HVDC system studies. In: Proceedings of the international conference AC and DC power transmission, pp 374–378

Chapter 5
Characteristics Analysis and Suppression of Onshore Valve-Side Fault

Abstract A valve-side ground fault in the converter station of an MMC-HVDC system presents a significant risk to power system security. This chapter examines the prevalent single-phase-to-ground (SPG) fault characteristics on the valve side of onshore converter stations within OWF-MMC-HVDC systems. It explores how these onshore SPG faults propagate and impact offshore HVDC transmission lines, offshore converter stations, and wind farms, all within the framework of MMC control principles. To mitigate these risks, the chapter proposes an overvoltage suppression strategy that utilizes zero-sequence voltage modulation. The strategy's impact on the system's transient stability is analyzed, and key parameters are designed with a balance between practical engineering considerations and system performance. The theoretical analysis of fault characteristics and the effectiveness of the proposed overvoltage suppression strategy are validated using the PSCAD/EMTDC simulation platform. The simulation results closely match the theoretical analysis, demonstrating the strategy's effectiveness in mitigating overvoltage issues.

5.1 Valve-Side SPG Fault Characteristics and Propagation Mechanism of MMCs

The structure of a typical OWF-MMC-HVDC system is depicted in Fig. 5.1, and this configuration is used in projects like the Rudong project in China and Helwin 2 in Europe [1]. The onshore converter station, also known as the GSMMC converter station, includes the MMC, converter transformer, transmission line, and grounding device. To limit short-circuit currents, a grounding device is connected in parallel with a star reactor on the valve side of the transformer, and it is grounded through a large resistance at the midpoint of the reactor [2]. To prevent zero-sequence currents, the transformer windings are configured in a delta/star (Δ/Y) arrangement in both the GSMMC and the WFMMC converter stations (offshore converter station) [3]. Both the GSMMC and WFMMC use HBMMC configurations [4]. This section focuses on the characteristics and

W. Yao et al., *Active Control of Large-Scale Offshore Wind Farms Connected Via VSC-HVDC*, Power Systems, https://doi.org/10.1007/978-981-97-9346-4_5

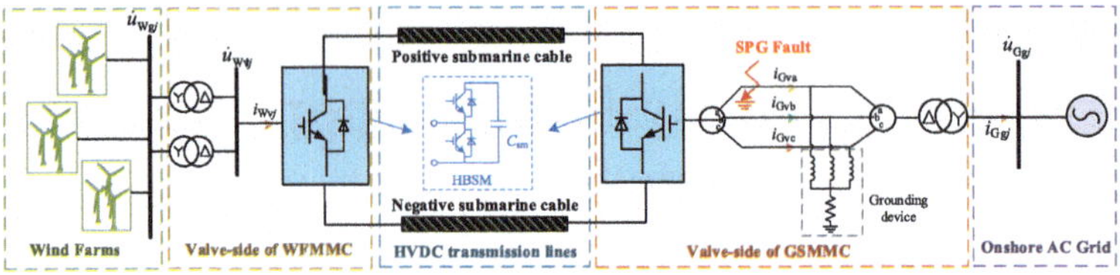

Fig. 5.1 Schematic diagram of the structure of a typical OWF-MMC-HVDC and valve-side SPG fault at the onshore converter station

propagation mechanisms of single-phase-to-ground (SPG) faults on the valve side of the GSMMC, with a brief overview of the valve-side SPG fault characteristics of the WFMMC.

5.1.1 *Valve-Side SPG Fault Characteristics of GSMMC and Impact on Onshore AC Grid*

The three-phase half-bridge MMC (HBMMC) topology is depicted in Fig. 5.2, where $\dot{u}_{vj}$ denotes the valve-side equivalent AC voltage of phase j (j = a, b, c). The parameters L_0 and R_0 represent the bridge arm inductance and bridge arm equivalent resistance, respectively. In the context of Fig. 5.2, if a single-phase-to-ground (SPG) fault occurs at the valve side of the GSMMC (using phase a as an example), the internal AC equivalent circuit of the converter station is illustrated in Fig. 5.3. Here, u_{Gvj} denotes the valve-side equivalent AC voltage of phase j for GSMMC, e_j represents the internal electrodynamic force of the bridge arm in phase j, Z_{arm} stands for the equivalent impedance of the bridge arm, and Z_f denotes the fault grounding

Fig. 5.2 Three-phase HBMMC topology

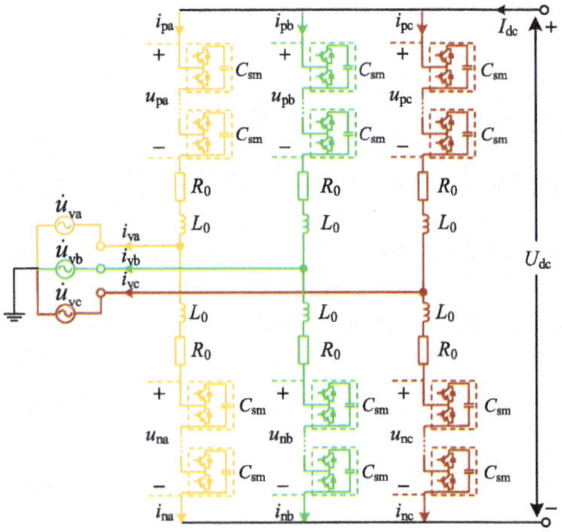

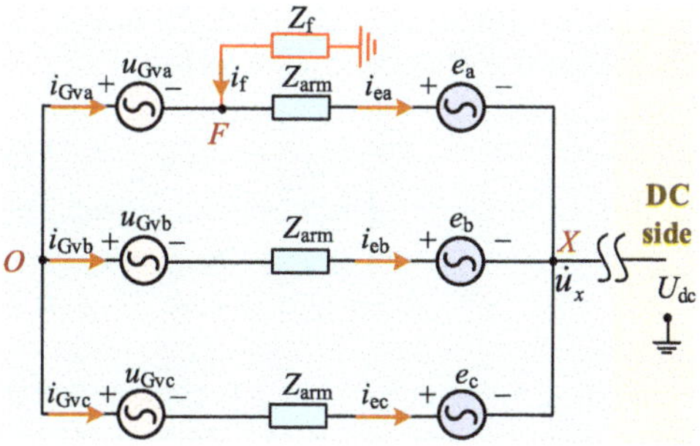

Fig. 5.3 Internal AC equivalent circuit after a valve-side SPG fault occurs of the GSMMC

impedance. Typically, the resistance in the grounding device is in the range of several thousand ohms, which can be approximated as an open circuit in the equivalent circuit and thus is not explicitly shown.

Under normal operating conditions, the AC equivalent current i_{Gvj} equals the electrodynamic force current i_{ej} ($i_{Gvj} = i_{ej}$). When analyzing AC side signals, the DC side can be treated as a DC current source set to 0. Referring to Fig. 5.3, the node current equations for nodes F, O, and X can be formulated using Kirchhoff's Current Law (KCL) as follows:

$$\begin{cases} \dot{i}_f + \dot{i}_{Gva} - \dot{i}_{ea} = 0 \\ \dot{i}_{Gva} + \dot{i}_{eb} + \dot{i}_{ec} = 0 \\ \dot{i}_{ea} + \dot{i}_{eb} + \dot{i}_{ec} = 0 \end{cases} \quad (5.1)$$

where, the fault current flowing through the fault impedance is determined to be zero ($i_f = 0$). This implies that no fault overcurrent is generated regardless of the value of the fault grounding impedance Z_f.

Since Z_f does not affect the fault current, it can be disregarded. The circuit depicted in Fig. 5.3 is expanded into Fig. 5.4. The voltage at node X ($\dot{u}_x$) can be calculated using Kirchhoff's current law (KCL):

$$\begin{aligned} &(\dot{u}_{Gvb} - \dot{u}_{Gva} - \dot{e}_b - \dot{u}_x)/Z_{arm} \\ &+(\dot{u}_{Gvc} - \dot{u}_{Gva} - \dot{e}_c - \dot{u}_x)/Z_{arm} \\ &= -(-\dot{e}_a - u_x)/Z_{arm} \\ &\Rightarrow \dot{u}_x = -\dot{u}_{Gva} \end{aligned} \quad (5.2)$$

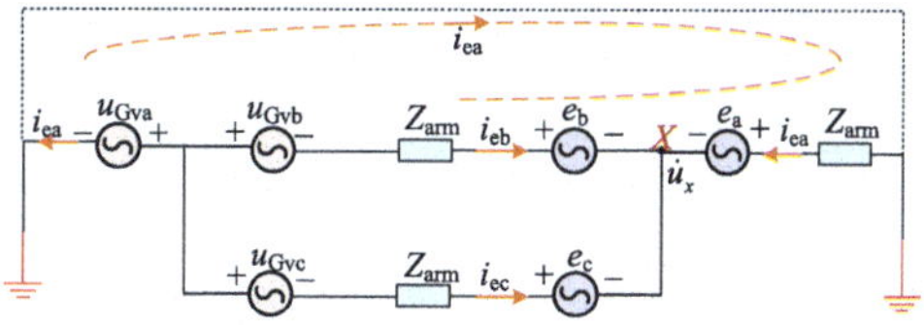

Fig. 5.4 The expansion diagram of the internal AC equivalent circuit after valve-side SPG fault of the GSMMC

Incorporating the results of (5.2) into (5.1), the current at the valve-side of each phase after the SPG fault is given by:

$$\begin{cases} \dot{i}_{\text{Gva}} = -(-\dot{e}_{\text{a}} - u_x)/Z_{\text{arm}} \\ \dot{i}_{\text{Gvb}} = (\dot{u}_{\text{Gvb}} - \dot{u}_{\text{Gva}} - \dot{e}_{\text{b}} - \dot{u}_x)/Z_{\text{arm}} \\ \dot{i}_{\text{Gvc}} = (\dot{u}_{\text{Gvc}} - \dot{u}_{\text{Gva}} - \dot{e}_{\text{c}} - \dot{u}_x)/Z_{\text{arm}} \end{cases}$$
$$\Rightarrow \dot{i}_{\text{Gv}j} = (\dot{u}_{\text{Gv}j} - \dot{e}_j)/Z_{\text{arm}} \tag{5.3}$$

Combining 5.3 with Fig. 5.4, it is evident that the current in each phase after the SPG fault remains the same as before the fault. Thus, the valve-side current for each phase, $i_{\text{Gv}j}$, is unaffected by the fault condition and remains consistent with the pre-fault current values.

The impact of SPG faults on AC voltage is analyzed as follows: When an SPG fault occurs, the voltage on the valve side of the GSMMC undergoes changes, as illustrated in Fig. 5.5. Specifically, the voltage in the faulted phase drops to 0, while the voltages in the non-faulted phases rise to the level of the line (phase-to-phase) voltage. Consequently, the line voltage remains consistent before and after the fault, maintaining stable voltage levels despite the occurrence of the SPG fault.

With the transformer winding configuration shown in Fig. 5.1, the relationship between the voltage and current at the valve side of the GSMMC and the onshore AC grid is depicted in Fig. 5.6. Since the valve-side current and line voltage of the GSMMC remain unaffected, the onshore AC voltage and current are also unchanged. Therefore, with the transformer windings arranged in Δ/Y configuration, the onshore AC grid is isolated from the effects of a valve-side SPG fault at the GSMMC. The power delivered to the onshore grid from the MMC-HVDC system remains stable after a valve-side SPG fault. Additionally, the offshore power delivery channel

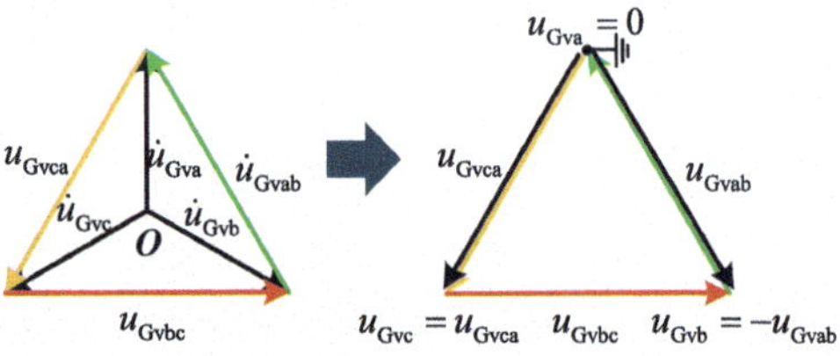

Fig. 5.5 Changes in the phase of the valve-side voltage in the GSMMC after the SPG fault

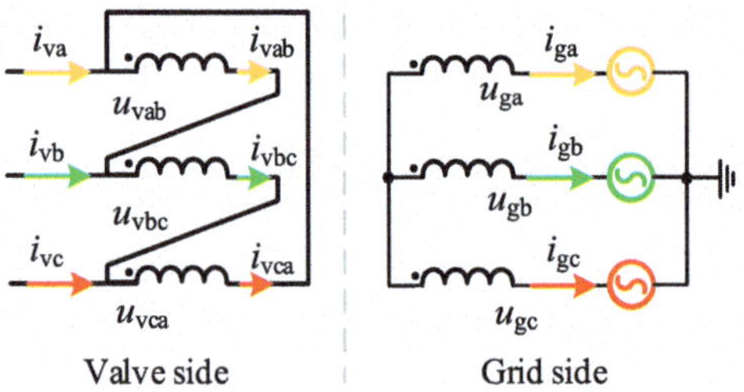

Fig. 5.6 The voltage and current relationship between the valve-side of GSMMC and the onshore AC grid

continues to function normally, with no excess power being directed into the SM capacitors, thereby preventing any risk of overvoltage in the SM capacitors during a valve-side SPG fault.

5.1.2 *Impact on HVDC Transmission Lines*

In Fig. 5.3, the valve-side SPG fault of the GSMMC will superimpose a voltage $\dot{u}_x$ on the internal electrodynamic force of each bridge arm phase $\dot{e}_j$. The fundamental equivalent circuit for a single phase of the MMC is depicted in Fig. 5.7.

The upper bridge arm voltage u_{pj}, down bridge arm voltage u_{nj}, upper bridge arm DC-side voltage u_{p}, and down bridge arm DC-side voltage u_{n} are described by Eqs. (5.4) and (5.5) before and after the fault.

$$\begin{cases} \dot{u}_{\mathrm{p}j} = U_{\mathrm{dc}}/2 - \dot{e}_j \\ \dot{u}_{\mathrm{n}j} = U_{\mathrm{dc}}/2 + \dot{e}_j \\ \dot{u}_{\mathrm{p}} = \dot{u}_{\mathrm{p}j} + \dot{e}_j \\ \dot{u}_{\mathrm{n}} = -\dot{u}_{\mathrm{n}j} + \dot{e}_j \end{cases} \tag{5.4}$$

$$\begin{cases} \dot{u}_{\mathrm{p}j} = U_{\mathrm{dc}}/2 - (\dot{e}_j + \dot{u}_x) \\ \dot{u}_{\mathrm{n}j} = U_{\mathrm{dc}}/2 + (\dot{e}_j + \dot{u}_x) \\ \dot{u}_{\mathrm{p}} = \dot{u}_{\mathrm{p}j} + \dot{e}_j \\ \dot{u}_{\mathrm{n}} = -\dot{u}_{\mathrm{n}j} + \dot{e}_j \end{cases} \tag{5.5}$$

Bring (5.2) into (5.5) and further simplify as follows,

$$\begin{cases} \dot{u}_{\mathrm{p}} = U_{\mathrm{dc}}/2 - \dot{u}_{\mathrm{Gva}} \\ \dot{u}_{\mathrm{n}} = -U_{\mathrm{dc}}/2 - \dot{u}_{\mathrm{Gva}} \end{cases} \tag{5.6}$$

where, U_{dc} is the pole-to-pole DC voltage.

On the DC side, a pole-to-ground voltage appears with an amplitude equal to the fault phase voltage and a phase difference of π relative to the original voltage of each pole. Equation (5.7) indicates that the pole-to-pole DC voltage remains constant.

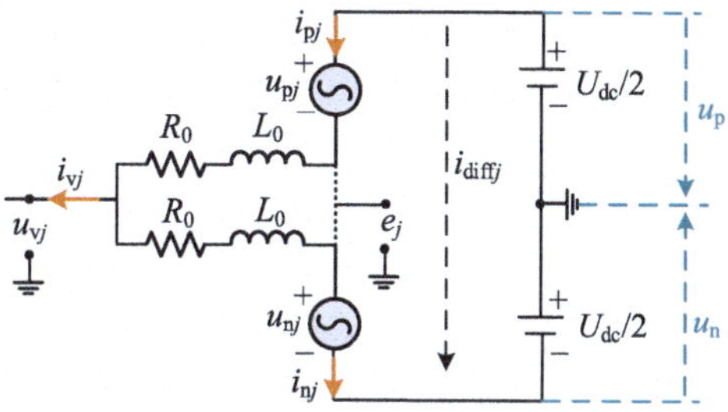

Fig. 5.7 The fundamental equivalent circuit for single phase of MMC

$$\dot{u}_{\mathrm{p}} - \dot{u}_{\mathrm{n}} = U_{\mathrm{dc}} \tag{5.7}$$

In (5.6), the amplitude of $\dot{u}_{\mathrm{Gva}}$ is related to the modulation ratio of MMC. The relationship between the modulation ratio m_{ac} and the amplitude of the single-phase AC voltage at the valve-side ($u_{\mathrm{Gv}j}^{\mathrm{amp}}$) can be written as

$$u_{\mathrm{Gv}j}^{\mathrm{amp}} = m_{\mathrm{ac}} U_{\mathrm{dc}}/2 \tag{5.8}$$

hence, the DC-side voltage amplitude can be expressed as (5.9).

$$\begin{cases} u_{\mathrm{p}}^{\mathrm{amp}} = U_{\mathrm{dc}}/2 + u_{\mathrm{Gva}}^{\max} \\ u_{\mathrm{n}}^{\mathrm{amp}} = \left|-U_{\mathrm{dc}}/2 - u_{\mathrm{Gva}}^{\max}\right| \end{cases} \tag{5.9}$$

When m_{ac} reaches its maximum value of 1, the pole-to-ground voltage amplitude of the offshore HVDC transmission line peaks at U_{dc}.

The symmetrical monopole design of MMC-HVDC ensures that the positive and negative currents in the offshore transmission line are balanced. Consequently, while the DC transmission power of each pole fluctuates with changes in the DC voltage, the total transmission power of the system remains unchanged.

5.1.3 Impact on WFMMC and Wind Farms

The characteristics of valve-side single-phase-to-ground (SPG) faults in the GSMMC are transmitted to the offshore converter station via the offshore HVDC transmission lines. Referring to Fig. 5.7, the AC voltage on the valve side of the WFMMC before the SPG fault can be described as (5.10),

$$\dot{u}_{\mathrm{Wv}j} = \dot{e}_{\mathrm{W}j} - (\dot{i}_{\mathrm{Wv}j} R_0 + L_0 \mathrm{d}\dot{i}_{\mathrm{Wv}j}/\mathrm{d}t)/2 \tag{5.10}$$

where, $\dot{e}_{\mathrm{W}j}$ is the internal electrodynamic force of the bridge arm for WFMMC and $\dot{i}_{\mathrm{Wv}j}$ is the valve-side AC current of WFMMC. Combining (5.2), (5.5) and (5.6), the valve-side AC voltage of WFMMC after SPG fault can be written as (5.11).

$$\begin{aligned} \dot{u}_{\mathrm{Wv}j} &= \dot{e}_{\mathrm{W}j} + \dot{u}_x - (\dot{i}_{\mathrm{Wv}j} R_0 + L_0 \mathrm{d}\dot{i}_{\mathrm{Wv}j}/\mathrm{d}t)/2 \\ &= \dot{e}_{\mathrm{W}j} - \dot{u}_{\mathrm{Gva}} - (\dot{i}_{\mathrm{Wv}j} R_0 + L_0 \mathrm{d}\dot{i}_{\mathrm{Wv}j}/\mathrm{d}t)/2 \end{aligned} \tag{5.11}$$

The impact of the GSMMC fault phase voltage, which includes a component with equal amplitude but opposite direction, is observed to be superimposed on the valve-side AC voltage of the offshore converter station. The specific effect of this component on the AC voltage of the converter station is analyzed below.

The influence of the GSMMC fault phase voltage on the valve-side voltage of the WFMMC is depicted in Fig. 5.8. Here, $\dot{u}_{\mathrm{GvF}}$ represents the fault phase voltage on the valve side of the GSMMC, $\phi_{\mathrm{W}xj}$ denotes the phase angle between phase j valve-side

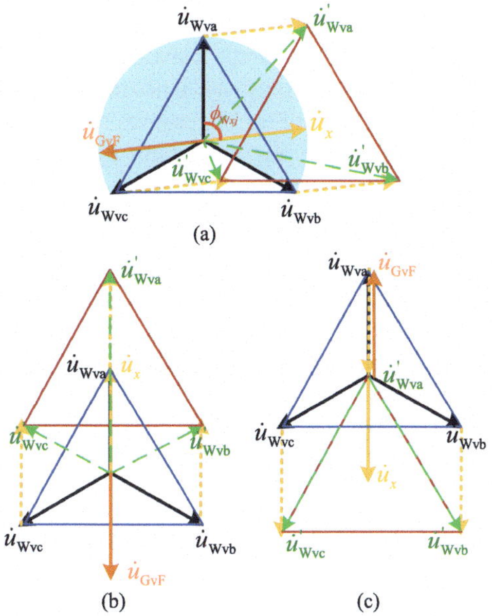

Fig. 5.8 The impact of the GSMMC fault phase voltage on the valve-side voltage of the WFMMC. **a** Scenario I: When $\dot{u}x$ does not coincide with or oppose the voltage of any phase. **b** Scenario II: Maximum overvoltage occurs when $\dot{u}x$ coincides with one phase voltage. **c** Scenario III: $\dot{u}_x$ is inverted with one phase voltage

voltage of the WFMMC and $\dot{u}_x$ before the SPG fault, and $\dot{u}'_{\mathrm{Wv}j}$ indicates the actual synthesized valve-side voltage of the WFMMC after the SPG fault.

Figure 5.8a depicts the general scenario. The valve-side phase voltage of the WFMMC is synthesized vectorially based on the amplitude and direction of $\dot{u}_x$. The resulting synthesized voltage is denoted as $\dot{u}'_{\mathrm{Wv}j}$, which can be computed in amplitude according to (12). When $|\phi_{\mathrm{W}xj}| < 2\pi/3$ (for instance, in the blue region corresponding to phase a), the amplitude $\left|\dot{u}'_{\mathrm{Wv}j}\right|$ of the newly synthesized voltage will exceed $|\dot{u}_{\mathrm{Wv}j}|$.

$$\left|\dot{u}'_{\mathrm{Wv}j}\right| = \sqrt{\left|\dot{u}_{\mathrm{Wv}j}\right|^2 + |\dot{u}_x|^2 + 2\left|\dot{u}_{\mathrm{Wv}j}\right| |\dot{u}_x| \cos\phi_{\mathrm{W}xj}} \tag{5.12}$$

In Fig. 5.8b, Scenario II illustrates a situation where $\dot{u}_x$ aligns with one phase voltage, leading to the maximum single-phase AC overvoltage at the valve-side of WFMMC. This scenario is undesirable and should be avoided.

Figure 5.8c shows Scenario III, where $\dot{u}_x$ opposes one phase voltage. Here, the valve-side voltage of that particular phase in WFMMC is minimized. If $\dot{u}_x$ matches the amplitude of the valve-side AC voltage of WFMMC, one phase voltage drops to zero, while the remaining two phase voltages rise to the line voltage.

Following this analysis, the impact of SPG faults on offshore wind farms is discussed. From Fig. 5.8, it is evident that the valve-side line voltage of WFMMC maintains both amplitude and phase integrity. Utilizing a transformer with a Δ/Y winding configuration, as depicted in Fig. 5.6, ensures the wind farm remains unaffected by SPG faults. Consequently, the wind farm does not trigger high or low voltage ride-through conditions. Furthermore, the AC current of WFMMC and the output current of the wind farm remain stable, as the DC current remains nearly unchanged following the SPG fault.

5.1.4 Valve-Side SPG Fault Characteristics of WFMMC

The valve-side SPG fault characteristics of the WFMMC are similar to those of the GSMMC for several key reasons:

- The winding configuration of the offshore converter transformer is the same as that of the onshore converter transformer. Therefore, the way valve-side SPG faults are transmitted to the grid side is consistent for both offshore and onshore converter stations.
- The structural design of the GSMMC and WFMMC is identical, as shown in Fig. 5.3. As a result, the transfer characteristics of valve-side SPG faults to the DC side are the same, whether the fault occurs at an offshore or onshore converter stations.

In conclusion, the transmission characteristics of valve-side SPG faults to both the grid side and DC side of the offshore converter station are the same as those of the onshore converter station.

5.2 Overvoltage Suppression Strategy Based on Zero-Sequence Voltage Modulation

As analyzed in Section II, a valve-side SPG fault in the GSMMC can cause oscillations in the DC line's pole-to-ground voltage. These oscillations can travel to the WFMMC, leading to valve-side AC overvoltage. Both DC line overvoltage and AC overvoltage can activate protection mechanisms at the converter station, potentially resulting in station blocking. In the case of a permanent SPG fault in the GSMMC, suppressing the overvoltage can extend the blocking duration and help mitigate the impact of overvoltage at the converter station. This allows the onshore system more time to address frequency issues caused by the offshore system disconnecting. If the SPG fault in the GSMMC is transient, overvoltage suppression reduces the risk of converter station blocking and facilitates fault ride-through for the MMC-HVDC system.

Table 5.1 The overvoltage capacity requirements of MMC-HVDC

Operating conditions	Tolerance time
$(1.02{\sim}1.15)U_{\text{dcN}}/2$	Long-term tolerance
$(1.15{\sim}1.5)U_{\text{dcN}}/2$	>5 min
$1.7U_{\text{dcN}}/2$	Tolerate 500 ms in blocking state
$(1.02{\sim}1.1)U_{\text{ac}j}$	Long-term tolerance
$(1.1{\sim}1.5)U_{\text{ac}j}$	>20 s
$1.5U_{\text{ac}j}$	>400 ms

Table 5.1 details the overvoltage capacity requirements for an MMC-HVDC project in China, where U_{dcN} represents the rated DC pole-to-pole voltage. This chapter introduces an overvoltage suppression strategy designed according to the principles outlined in Table 5.1.

5.2.1 Design of DC Overvoltage Suppression Strategy

Following an SPG fault, the sequence equivalent circuit from the onshore AC grid to the fault point is depicted in Fig. 5.9. The relationship between the sequential current and voltage in this circuit is described as follows:

$$\dot{i}_{\text{a}}^{+} = \dot{i}_{\text{a}}^{-} = \dot{i}_{\text{a}}^{0} = 0 \tag{5.13}$$

$$\begin{cases} \dot{u}_{\text{a}}^{-} = 0 \\ \dot{u}_{\text{a}}^{+} = -\dot{u}_{\text{a}}^{0} \end{cases} \tag{5.14}$$

If the positive-sequence voltage and zero-sequence voltage are equal in amplitude and opposite in direction, then $\dot{u}_x = \dot{u}_{\text{a}}^{0}$. Consequently, detecting the valve-side

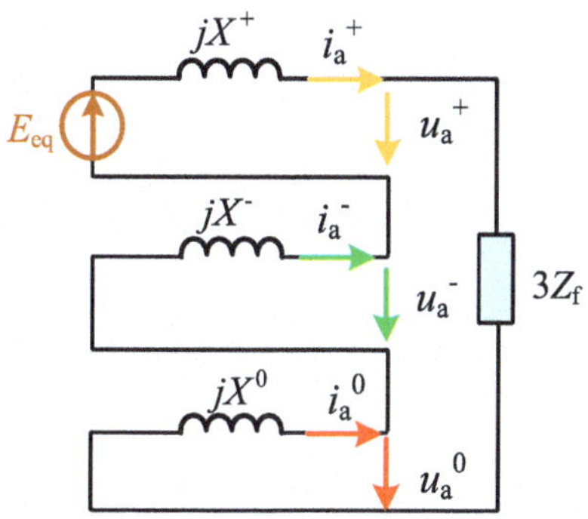

Fig. 5.9 The sequence equivalent circuit from the onshore AC grid to the fault location

zero-sequence voltage of GSMMC allows for obtaining the superimposed DC-side oscillation $\dot{u}_x$.

Before the SPG fault, the voltage reference values of the upper and down bridge arms of GSMMC are shown in (5.15),

$$\begin{cases} \dot{u}_{\mathrm{p}j\_\mathrm{ref}} = U_{\mathrm{dc}}/2 - \dot{e}_j - \dot{u}_{\mathrm{diff}\,j} \\ \dot{u}_{\mathrm{n}j\_\mathrm{ref}} = U_{\mathrm{dc}}/2 + \dot{e}_j - \dot{u}_{\mathrm{diff}\,j} \end{cases} \tag{5.15}$$

where, $\dot{u}_{\mathrm{diff}\,j}$ represents the unbalanced voltage caused by circulating currents within the converter [5].

The analysis from Figs. 5.3 and 5.7 indicates that $\dot{e}_j$ is superimposed with $\dot{u}_x$ following the SPG fault. Furthermore, since $\dot{u}_x = \dot{u}_{\mathrm{a}}^0$, the voltage reference values for the upper and lower bridge arms of GSMMC can be adjusted according to (5.16) after the SPG fault.

$$\begin{cases} \dot{u}_{\mathrm{p}j\_\mathrm{ref}} = U_{\mathrm{dc}}/2 - (\dot{e}_j + \dot{u}^0) - \dot{u}_{\mathrm{diff}\,j} \\ \dot{u}_{\mathrm{n}j\_\mathrm{ref}} = U_{\mathrm{dc}}/2 + (\dot{e}_j + \dot{u}^0) - \dot{u}_{\mathrm{diff}\,j} \end{cases} \tag{5.16}$$

From (5.16), it is evident that the inputs to the SMs can be modulated by the zero-sequence voltage $\dot{u}^0$, effectively suppressing overvoltage. The structure of the proposed overvoltage suppression strategy is illustrated in Fig. 5.10, where the blue segment represents the overvoltage suppression control based on the zero-sequence voltage. The remainder of the structure comprises the standard MMC controller [6]. The d-axis of the AC outer loop employs DC voltage control [7]. S_{SM} denotes the SM modulation signal of MMC. The parameter k acts as a constraint factor, limiting the amplitude of the zero-sequence voltage input to the SM modulation link. The voltage reference values for the upper and lower bridge arms of GSMMC, considering the influence of k, can be expressed as:

$$\begin{cases} \dot{u}_{\mathrm{p}j\_\mathrm{ref}} = U_{\mathrm{dc}}/2 - (\dot{e}_j + k\dot{u}^0) - \dot{u}_{\mathrm{diff}\,j} \\ \dot{u}_{\mathrm{n}j\_\mathrm{ref}} = U_{\mathrm{dc}}/2 + (\dot{e}_j + k\dot{u}^0) - \dot{u}_{\mathrm{diff}\,j} \end{cases} \tag{5.17}$$

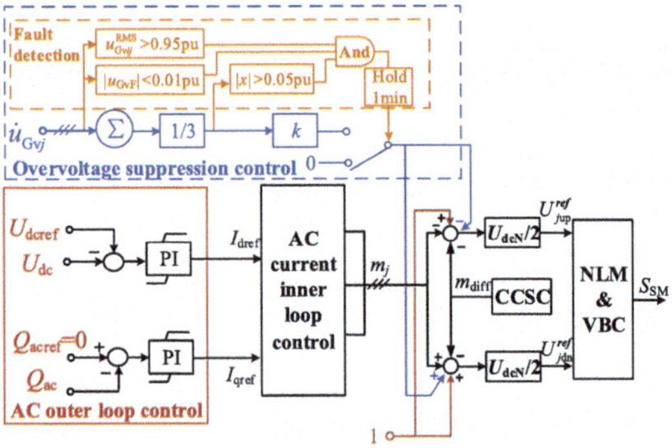

Fig. 5.10 Overvoltage suppression strategy based on zero-sequence voltage modulation for GSMMC

The fault detection module, highlighted in the orange area, uses the asymmetric fault detection method from commutation failure prevention (CFPREV) [8]. An SPG fault is confirmed when the following three conditions are met simultaneously:

1.The zero-sequence voltage exceeds 0.05 pu. 2.One phase voltage drops to nearly zero (e.g., below 0.01 pu). 3.The root mean square (RMS) value of the line voltage remains stable (e.g., above 0.95 pu).

To avoid false triggers due to noise or detection errors, a hysteresis block is placed after the voltage detection block. Given that the AC system operates at a rated frequency of 50 Hz, and the phase voltage typically dips below 0.01 pu for less than 1 ms during a fault, a 1 ms delay is applied before confirming the fault. An SPG fault is only verified if the fault phase voltage stays below 0.01 pu for more than 1 ms.

Once the fault is detected, the overvoltage suppression control is activated and remains engaged for 1 min. After this period, detection continues, and if the fault conditions persist, the suppression control remains active.

5.2.2 Influence Mechanisms of Overvoltage Suppression Strategy on System Transient Processes

Zero-sequence voltages, influenced by transient surges or drops following a fault, can significantly impact MMC modulation. If these voltages are directly superimposed on the MMC modulation link, it can lead to instantaneous mass input or removal of SMs, affecting system stability. The adverse effects of this control strategy on system transients can be outlined in a four-step process, as depicted in Fig. 5.11.

Step I: An SPG fault triggers valve-side zero-sequence voltage u_G^0, which propagates to the modulation link. This zero-sequence component induces transient mass input or removal of SMs.

Step II: The imbalance in the input and removal of upper and lower bridge arm SMs alters the total number of active SMs.

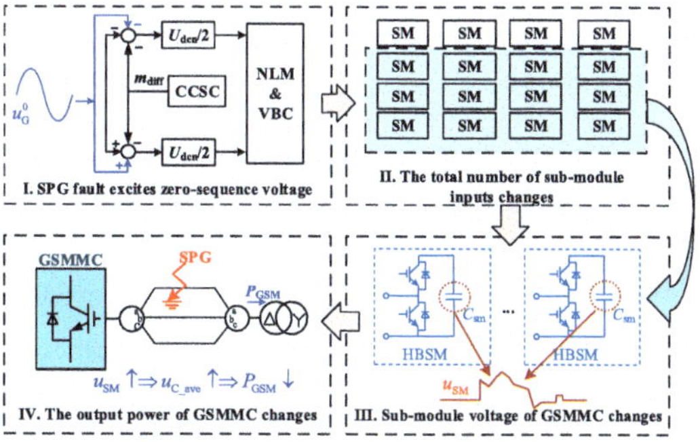

Fig. 5.11 Mechanisms of overvoltage suppression strategy on system transient processes

Step III: With GSMMC employing constant DC voltage control, changes in the total number of SMs affect internal energy (e.g., capacitor voltage u_{SM}) within the SMs.

Step IV: Utilizing voltage balancing control (VBC) [9], changes in the voltage of some SMs influence others. For instance, an increase in SM voltage raises the average capacitance voltage (u_{Cave}) of internal GSMMC SMs. This results in increased absorption of output power from wind farms by GSMMC, thereby reducing power delivered to the onshore AC grid.

Moreover, excessive zero-sequence voltage compensation can lead to overmodulation issues, distorting zero-sequence voltages and degrading modulation quality. Hence, careful adjustment of the compensation factor k is crucial to prevent these complications.

5.2.3 Determination of the Key Parameter k

According to (5.16) and Fig. 5.10, the parameter k directly influences the effectiveness of overvoltage suppression. As described in subsection III-B, a higher value of k enhances overvoltage suppression but also exacerbates adverse effects on system transients. Therefore, selecting an appropriate value for k involves balancing overvoltage suppression capability with the need to mitigate drastic changes in electromagnetic quantities during transient events.

From Table 5.2, it is established that maintaining DC pole-to-ground overvoltage suppression below $1.5U_{dcN}/2$ ensures extended tolerance time for MMC-HVDC. During normal operation when the DC-side voltage equals the rated DC pole-to-pole voltage U_{dcN}, the permissible range for k is determined by combining (5.8), as indicated in (5.19).

$$\begin{aligned} &(1 + m_{ac} - km_{ac})U_{dcN}/2 \leq 1.5U_{dcN}/2 \\ &\Rightarrow k \geq 1 - 0.5/m_{ac} \end{aligned} \tag{5.18}$$

In general, $m_{ac} < 1$. To prevent excessive suppression that could lead to undesirable transient processes, this chapter aims to limit the pole-to-ground DC overvoltage suppression to $1.5U_{dcN}/2$, which corresponds to setting $k = 1 - 0.5/m_{ac}$.

It's important to note that when $k = 1 - 0.5/m_{ac}$, the overvoltage suppression capability is constrained, particularly if the MMC is overmodulated. In such cases, increasing k slightly above $1 - 0.5/m_{ac}$ becomes necessary to ensure effective overvoltage suppression.

The value of k primarily influences overvoltage tolerance and overmodulation tolerance. Projects with weaker overvoltage tolerance require a higher k to manage overmodulation effectively, which can otherwise lead to system instability. Conversely, projects with strong overvoltage tolerance can maintain a lower k value that satisfies both overvoltage and overmodulation acceptance criteria. This balanced

Table 5.2 Main parameters of MMC-HVDC test system

System parameters	Value
Rated DC pole-to-ground voltage U_{dc0}	±320 kV
Rated DC pole-to-pole voltage U_{dcN}	640 kV
Rated DC transmission power P_{dc0}	1200 MW
Number of individual bridge arm SMs N	160
Rated voltage MMC SM U_{SM}	2 kV
Rated power of wind farms P_{WF0}	1200 MW
Onshore converter transformer rated ratio K_{GSN}	230 kV/338 kV
Offshore converter transformer rated ratio K_{WFN}	230 kV/338 kV
Modulation ratio of GSMMC m_{ac}	0.86

choice of k ensures that overvoltage remains within acceptable limits across varying operational conditions in projects with high overvoltage tolerance.

5.3 Simulation Verification

The OWF-MMC-HVDC model depicted in Fig. 5.1 is implemented using PSCAD/EMTDC, with detailed system parameters provided in Table 5.2. Below, separate verification of fault mechanism analysis and the proposed overvoltage suppression strategy is conducted. Each positive and negative submarine cable transmits 600 MW under normal operating conditions.

5.3.1 Verification of Valve-Side SPG Fault Characteristics and Propagation Mechanism of GSMMC

Simulation setup: The GSMMC of the MMC-HVDC system experiences a valve-side metallic SPG fault occurring at 3.0 s.

5.3.1.1 Valve-Side SPG Fault Characteristics of GSMMC

The impact of the valve-side SPG fault on the GSMMC is first analyzed and verified, focusing on its effects on the internal components of the onshore converter station and the AC grid side. The phase-to-ground voltage and current on the valve-side of the GSMMC (u_{Gvj} and i_{Gvj}), the phase voltage on the AC grid side (u_{Ggj}), and the output power of the GSMMC before and after the fault are illustrated in Fig. 5.12.

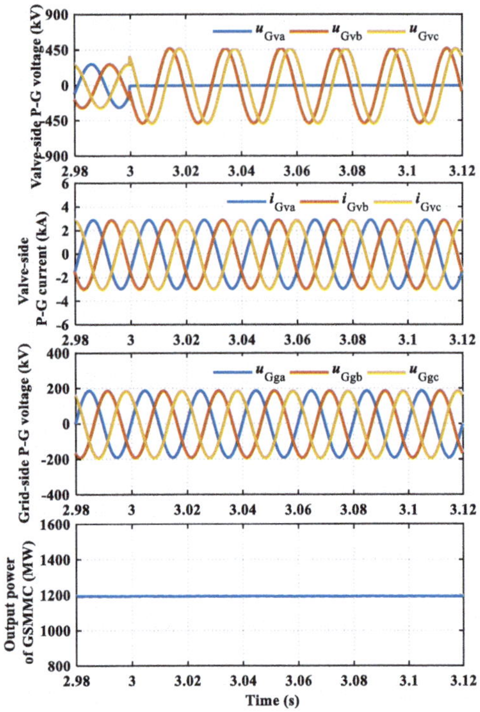

Fig. 5.12 Valve-side SPG fault characteristics of GSMMC and impact on onshore AC grid

According to Fig. 5.12, after the SPG fault occurs, the valve-side fault phase voltage drops to 0, while the non-fault phase voltage rises to the line voltage. No overcurrent is observed following the fault. The grid-side phase voltage remains unchanged, indicating that the fault on the valve-side does not affect the line voltage when the converter transformer windings are configured in delta-star (Δ/Y). Additionally, the output power of the GSMMC remains constant, confirming that the AC grid is unaffected by the valve-side SPG fault. The system exhibits rapid recovery once the SPG fault is cleared.

5.3.1.2 Valve-Side SPG Faults Propagation Mechanism from GSMMC to HVDC Transmission Lines

The valve-side zero-sequence voltage of GSMMC (u_G^0), the increment in HVDC transmission line pole-to-ground voltage (Δu), positive and negative DC voltages, DC current, DC transmission power, and average voltage of SM capacitors of GSMMC and WFMMC (U_{GSSM} and U_{WFSM}) before and after the SPG fault occurrence are depicted in Fig. 5.13.

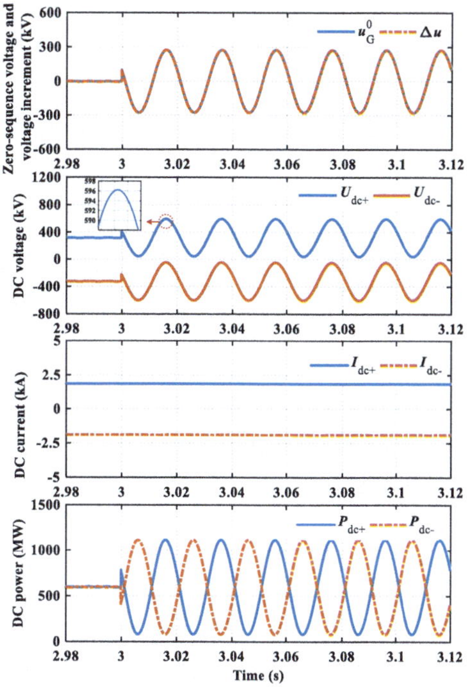

Fig. 5.13 Characteristics of offshore HVDC transmission lines before and after SPG fault

The results from Fig. 5.13 clearly demonstrate that the zero-sequence voltage u_G^0 precisely matches the increase in the DC pole-to-ground voltage Δu. The peak value of the DC pole-to-ground voltage reaches approximately 596 kV after the zero-sequence voltage is superimposed. The calculated $m_{ac} = 0.86$ from (5.8) matches the preset value accurately. Furthermore, the DC current remains stable, and the DC power to ground oscillates with the voltage, while the total transmitted power remains constant at 1200 MW. The average submodule capacitor voltages of both the GSMMC and WFMMC exhibit minimal change before and after the fault. Importantly, the valve-side SPG fault does not cause overvoltage issues in the bridge arms. These findings confirm the correctness of the fault propagation mechanism analysis.

5.3.1.3 Valve-Side SPG Faults Propagation Mechanism from GSMMC to WFMMC and Wind Farms

Following the propagation of the valve-side SPG fault from the GSMMC to the WFMMC through the HVDC transmission line, Fig. 5.14 illustrates the phase-to-ground voltage, line-to-line (P-P) voltage at the valve-side of the WFMMC, and the phase voltage at the wind farm side.

The valve-side voltage of the WFMMC exhibits changes in amplitude and phase due to the superposition of the zero-sequence voltage transmitted from the GSMMC.

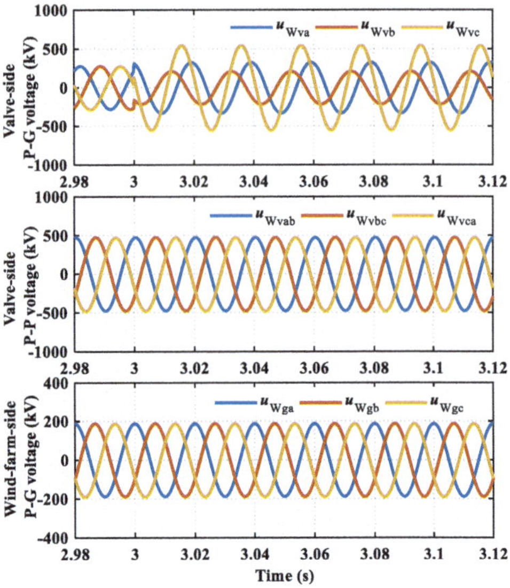

Fig. 5.14 Characteristics of WFMMC and wind farms before and after SPG fault

However, the line voltage at the WFMMC valve-side remains unchanged. The voltage experienced by the offshore wind farms also remains unaffected, especially in configurations where the windings of the converter transformer are in a Δ/Y configuration. Importantly, the valve-side SPG fault in the GSMMC does not induce low or high voltage ride-through events at the wind farms.

Three simulation scenarios, as depicted in Fig. 5.8, are established to investigate the correlation between the zero-sequence voltage phase following an SPG fault and the phase voltage amplitude at the valve-side of the WFMMC. The relationship between the generated zero-sequence voltage post-fault and the valve-side phase voltage of the WFMMC is illustrated in Fig. 5.15.

In Fig. 5.15a, it is observed that the zero-sequence voltage causes an increase in voltage for two phases while decreasing the voltage of the remaining phase. When the zero-sequence voltage coincides with phase A voltage, as shown in Fig. 5.15b, the amplitudes of the two phases' voltages are directly superimposed, resulting in maximum overvoltage in the WFMMC. Conversely, Fig. 5.15c shows the scenario opposite to Fig. 5.15b; here, the equal rated valve-side voltages of the rectifier and inverter stations of the MMC-HVDC system cause the phase A valve-side voltage in the WFMMC to be offset to zero by the zero-sequence voltage. The simulation results are consistent with the theoretical analysis.

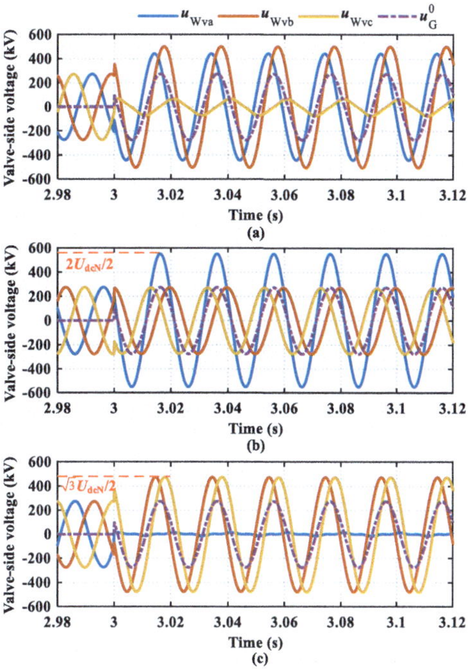

Fig. 5.15 The impact of the GSMMC fault phase voltage on the valve-side voltage of the WFMMC. **a** Scenario I: When $\dot{u}x$ does not coincide with or oppose the voltage of any phase. **b** Scenario II: Maximum overvoltage occurs when $\dot{u}x$ coincides with one phase voltage. **c** Scenario III: $\dot{u}_x$ is inverted with one phase voltage

5.3.2 *Valve-Side SPG Fault Characteristics of WFMMC*

Simulation setup: At 3.0 s, the WFMMC of the MMC-HVDC system experiences a valve-side metallic SPG fault. Assuming the fault does not lead to system shutdown, Fig. 5.16 illustrates the valve-side voltage, grid-side voltage, DC voltage, and valve-side zero-sequence voltage (Δu_{Wv}^0) of the WFMMC.

Comparing Figs. 5.12, 5.13, and 5.16, it is clear that both the offshore and onshore converter stations exhibit identical valve-side SPG fault characteristics.

5.3.3 *Validation of Overvoltage Suppression Strategy*

5.3.3.1 Effectiveness Validation

Combining (5.19) with Table 5.2, the constraint coefficient k is calculated to be $k \geq 0.419$. The effects of the proposed overvoltage suppression strategy on the DC pole-to-ground voltage, output power of GSMMC, and internal energy of GSMMC for different values of k are shown in Fig. 5.17.

As illustrated in Fig. 5.17, the proposed strategy effectively suppresses overvoltage when k is set above 0. The higher the value of k, the better the overvoltage suppression.

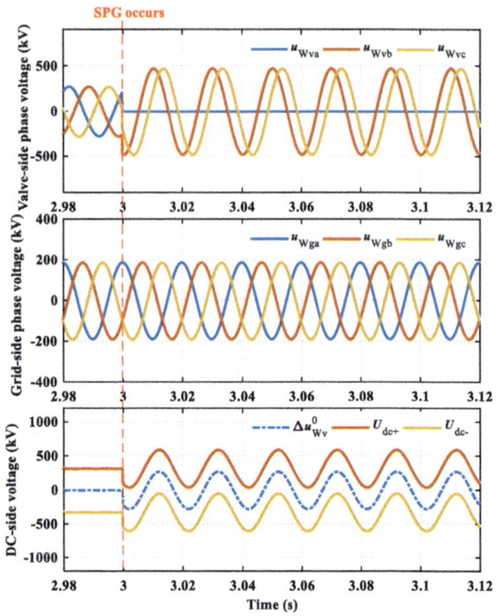

Fig. 5.16 Voltage characteristics of valve-side SPG fault of WFMMC

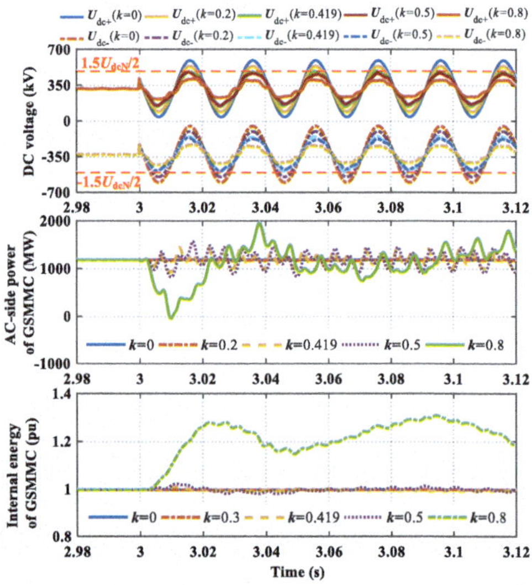

Fig. 5.17 The effect of the proposed strategy on the overvoltage suppression effect and on the output power of GSMMC for different values of *k*

However, increasing the value of *k* also intensifies the internal energy variation of the GSMMC, reflected in the capacitor voltage fluctuations of the SMs, which limits the output power of the GSMMC.

To emphasize the significance of the proposed overvoltage suppression strategy, two schemes are compared: Conventional Scheme: This scheme relies on overvoltage protection to mitigate prolonged DC overvoltage shocks. According to Table 5.1, it

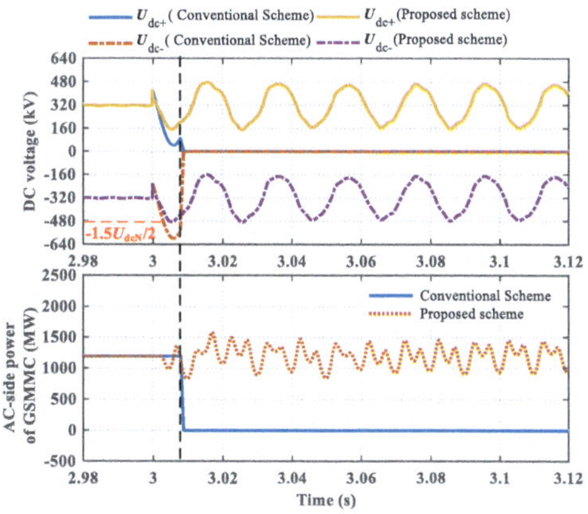

Fig. 5.18 Comparison of the conventional scheme and the proposed scheme

assumes that DC pole-to-ground overvoltages exceeding 1.5 pu and lasting longer than 5 ms will activate the DC overvoltage protection. Proposed Scheme: This scheme implements the proposed overvoltage suppression strategy, where the parameter k is set to 0.5 based on the analysis presented in Eq. (5.19) and illustrated in Fig. 5.17.

Simulation Scenario: At 3.0 s, the GSMMC in the MMC-HVDC system experiences a valve-side metallic SPG fault. The DC voltage and AC side power of the GSMMC are depicted in Fig. 5.18.

In Fig. 5.18, the conventional scheme mitigates continuous DC overvoltage shocks through overvoltage protection. However, when the DC voltage drops to 0, power transmission from offshore to the onshore grid is interrupted. In contrast, with the proposed scheme, the peak DC overvoltage is immediately suppressed upon fault occurrence, thereby preventing overvoltage protection activation. This ensures uninterrupted power transmission and enhances the utilization of offshore wind power.

In Fig. 5.17, when $k = 0.2$, the DC voltage remains stable without distortion, and the output power of the GSMMC is nearly constant. However, as k increases, voltage distortion begins to occur. This phenomenon can be explained by examining the number of upper bridge arm SMs and the bridge arm voltage for each phase at different values of k, as shown in Fig. 5.19.

At $k = 0.2$, the number of SMs in the bridge arm is within the modulation limits, and the bridge arm voltage is not distorted, with $m_{\mathrm{ac}} = 0.87$. For higher values of k, overmodulation occurs, leading to increased distortion in the bridge arm voltage. The degree of distortion rises with the increase in k. This overmodulation limits the overvoltage suppression capability of the proposed strategy, preventing the overvoltage from reaching $1.5U_{\mathrm{dcN}}/2$ at $k = 0.419$ in Fig. 5.17. It is only when k increases to 0.5 that the overvoltage suppression achieves the desired level, aligning with the theoretical analysis.

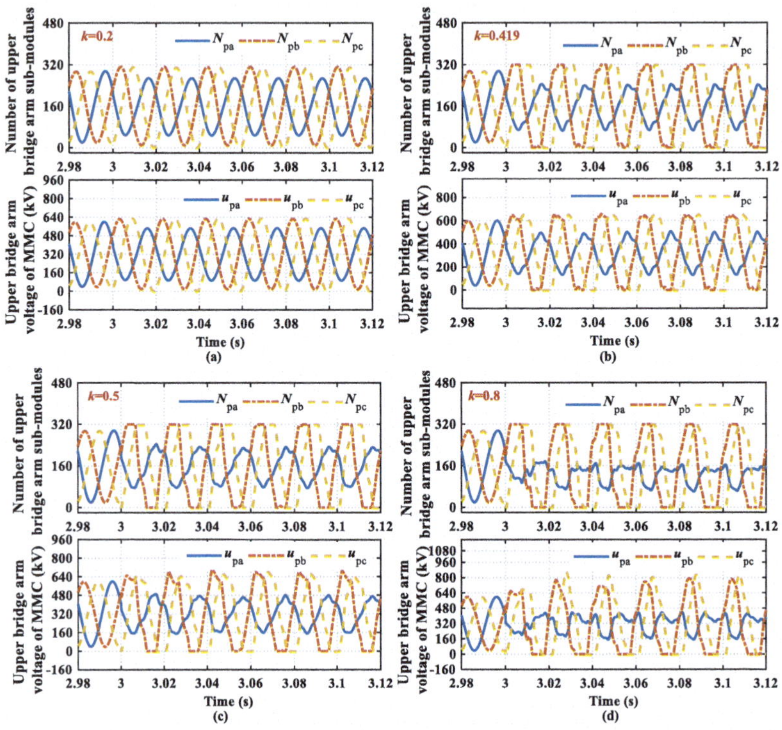

Fig. 5.19 The number of upper bridge arm SMs and the bridge arm voltage for each phase at different values of k. **a** $k = 0.2$. **b** $k = 0.419$. **c** $k = 0.5$. **d** $k = 0.8$

To assess the impact of overmodulation on the system during overvoltage suppression, the total harmonic distortion (THD) rate of the bridge arm voltage in the fault phase is evaluated, following the THD definition from [10], as follows:

$$\mathrm{THD_{fup}} = \frac{\sqrt{\sum_n \left|U_{\mathrm{fup}\_n}\right|^2}}{U_{\mathrm{fup}\_1}} \cdot 100\% \tag{5.19}$$

where, U_{fup_n} is the nth harmonic amplitude of the upper bridge arm of the fault phase, n is the harmonic order, with $n \neq 1$, and U_{fup_1} is the fundamental amplitude of the upper bridge arm of the fault phase.

Using fast Fourier transform (FFT) analysis, the harmonic amplitudes of the bridge arm voltage for the fault phase at various k values were obtained. The THDfup values were calculated and are illustrated in Fig. 5.20 [11]. The analysis was based on the measured bridge arm voltage of the fault phase over five power frequency cycles following the SPG fault. The results show that as the k value increases, $\mathrm{THD_{fup}}$ also rises, indicating greater voltage distortion. Thus, $\mathrm{THD_{fup}}$ serves as an indicator of the extent of overmodulation.

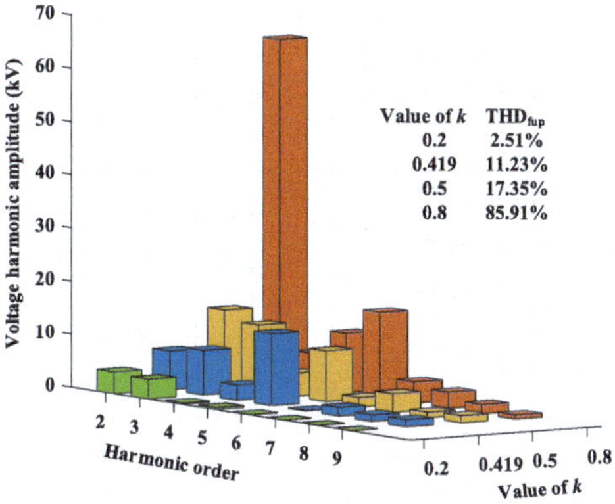

Fig. 5.20 The amplitudes of different harmonic orders of the bridge arm voltage on the fault phase for different values of k

5.3.3.2 Influence Mechanisms of Overvoltage Suppression Strategy on System Transient Processes

In Fig. 5.17, significant oscillations in the output power of the GSMMC are evident at $k = 0.8$. Figure 5.21 further substantiates this observation by examining the mechanism behind these pronounced transient power fluctuations at $k = 0.8$.

It can be observed from Fig. 5.21 that the mechanism of the overvoltage suppression strategy's influence on the system's transient process unfolds in four distinct steps. Step I: Following the fault occurrence, there is a sudden change in the zero-sequence voltage. Step II: Due to the high value of k and the proposed control strategy, the number of SMs added or removed by the upper and lower bridge arms changes abruptly. (Here, phase A is illustrated, with reference to Fig. 5.19 for other phases.) This sudden alteration causes an imbalance in the number of SMs between the upper and lower bridge arms and leads to a change in the total count of SMs. Step III: As the total number of SM inputs decreases, the capacitor voltage of the input SMs increases (indicating an increase in SM energy) to maintain a constant DC voltage. Further, under voltage balancing control, the SM capacitors within the GSMMC absorb output power from the wind farms, causing their voltages to collectively rise and thus increasing the total energy within the GSMMC. Step IV: Consequently, the output power of the GSMMC decreases as the output from the wind farms is redirected to elevate the SM capacitor voltages. The simulation results presented in Fig. 5.10 validate the accuracy of these influence mechanisms of the overvoltage suppression strategy on the system's transient processes.

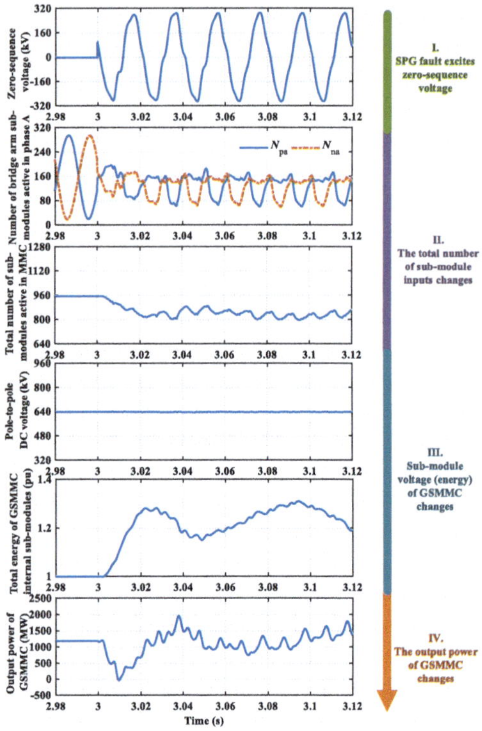

Fig. 5.21 The mechanism of the large oscillations in the transient power

5.4 Summary

This study examines the characteristics and propagation mechanisms of onshore valve-side SPG faults in OWF-MMC-HVDC systems and proposes an overvoltage suppression strategy to address these faults. Due to the unique grounding method used in OWF-MMC-HVDC systems, SPG faults do not cause widespread overcurrent issues. Within onshore converter stations, an SPG fault results in zero voltage for the faulted phase, an increase in the non-faulted phase voltages up to line voltage levels, and unchanged line voltages. The onshore AC grid remains unaffected by SPG faults because of the transformer winding configurations.

In offshore HVDC transmission lines, an SPG fault causes the fault-induced zero-sequence voltage to be superimposed onto the original pole-to-ground DC voltage, leading to DC overvoltage, while the pole-to-pole voltage remains stable. At offshore converter stations, the fault-induced zero-sequence voltage is similarly superimposed on each phase at the valve side, potentially causing AC overvoltagełespecially when the zero-sequence voltage aligns with one phase. However, line voltages at offshore converter stations remain unaffected. Offshore wind farms are protected from SPG faults when the converter transformer windings are configured in a (Δ/Y) configuration.

The study proposes an overvoltage suppression strategy based on zero-sequence voltage modulation, focusing on the influence of control parameters on system transient behavior. These parameters are calibrated according to the project's overvoltage tolerance to ensure both effective overvoltage suppression and system stability during transients. Simulation results validate the fault mechanism analysis and demonstrate the effectiveness of the proposed overvoltage suppression strategy.

References

1. Pan E, Yue B, Li X, Zhao Z, Zhu Q (2020) Integration technology and practice for long-distance offshore wind power in China. Energy Convers. Econ. 1(1):4–19. https://doi.org/10.1049/enc2.12003
2. Yang B, Liu B, Zhou H, Wang J, Yao W, Wu S, Shu H, Ren Y (2022) A critical survey of technologies of large offshore wind farm integration: Summarization, advances, and perspectives. Prot Control Mod Power Syst (in press)
3. Li G, Liang J, Ma F, UgaldeLoo CE, Liang H (2019) Analysis of single-phase-to-ground faults at the valve-side of HB-MMCs in HVDC systems. IEEE Trans Ind Electron 66(3):2444–2453. https://doi.org/10.1109/TIE.2018.2829666
4. Ji K, Tang G, Pang H, Yang J (2020) Impedance modeling and analysis of MMC-HVDC for offshore wind farm integration. IEEE Trans Power Deliv 35(3):1488–1501. https://doi.org/10.1109/TPWRD.2019.2946450
5. Xiang W, Lin W, Xu L, Wen J (2018) Enhanced independent pole control of hybrid MMC-HVdc system. IEEE Trans Power Deliv 33(2):861–872. https://doi.org/10.1109/TPWRD.2017.2715040
6. Liu Z, Zhao J (2021) Disturbance interaction analysis and suppression strategy of MMC-HVDC systems considering sub-module capacitor voltage ripples. IEEE Trans Power Syst 36(1):235–247. https://doi.org/10.1109/TPWRS.2020.3006234
7. Meng P, Xiang W, Chi Y, Wang Z, Lin W, Wen J (2022) Resilient DC voltage control for islanded wind farms integration using cascaded hybrid HVDC system. IEEE Trans Power Syst 37(2):1054–1066. https://doi.org/10.1109/TPWRS.2021.3106972
8. Zhou H, Yao W, Ai X, Li D, Wen J, Li C (2022) Comprehensive review of commutation failure in HVDC transmission systems. Electr Power Syst Res 205:107768. https://doi.org/10.1016/j.epsr.2021.107768
9. Zhou H, Yao W, Zhou M, Ai X, Wen J, Cheng S (2023) Active energy control for enhancing AC fault ride-through capability of MMC-HVDC connected with offshore wind farms. IEEE Trans Power Syst 38(3):2705–2718. https://doi.org/10.1109/TPWRS.2022.3179443
10. Matas J, Martin H, de la Hoz J, Abusorrah A, Al-Turki Y, Alshaeikh H (2020) A new THD measurement method with small computational burden using a SOGI-FLL grid monitoring system. IEEE Trans Power Electron 35(6):5797–5811. https://doi.org/10.1109/TPEL.2019.2953926
11. Wang L, Yao W, Xiong Y, Shi Z, Wen J (2022) Commutation failure analysis and prevention of UHVDC system with hierarchical connection considering voltage harmonics. IEEE Trans Power Deliv 37(4):3142–3154. https://doi.org/10.1109/TPWRD.2021.3124096

Chapter 6
Characteristics Analysis and Suppression of DC Submarine Cable Disconnection Fault

Abstract This chapter presents a Multi-Level Coordinated Control (MLCC) scheme specifically designed for symmetrical bipolar OWF-MMC-HVDC systems. The MLCC scheme utilizes the energy control capabilities of MMCs to address DC overcurrents resulting from submarine cable failures. The scheme is organized into four control levels, each targeting a specific aspect of system response: Offshore Fault Pole MMC: When a DC submarine cable breakage is detected, the fault pole MMC at the offshore converter station begins passive power absorption from the wind farms using its internal SM capacitors. This provides a critical time buffer for subsequent control actions to be implemented. Onshore Non-Faulty Pole MMC: At the second level, the non-faulty pole MMC at the onshore converter station actively adjusts DC voltage to suppress potential DC overcurrents. Offshore Non-Fault Pole MMC: The third level involves the non-fault pole MMC at the offshore converter station, which dynamically balances DC power flow based on real-time power input from the wind farms. Offshore Wind Farms: The fourth level focuses on the offshore wind farms, which reduce their power output to bring submarine cable transmission parameters back within acceptable limits. The effectiveness and accuracy of the proposed MLCC scheme across various scenarios are rigorously validated through PSCAD/EMTDC simulations.

6.1 Structure and Fault Characteristics of Symmetrical Bipolar MMC-HVDC

The structure of the symmetrical bipolar OWF-MMC-HVDC system is shown in Fig. 6.1. The wind farms use direct-drive permanent magnet synchronous generators (PMSGs), with their rated output matched to the WFMMC station's rated transmission power. Each PMSG is connected to a back-to-back full power converter, which includes a rotor-side converter (RSC) and a grid-side converter (GSC). The RSC controls the wind turbine's active power output, while the GSC maintains a stable DC voltage [1]. Due to varying wind conditions, the actual power output from the wind farms often falls below their rated capacity. To allow the non-faulty pole to

W. Yao et al., *Active Control of Large-Scale Offshore Wind Farms Connected Via VSC-HVDC*, Power Systems, https://doi.org/10.1007/978-981-97-9346-4_6

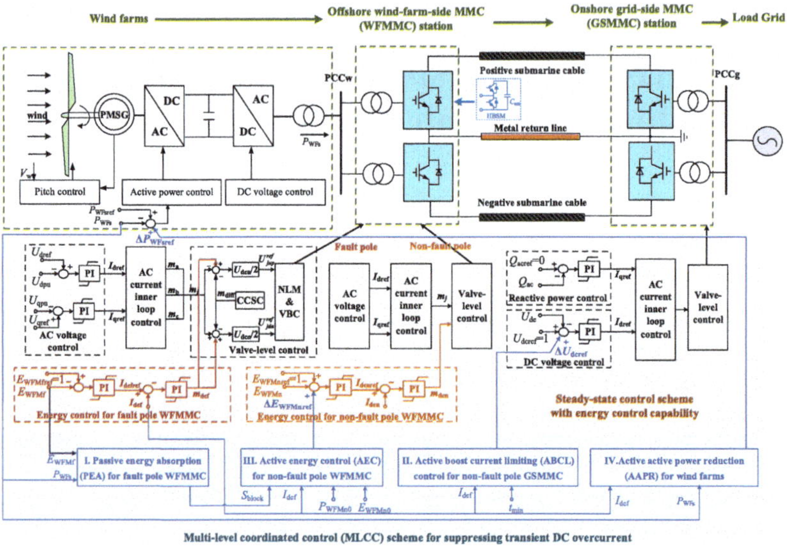

Fig. 6.1 The system structure and control scheme of the symmetrical bipolar OWF-MMC-HVDC

transmit more power from the wind farms after a fault is cleared, each wind farm outlet is connected to a common busbar.

The structure of the symmetrical bipolar MMC-HVDC is depicted in Fig. 6.1. Both offshore and onshore converter stations utilize Half-Bridge Modular Multilevel Converters (HBMMCs). Each polarity (positive and negative) is controlled independently by its own MMC. Typically, offshore platforms do not have grounding poles due to environmental regulations, so the positive and negative MMCs are grounded via a metallic return line connected to the onshore converter station.

The schematic diagram illustrating power exchange between the AC side and DC side of a single MMC is shown in Fig. 6.2. Here, P_{dc} and P_{ac} denote the input power on the DC side and the output power on the AC side of the MMC, respectively. U_{dc} and I_{dc} represent the unipolar DC voltage and DC current, respectively. All electrical quantities are standardized, and the arrow direction indicates the positive convention.

The AC side power of a single WFMMC is half of the real-time output power of wind farms, denoted as P_{WFs}. Therefore, P_{ac} and P_{dc} can be expressed as follows:

Fig. 6.2 AC and DC side power of a single WFMMC

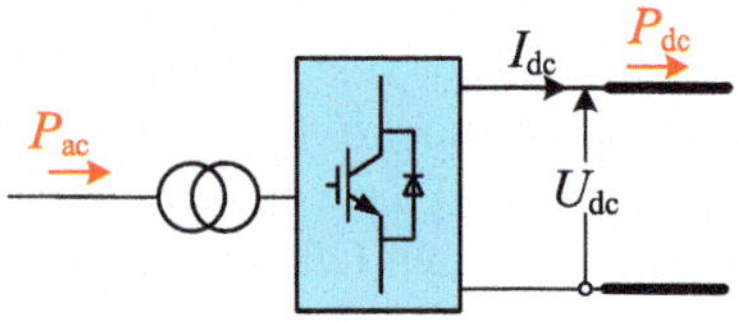

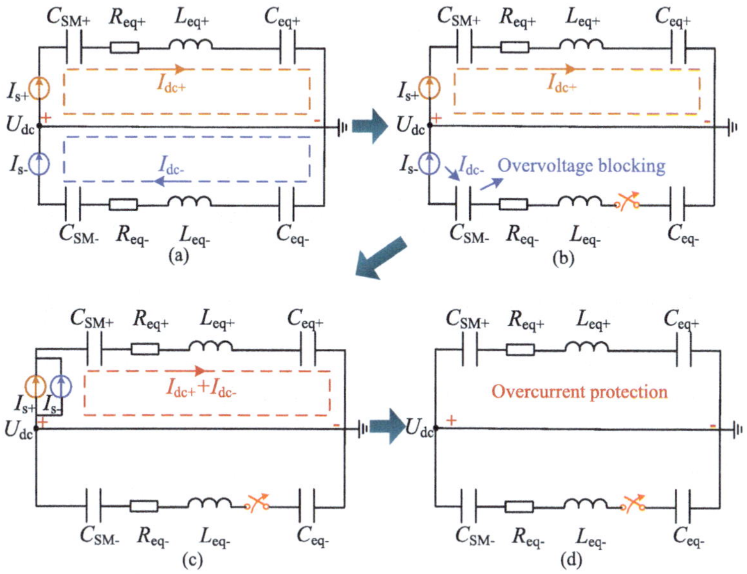

Fig. 6.3 Evolutionary process of single-pole breakage faults in DC cables with symmetrical bipolar MMC-HVDC. **a** Stage I: Non-fault state. **b** Stage II: Fault initial state. **c** Stage III: Fault pole blocking state. **d** Stage IV: Bipolar blocking state

$$\begin{cases} P_{\text{ac}} = \frac{1}{2} P_{\text{WFs}} \\ P_{\text{dc}} = U_{\text{dc}} I_{\text{dc}} \end{cases} \tag{6.1}$$

The symmetrical bipolar MMC-HVDC equivalent circuit is illustrated in Fig. 6.3. Here, $C_{\text{SM}+}$ and $C_{\text{SM}-}$ represent the total capacitances of the SM bridge arms for the positive and negative MMCs within the offshore converter station. $R_{\text{eq}+}$, $R_{\text{eq}-}$, $L_{\text{eq}+}$, and $L_{\text{eq}-}$ denote the equivalent resistance and equivalent inductance of the positive and negative bipolar system, respectively. $C_{\text{eq}+}$ and $C_{\text{eq}-}$ refer to the equivalent capacitances of the positive and negative bipolar system, excluding $C_{\text{SM}+}$ and $C_{\text{SM}-}$.

As shown in Fig. 6.3, the progression of a single-pole breakage fault in symmetrical bipolar MMC-HVDC systems unfolds through four distinct stages

- **Stage I: Non-faulty state**. As depicted in Fig. 6.3a, in this stage, both the positive and negative bipolar systems transmit half of the power generated by the offshore wind farms. The operation of both systems remains stable. The positive and negative currents can be expressed as per (6.1).

$$I_{\text{dc}+} = I_{\text{dc}-} = \frac{P_{\text{WFs}}}{2U_{\text{dc}}} \tag{6.2}$$

- **Stage II: Fault initial state**. In Fig. 6.3b, during a single-pole breakage fault in the submarine cable (specifically the negative pole), the power transmission capability of the negative pole is disrupted. Wind farm output continues to charge the submodule capacitor in the negative MMC. Ignoring current fluctuations and combining this with (6.2), the voltage across the submodule capacitance during charging (U_{SM-}) can be expressed using (6.3). Here, t_{ch} denotes the charging duration, and U_{SM0-} represents the initial voltage across the submodule capacitance before charging.

$$U_{SM-} = \frac{I_{dc-} \cdot t_{ch}}{C_{SM-}} + U_{SM0-} = \frac{P_{WFs} \cdot t_{ch}}{2U_{dc}C_{SM-}} + U_{SM0-} \tag{6.3}$$

- **Stage III: Fault pole blocking state**. In Fig. 6.3c, the negative MMC enters a blocked state when the voltage across its SM capacitor exceeds the protection threshold due to continuous charging. Consequently, the power from the wind farms, initially transmitted through the negative terminal, shifts to the positive pole, leading to an increase in positive current. The total positive current (I_{Tol}) can be mathematically described as

$$I_{Tol} = I_{dc+} + I_{dc-} = \frac{P_{WFs}}{U_{dc}} \tag{6.4}$$

- **Stage IV: Bipolar blocking state**. In Fig. 6.3d, if the peak or duration of the positive overcurrent exceeds the protection threshold, the positive MMC becomes blocked, resulting in the complete de-operation of the MMC-HVDC system.

Based on the analysis above, it's evident that the WFMMC lacks control over the power input from wind farms to the HVDC system. During a single-pole breakage fault, if the non-faulty pole experiences overcurrent, the MMC-HVDC system will cease operation entirely. Thus, the offshore converter station's ability to control energy also enables it to suppress overcurrent issues.

6.2 Steady-State Control Scheme and Transient MLCC Scheme for MMC-HVDC

6.2.1 MMC-HVDC Steady-State Control Scheme With Energy Control Capability

The single converter employs an HBMMC topology, depicted in Fig. 5.2. Each SM capacitor, with capacitance C_{sm} [2], serves as the primary controllable energy storage element. Considering only SM capacitance, the rated energy storage capacity of a single MMC is

$$E_{MMC0} = \frac{1}{2} \times C_{sm} \times U_{sm}^2 \times 6N \tag{6.5}$$

where, U_{sm} denotes the rated capacitor voltage of a single SM and N represents the number of SMs in a single bridge arm.

As depicted in Fig. 6.2, the correlation between the internal energy of WFMMC and the power on the AC and DC sides can be formulated as:

$$\frac{\mathrm{d}E_{\text{WFM}}}{\mathrm{d}t} = \frac{S_N}{E_{\text{MMC0}}} \cdot (P_{\text{ac}} - P_{\text{dc}}) \tag{6.6}$$

where, S_{N} represents the base value of energy, typically set to the rated power output of wind farms (P_{WF0}).

From (6.1) and (6.6), the unipolar DC current can be expressed as:

$$I_{\text{dc}} = -\frac{\mathrm{d}E_{\text{WFM}}}{\mathrm{d}t} \cdot \frac{E_{\text{MMC0}}}{S_N U_{\text{dc}}} + \frac{P_{\text{WFs}}}{2U_{\text{dc}}} \tag{6.7}$$

Hence, the relationship between DC current control and energy control can be established as:

$$I_{\text{dcref}} = k_{\text{p}}(E_{\text{WFMref}} - E_{\text{WFM}}) + k_{\text{i}} \int (E_{\text{WFMref}} - E_{\text{WFM}})\mathrm{d}t \tag{6.8}$$

where, E_{WFMref} and I_{dcref} represent the reference values for WFMMC internal energy and DC current, respectively. The parameters k_{p} and k_{i} denote the proportional and integral gains of the PI controller.

Therefore, as depicted in Fig. 6.1, the AC/DC current dual control strategy for the half-bridge WFMMC is proposed based on the hybrid MMC control strategy [3]. This dual control strategy comprises an AC current control loop and a DC current control loop. The AC current loop employs AC voltage control to maintain a stable grid-connected voltage for wind farms. The DC current loop utilizes energy control as per (6.8), enabling WFMMCs to regulate the power output from wind farms to the DC submarine cables.

Regarding the control strategy for the onshore GSMMC, shown in Fig. 6.1, conventional current vector control is applied [4], where the d-axis and q-axis of AC current control manage DC voltage and reactive power, respectively [5].

6.2.2 MLCC Scheme for Transient DC Overcurrent Suppression

To effectively mitigate overcurrents caused by single-pole submarine cable breakage, an MLCC scheme is proposed. Illustrated in Fig. 6.4, this MLCC scheme operates across four primary levels: passive energy absorption (PEA) for the fault pole WFMMC, active boost current limiting (ABCL) control for the non-fault pole GSMMC, active energy control (AEC) for the non-fault pole WFMMC, and active power reduction (AAPR) for wind farms.

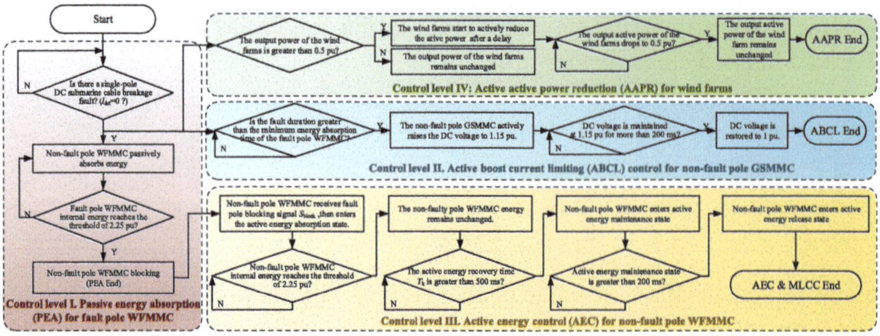

Fig. 6.4 Flowchart of the algorithm for the MLCC scheme

6.3 Control Structure and Characteristics of MLCC

The MLCC employs a four-level structure to enhance DC overcurrent suppression by coordinating each level through control logic and response timing. This section will provide a detailed analysis of the control scheme and timing for each level.

6.3.1 Control Level I: PEA for Fault Pole WFMMC

Analyzing Fig. 6.3 reveals that following a single-pole breakage fault in the submarine DC cable, the wind farm initially charges the fault pole WFMMC until the protection threshold is reached. Therefore, the short-term power absorption capability of the fault pole WFMMC can be utilized to delay the activation of other control measures.

In the case study of the Xiamen symmetrical bipolar MMC-HVDC project in China, Table 6.1 presents the protection settings for the MMC SM. Table 6.1 illustrates that under the condition of avoiding SM overvoltage protection activation, the average values of single module capacitor voltage and SM capacitor voltage maintain safety margins of 0.7 pu and 0.6 pu, respectively.

To ensure operational stability, additional voltage margins must be reserved for the non-fault pole WFMMC to prevent voltage overshoots that could trigger SM protection during internal energy rises. Similarly, the fault pole WFMMC requires an

Table 6.1 Voltage protection threshold and action of the SM

Protection type	Overvoltage protection of single module capacitor	Average value protection of capacitor voltage of all SMs
Threshold	1.7 pu	1.6 pu
Action	Bypass + Blocking	Bypass + Blocking

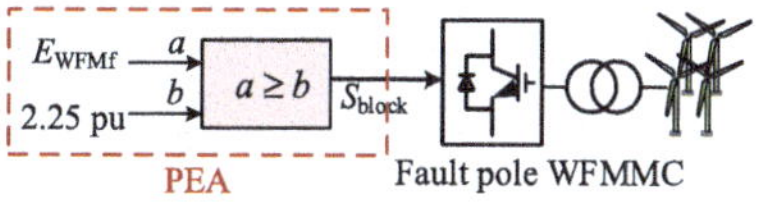

Fig. 6.5 PEA controller for fault pole WFMMC

additional voltage margin to prevent overvoltage shocks during its blocking operation. Hence, both fault and non-fault WFMMCs have their upper voltage limits set at 1.5 pu. Each WFMMC has the capacity to store extra energy beyond its rated energy storage:

$$\Delta E_{\mathrm{WFM}} = \frac{1}{2} \times C_{\mathrm{sm}} \times (1.5^2 - 1) \times U_{\mathrm{sm}}^2 \times 6N = 1.25 E_{\mathrm{MMC0}} \tag{6.9}$$

The passive energy absorption (PEA) controller for the fault pole WFMMC is depicted in Fig. 6.5. Once the internal energy of the fault pole WFMMC (E_{WFMf}) surpasses 2.25 pu, it initiates active blocking and sends a blocking signal (S_{block}) to the non-fault pole WFMMC. The fault pole WFMMC reaches its protection threshold for internal energy in the shortest time when wind farms are operating at rated power (P_{WF0}). The minimum energy rise time is:

$$t_{\min} = \frac{\Delta E_{\mathrm{WFM}}}{0.5 P_{\mathrm{WF0}}} \tag{6.10}$$

6.3.2 *Control Level II: ABCL for Non-Fault Pole GSMMC*

When transmitting a specified power through the DC submarine cable, raising the DC voltage can effectively decrease the DC current. Thus, the active boost current limiting (ABCL) control scheme is proposed for the non-fault pole GSMMC. The controller configuration is illustrated in Fig. 6.6.

Upon detecting a sudden drop in the fault pole current (I_{dcf}) to zero, indicating a fault occurrence, the non-fault pole GSMMC initiates a delay of $t_{\min}$ before increasing the DC voltage to the ABCL level within 5 ms. This ABCL level is maintained for a specific duration, after which the DC voltage returns to 1 pu.

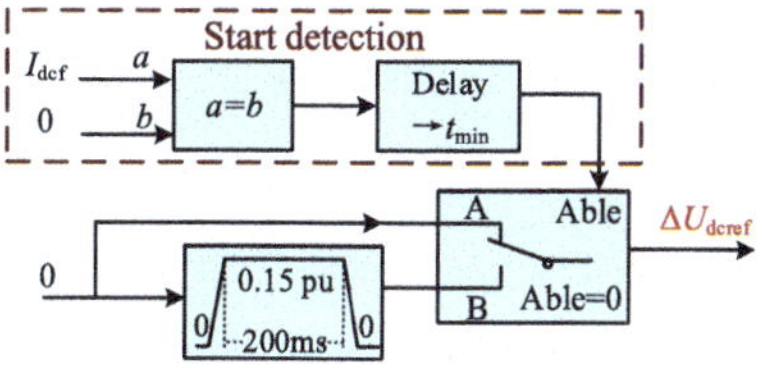

Fig. 6.6 ABCL controller for non-fault pole GSMMC

Table 6.2 Single-pole overvoltage and overcurrent capability of MMC

Operating conditions	Tolerance time
$(1.02\sim1.15)U_{dc0}$	Long-term tolerance
$(1.15\sim1.6)U_{dc0}$	>5 min
$1.7U_{dc0}$	Tolerate 500 ms in blocking state
$(1.02\sim1.15)I_{dc0}$	Long-term tolerance
$(1.15\sim1.3)I_{dc0}$	>1 s
$2I_{dc0}$	>10 ms
$3I_{dc0}$	>2 ms

In an MMC-HVDC project in China, the DC line is required to sustain an overvoltage of 1.15 pu for at least 5 min, with a robust tolerance against strong overcurrent that lasts from milliseconds to a hundred milliseconds, as detailed in Table 6.2. Considering the time required for overcurrent suppression and the line's voltage tolerance, the ABCL level is set at 1.15 pu and maintained for 200 ms.

6.3.3 Control Level III: AEC for Non-Fault Pole WFMMC

When the fault pole WFMMC is blocked, all wind farm power output is channeled through the non-fault pole. The non-fault pole WFMMC leverages the SM capacitance voltage margin to precisely regulate the injection of wind power into the DC cable. The control scheme for this purpose is illustrated in Fig. 6.7.

AEC consists of three stages: initiation detection, active energy absorption, and active energy maintenance and release.

During initiation detection, AEC initiates upon detecting both zero fault pole current and receipt of a fault pole blocking signal.

For active energy absorption, the non-fault pole WFMMC dynamically adjusts its energy absorption rate to maximize surplus power absorption while preventing over-absorption. This adjustment is based on the difference between the actual power received from wind farms (P_{WFs}) and the rated power of the non-fault pole (P_{WFMn0}). The precise relationship governing this adjustment is detailed in Eq. (6.11).

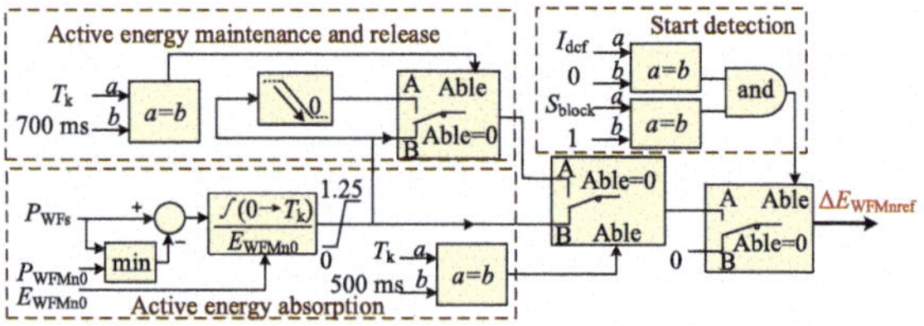

Fig. 6.7 AEC controller for non-fault pole WFMMC

$$\Delta E_{\mathrm{WFMnref}} = \begin{cases} \frac{\int_0^{T_k} (P_{\mathrm{WFs}} - P_{\mathrm{WFMn0}})dt}{E_{\mathrm{WFMn0}}}, P_{\mathrm{WFs}} > P_{\mathrm{WFMn0}} \\ 0, P_{\mathrm{WFs}} \leq P_{\mathrm{WFMn0}} \end{cases} \tag{6.11}$$

where, E_{WFMnref} represents the energy reference value for the non-fault pole WFMMC, T_k denotes the duration of the AEC process, and E_{WFMn0} signifies the rated energy capacity of the non-fault pole WFMMC. The entire duration of the active energy absorption process is dynamically adjusted based on the startup delay and the rate at which power decreases from wind farms. Typically, wind farms can scale down their output power from rated to 0 within milliseconds to seconds [6]. It is assumed that wind farms can reduce their output to 0.5 pu within 200 ms after activating power reduction control. Therefore, the active energy absorption stage is set to last 500 ms to mitigate the impact of wind farm startup delays and active power reduction rates.

Upon completing the active energy absorption stage, the AEC transitions into a 200 ms energy maintenance phase to stabilize the system. Subsequently, the non-fault pole WFMMC actively supplies power to the DC submarine cable until its internal energy is restored to 1 pu.

6.3.4 *Control Level IV AAPR for Wind Farms*

The active power reduction controller for wind farms is illustrated in Fig. 6.8, where P_{WFsref} denotes the reference value for the output active power of the wind farms. Upon detecting a fault at the offshore converter station, such as through the fault pole current I_{dcf}, a signal is sent to the wind farms to initiate active power reduction. The wind farms adjust their power output to 0.5 pu after a communication delay of several tens of milliseconds. According to [6], the rate at which wind farm power decreases is 0.0025 pu per millisecond. If the wind farm's output power does not reach 0.5 pu at the time of the fault detection, the power reduction signal is not activated.

6.3.5 *Timing Characteristics of MLCC in Multiple Scenarios*

The timing characteristics of MLCC are depicted in Fig. 6.9, where t_1 marks the fault onset moment (PEA start), t_2 denotes the ABCL activation moment, t_3 signifies the

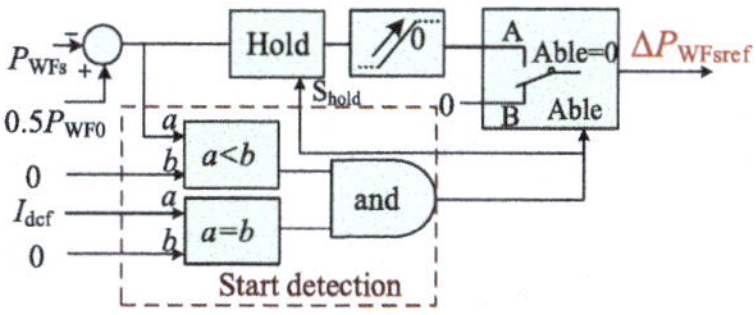

Fig. 6.8 AAPR controller for wind farms

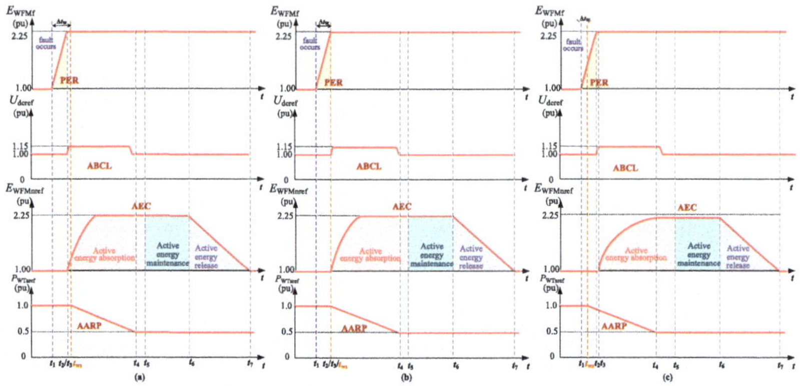

Fig. 6.9 Timing characteristics of the MLCC in three different scenarios. **a** Scenario 1: The wind farm start-up delay Δt_w is greater than the fault pole WFMMC's minimum energy rise time $t_{\min}$. **b** Scenario 2: The wind farm start-up delay Δt_w is equal to the fault pole WFMMC's minimum energy rise time $t_{\min}$. **c** Scenario 3: The wind farm start-up delay Δt_w is less than the fault pole WFMMC's minimum energy rise time $t_{\min}$

AEC initiation time, t_{ws} indicates the AAPR activation time, t_4 denotes the conclusion of AARP, t_5 marks the end of the active energy absorption stage, t_6 marks the end of the active energy maintenance stage, and t_7 indicates the end of active energy release (MLCC completion moment). The activation of AAPR in wind farms depends on communication from WFMMC. MLCC characteristics vary based on different communication delay scenarios.

The collaboration among fault pole WFMMC, non-fault pole WFMMC, non-fault pole GSMMC, and wind farms within the MLCC scheme is further elucidated below based on Fig. 6.9.

In scenario 1, the start-up delay of the wind farm Δt_w is greater than the minimum energy rise time of the fault pole WFMMC $t_{\min}$. Timing of the MLCC can be divided into 4 stages as follows.

Stage I ($t < t_1$): The system is in a state of equilibrium, with the internal energy of both the fault and non-fault poles stabilised at a value of 1 pu. The DC voltage reference value is 1 pu.

Stage II ($t_1 \leq t \leq t_2$ or t_3): At moment t_1, the fault occurs. The fault pole WFMMC passively absorbs the power output from wind farms and its internal energy E_{WFMf} rises. When the wind farm outputs the rated active power, the internal energy of the fault pole WFMMC reaches the blocking threshold of 2.25 pu at $t_1+t_{\min}$. This is the moment when the ABCL is activated and the non-fault pole WFMMC receives the blocking signal to start the AEC as well. Hence, the t_2 and t_3 moments coincide.

Stage III ($t_3 < t < t_5$): As a consequence of the communication delay, the wind farms commence the AARP stage during this phase. At this point in time, designated as t_4, the output power from the wind farms is reduced to a level of 0.5 pu. The slope of the active energy absorption stage is adjusted by the AEC in accordance with (6.11). Upon reaching a value of 2.25 pu, the E_{WFMnref} does not increase any further.

It is notable that, should the wind farm output power be less than the rated power of the non-fault pole DC submarine cable at the time of a fault, the AARP will not be activated by the wind farm in question, nor will the AEC.

Stage IV ($t_5 \leq t \leq t_7$): At this point, designated as t_5, the active energy absorption stage of the AEC reaches its conclusion. This is followed by a 200-millisecond active energy retention phase, which is then succeeded by active energy release. At the present moment, t_7, the MLCC has reached its end. The symmetric bipolar MMC-HVDC now operates in a new steady state, which is characterised by the absence of any bipolar operation and instead features solely monopolar operation.

It is possible to divide scenarios 2 and 3 into the four stages of scenario 1. In contrast, in scenario 2, the times t_2, t_3, and t_w overlap as a consequence of the equality of the time differences:t_w=t_{min}. In scenario 3, the PER operates for a longer period than the minimum time threshold, due to the ability of the AAPR to commence functioning rapidly. Furthermore, the internal energy of the non-fault pole WFMMC may not reach the requisite threshold level by the time the wind farms complete the AARP.

6.4 Simulation Verification

In reference to the parameters of the Xiamen OWF-MMC-HVDC project in China, the primary parameters of the offshore wind power system, as illustrated in Fig. 6.1, are presented in Table 6.3. When the output power of wind farms is rated power, the issue of a power surplus resulting from a DC submarine cable fault represents a significant challenge. Accordingly, the output power of the wind farms in the event of a fault is set to be equal to P_{WF0}. It can be calculated from Eqs. (6.9) and (6.10) that t_{min} is equal to 50 ms. The energy release slope $k_{re} = -0.002$ pu/ms.

Table 6.3 Main parameters of offshore wind power test system

System parameter	Number
Rated DC voltage U_{dc0}	±320 kV
Rated DC transmission power P_{dc0}	±600 MW
MMC SM capacitance C_{sm}	12500 μF
Number of individual bridge arm SMs N	160
Rated voltage MMC SM U_{sm}	2 kV
Rated power of wind farms P_{WF0}	1200 MW
Base value of energy S_N	1200 MW

6.4.1 Verification of Fault Characteristics

Initially, validate the fault characteristics of single-pole submarine cable breakage in the symmetrical bipolar MMC-HVDC system. Simulation parameters include a fault occurring on the negative submarine cable at 3 s, with no configuration of MLCC or cable overcurrent protection. Figures depict the positive and negative power transmission of the DC submarine cable (P_{dc+} and P_{dc-}), DC currents (I_{dc+} and I_{dc-}), and the internal energy variations of the positive and negative WFMMC (E_{WFM+} and E_{WFM-}), as illustrated in Fig. 6.10.

- Before 3 s, the system operates stably, validating the correctness of the MMC-HVDC steady-state control strategy depicted in Fig. 6.1.
- Following the fault occurrence, the negative WFMMC absorbs output power from the wind farm passively, resulting in an increase in its internal energy. No DC overcurrent occurs during this period.
- Upon exceeding a predefined threshold, the internal energy of the negative WFMMC triggers its blocking. The power initially destined for the negative WFMMC transfers to the positive WFMMC, causing an overcurrent in the positive DC cable. The peak positive overcurrent reaches 2.15 pu. Based on Table 6.2, there is a significant likelihood of triggering the positive WFMMC to block due to overcurrent during this phase.

6.4.2 Verification of the MLCC Scheme

The performance of the MLCC is analyzed and validated under different AARP start-up delays for wind farms. In the simulation, a break fault occurs on the negative

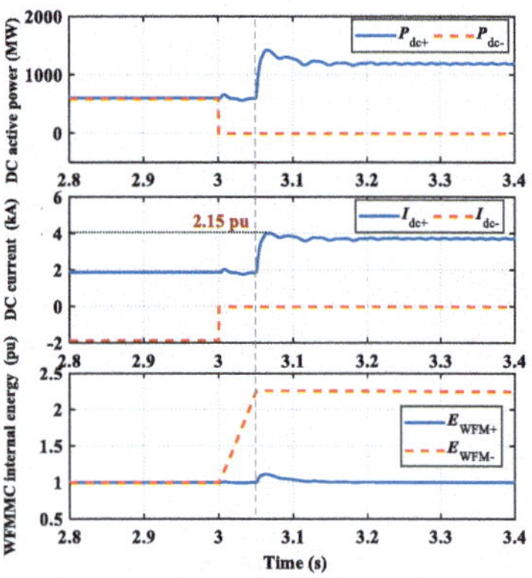

Fig. 6.10 Characteristics of single-pole submarine cable breakage

submarine cable at 3 s, with MLCC applied to the MMC-HVDC system. The AARP for wind farms is activated at 30 ms, 50 ms, and 80 ms after the fault. The simulation results are shown in Fig. 6.11.

- In the scenarios depicted in Fig. 6.11a, b, and c, each control link within the MLCC operates effectively, consistent with the characteristics shown in Fig. 6.10. Following the fault occurrence, the negative WFMMC promptly absorbs wind farm output power based on the PEA. Subsequently, the positive GSMMC and WFMMC activate ABCL and AEC, respectively, to suppress positive overcurrents. Wind farm output power is maintained above 0.5 pu after dropping.
- With decreasing start-up delay of AARP, the capability of MLCC to suppress overcurrents gradually improves. At $\Delta t_w = 80$ ms, the peak positive overcurrent is 1.45 pu; at $\Delta t_w = 50$ ms, it is 1.33 pu; and at $\Delta t_w = 30$ ms, it reduces to 1.24 pu. In all scenarios, the overcurrent peak is notably lower compared to scenarios without MLCC (2.15 pu).
- As shown in Fig. 6.11a and b, when the internal energy of the positive WFMMC reaches the threshold, a second overcurrent peak occurs due to the ongoing AARP process of wind farms. Figure 6.11c demonstrates that if the AARP process concludes before the internal energy of the positive WFMMC reaches the threshold, a second current peak does not occur.

Six different schemes are configured to evaluate the current suppression capabilities of the MLCC, as detailed in Table 6.4. Scheme I employs conventional wind farm power reduction control, considering the possibility of converter station blocking [6]. The AEC-CAR scheme uses Active Energy Control with a constant absorption rate, ensuring that during energy absorption, the rate matches half of the wind farms' rated power output, thereby maintaining rated power transmission at the non-fault pole. The proposed MLCC incorporates a variable absorption rate AEC (AEC-VAR), adjusting energy absorption based on the wind farm's output power.

The simulation is set up to simulate a break fault on the negative submarine cable at 3 s, with the MMC-HVDC system evaluating positive overcurrent suppression across these six schemes. The results are presented in Fig. 6.12. A comparative analysis of four key characteristics among the schemes is shown in Fig. 6.13. The criterion for

Table 6.4 Setting of six comparison schemes

Scheme number	Program content
Scheme I	PER + AARP
Scheme II	PER + AARP + ABCL
Scheme III	PER + AARP + AEC-CAR
Scheme IV	PER + AARP + AEC-VAR
Scheme V	PER + AARP + ABCL + AEC-CAR
Scheme VI	PER + AARP + ABCL + AEC-VAR (MLCC)

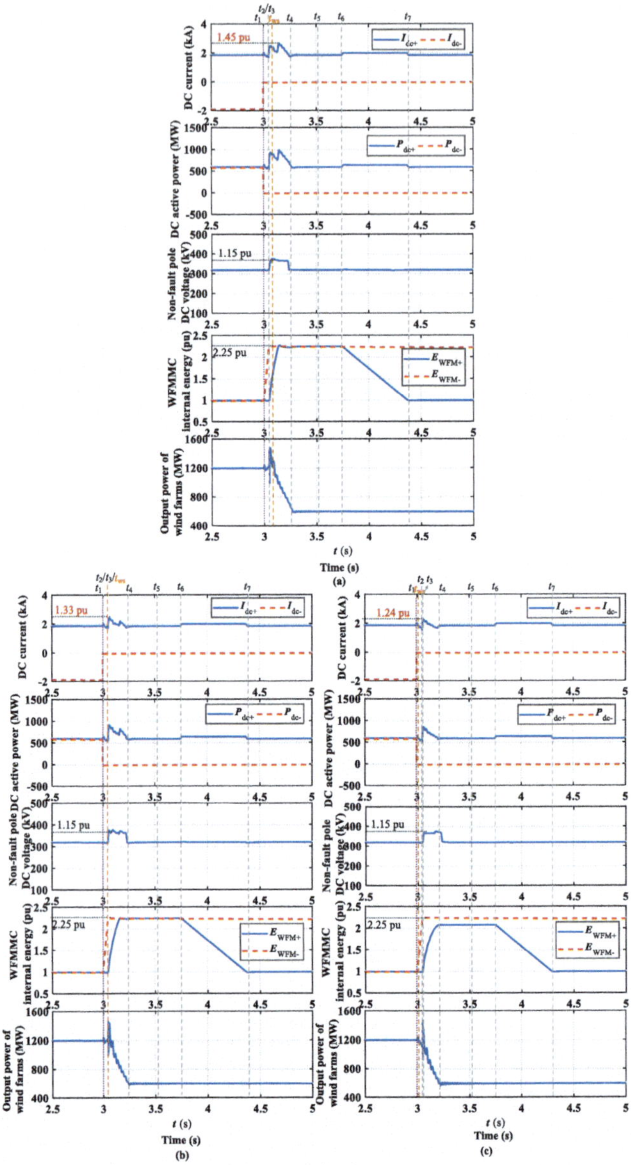

Fig. 6.11 MLCC characteristics at different start-up delays of AARP for wind farms when the minimum energy rise time of the fault pole WFMMC is 50 ms ($t_{\min} = 50$ ms). **a** The start-up delay of AARP is 80 ms ($\Delta t_{\mathrm{w}} = 80$ ms). **b** The start-up delay of AARP is 50 ms ($\Delta t_{\mathrm{w}} = 50$ ms). **c** The start-up delay of AARP is 30 ms ($\Delta t_{\mathrm{w}} = 30$ ms)

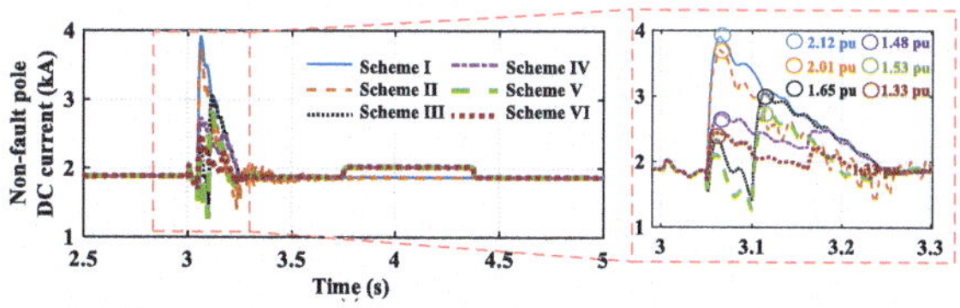

Fig. 6.12 Comparison of fault current characteristics under six DC overcurrent suppression schemes

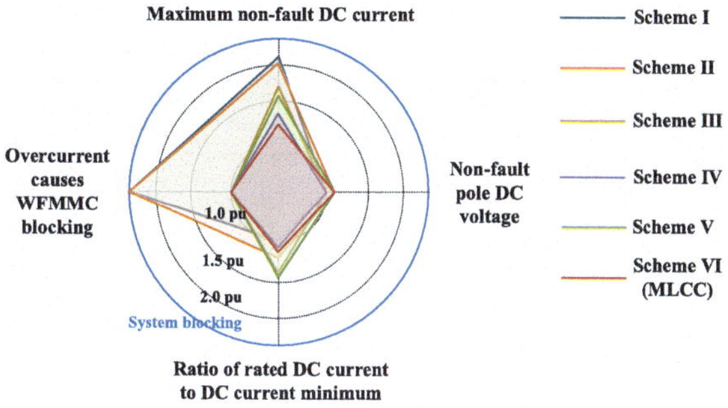

Fig. 6.13 A comprehensive performance comparison of six DC overcurrent suppression schemes

blocking the non-fault pole WFMMC is triggered if overcurrent amplitudes exceed 2 pu for more than 10 ms.

It can be found from Figs. 6.12 *and* 6.13*:*

- Scheme VI (MLCC) demonstrates the highest capability in suppressing overcurrents.
- Relying solely on wind farms and GSMMC to suppress overcurrents can lead to frequent blocking of the non-fault pole WFMMC. The proposed AEC for the non-fault pole WFMMC significantly enhances overcurrent suppression effectiveness.
- In Schemes III and V, the AEC-CAR consistently operates during the active energy absorption stage at a rate that matches the rated output power of wind farms. During AARP periods, this can lead to excessive absorption of wind farm output power, resulting in low DC current.
- The AEC-CAR causes rapid buildup of internal energy in the non-fault pole WFMMC to threshold levels. During this phase, insufficient reduction in wind farm output power through AARP processes leads to significant secondary current peaks.

6.4.3 Characteristics of MLCC Under Different Output Power of Wind Farms

Maintaining constant rated conditions for offshore wind farms is inherently challenging. The following analysis evaluates the performance of the MLCC under varying wind farm output power levels.

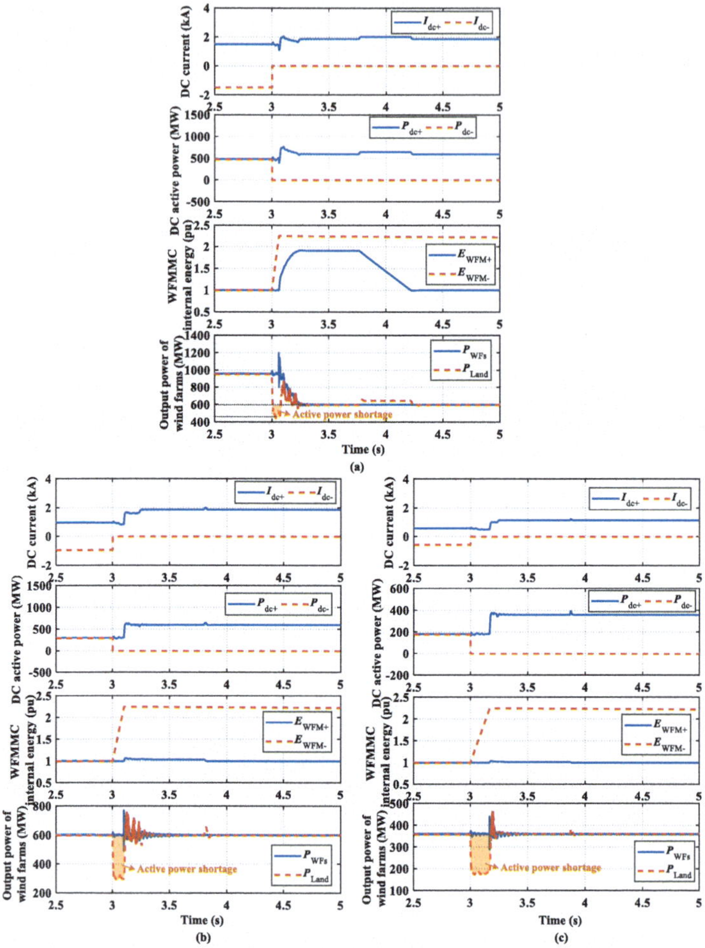

Fig. 6.14 The characteristics of MLCC under different output power of wind farms. **a** The output power of wind farms is 0.8 pu (960 MW). **b** The output power of wind farms is 0.5 pu (600 MW). **c** The output power of wind farms is 0.3 pu (360 MW)

In the simulation, a break fault occurs on the negative submarine cable at 3 s. Wind farm outputs are set to 0.8 pu (960 MW), 0.5 pu (600 MW), and 0.3 pu (360 MW). The system's response under MLCC operation is shown in Fig. 6.14, where P_{Land} represents the active power transmitted by the MMC-HVDC to the onshore AC grid.

Key observations from Fig. 6.14*:*

- When the wind farm output is below its rated capacity, the DC current in the submarine cable before the fault also remains below its rated level. After the fault, the MLCC effectively mitigates overcurrent on the non-fault pole.

- As indicated in Fig. 6.14b and c, the AEC ceases to absorb power from the wind farm when its output is at or below half of its rated power. Under these conditions, where no overcurrent occurs, the full output of the wind farm is successfully transmitted to the onshore AC grid.
- After the fault, the PEA link of the faulted pole WFMMC absorbs half of the wind farm's active power, which is then rerouted to the non-fault pole. This causes the power transmitted from the non-fault pole WFMMC to temporarily drop below the rated capacity of the single-pole submarine cable, resulting in a brief reduction in power delivered to the onshore grid.
- A potential solution to this issue is for the faulted pole WFMMC to automatically block itself if it detects that the active power sampled just before the fault exceeds half of the transmission capacity of the single-pole submarine cable. This would prevent the PEA process from continuing at the faulted pole after the fault occurs.

This analysis confirms the accuracy and effectiveness of the proposed MLCC, demonstrated through the coordination of PEA for the faulted pole WFMMC, ABCL for the non-fault pole GSMMC, AEC for the non-fault pole WFMMC, and AARP for the wind farms. In practical applications, PEA, AEC, and ABCL can be further supported by additional measures, such as installing energy-dissipating devices at offshore converter stations or selectively disconnecting wind farm units. These strategies can provide additional time or reduce capacity costs when implementing alternative solutions.

6.5 Summary

This chapter introduces a Multi-Level Coordinated Control (MLCC) scheme for symmetrical bipolar OWF-MMC-HVDC systems. The MLCC scheme utilizes the energy control capabilities of MMCs to manage DC overcurrents caused by single-pole submarine cable breakage. It coordinates PEA for the fault pole WFMMC, AEC for the non-fault pole WFMMC, ABCL for the non-fault pole GSMMC, and AARP for offshore wind farms to collectively suppress DC overcurrent at the non-fault pole, ensuring a smooth transition to a new steady-state operation of the MMC-HVDC system. The simulation results lead to several key findings:

- The proposed steady-state control strategy for MMC-HVDC ensures stable system operation.
- MLCC reduces the peak DC current in the non-fault pole from 2.15 pu to approximately 1.3 pu following a single-pole submarine cable breakage.
- AEC for the non-fault pole WFMMC significantly contributes to the suppression of DC overcurrent.
- MLCC effectively prevents MMC-HVDC system outages due to overcurrent.
- MLCC optimizes power transmission through non-fault pole submarine cables for wind farms operating below rated power while maintaining effective DC overcurrent suppression.

The complexity of the control logic in MLCC may impact the reliability of the MMC controller. In practical applications, ABCL or AEC can be simplified according to engineering overcurrent tolerance. Future research could explore the application of MLCC in DC grids to efficiently coordinate multiple converters for DC overcurrent suppression.

References

1. Li W, Zhu M, Chao P, Liang X, Xu D (2020) Enhanced FRT and postfault recovery control for MMC-HVDC connected offshore wind farms. IEEE Trans Power Syst 35(2):1606–1617. https://doi.org/10.1109/TPWRS.2019.2944940
2. Saad H, Guillaud X, Mahseredjian J, Dennetire S, Nguefeu S (2015) MMC capacitor voltage decoupling and balancing controls. IEEE Trans Power Deliv 30(2):704–712. https://doi.org/10.1109/TPWRD.2014.2338861
3. Xiang W, Lin W, Xu L, Wen J (2018) Enhanced independent pole control of hybrid MMC-HVdc system. IEEE Trans Power Deliv 33(2):861–872. https://doi.org/10.1109/TPWRD.2017.2715040
4. Zhou H, Yao W, Ai X, Zhang J, Wen J, Li C (2022) Coordinated power control of electrochemical energy storage for mitigating subsequent commutation failures of HVDC. Int J Electr Power Energy Syst 134:107455. https://doi.org/10.1016/j.ijepes.2021.107455
5. Fan X, Shu J, Zhang B (2018) IEEE Trans Power Syst 33(4):4625–4633. https://doi.org/10.1109/TPWRS.2018.2792008
6. Deng F, Chen Z (2013) Operation and control of a DC-Grid offshore wind farm under DC transmission system faults. IEEE Trans Power Deliv 28(3):1356–1363. https://doi.org/10.1109/TPWRD.2013.2261561

Chapter 7
Analysis and Mitigation Strategies of Fault Characteristics on the Offshore Wind Farm Side

Abstract Modular Multilevel Converter-based High Voltage Direct Current (MMC-HVDC) systems are increasingly recognized as the ideal solution for transmitting power over long distances from large-scale offshore wind farms (OWFs). However, offshore AC asymmetric faults pose significant fault ride-through (FRT) challenges for both OWFs and MMC-HVDC systems, particularly due to the involvement of power electronic converters on both sides of the fault. This chapter introduces a Multi-Stage Sequential Network Energy Control (MSNEC) strategy to tackle these challenges. The approach begins by clarifying the constraint relationships within sequence networks during AC asymmetric faults and explores how to maximize the energy control capacity of the WFMMC, focusing on energy transformation and transfer. The strategy reduces fault phase current amplitude by injecting submodules into the faulted phase while simultaneously introducing additional submodules into the non-faulted phase to achieve a series voltage-dividing effect, actively preventing overvoltage in the non-faulted phase. The MSNEC strategy thoroughly addresses FRT requirements across different stages: fault initiation, steady state, and recovery. It combines sequential network control, energy management, and voltage support from the converter station. The approach achieves AC asymmetric FRT through four key aspects: power equivalence, negative sequence suppression, energy control, and voltage stabilization. A simulation model of OWFs connected to the onshore grid via MMC-HVDC, implemented in PSCAD/EMTDC, validates the effectiveness of MSNEC. The results show that MSNEC not only enhances the AC asymmetric FRT capability of OWFs and WFMMC but also promotes efficient renewable energy utilization and improves the economic feasibility of OWF grid connections.

7.1 Sequential Network Mechanism Analysis of Offshore AC Asymmetric Faults

The topology of OWFs connected to an onshore AC grid via MMC-HVDC, along with the MSNEC approach, is shown in Fig. 7.1. The OWFs use direct-drive permanent magnet synchronous generators (PMSGs) that are rated to match the capacity

W. Yao et al., *Active Control of Large-Scale Offshore Wind Farms Connected Via VSC-HVDC*, Power Systems, https://doi.org/10.1007/978-981-97-9346-4_7

of the WFMMC. Each PMSG is linked to a back-to-back power electronic converter comprising a RSC and a GSC. Both the WFMMC and GSMMC are built using half-bridge sub-modules. The OWFs connect to the WFMMC through an offshore AC bus, as described in [1]. Since converters are present at both ends of the offshore AC section, the fault characteristics manifest as power supplies controlled by electronics on both sides. Therefore, effective fault ride-through (FRT) necessitates coordinated FRT between the OWF and WFMMC.

This chapter outlines the structure of MMC-HVDC integrated offshore wind farms and the control strategies they employ. The WFMMC uses a combination of Negative Sequence Current Control (NSCC) and Energy Control (EC) to regulate the AC modulation ratio, while the GSC applies internal loop current limiting to manage fault current and provide reactive power compensation during faults. The lower part of Fig. 7.1 depicts the multi-stage control of the proposed MSNEC, addressing different phases of the FRT process.

In the fault identification phase, a local fault detection device recognizes the fault occurrence at time t_0 by detecting changes in the d-axis voltage amplitude and the currents in the three ABC phases. After a 20 ms delay, the system enters the FRT control phase. During this phase, the WFMMC's adaptive EC and positive and negative sequence separation control are activated, along with the OWF's reactive

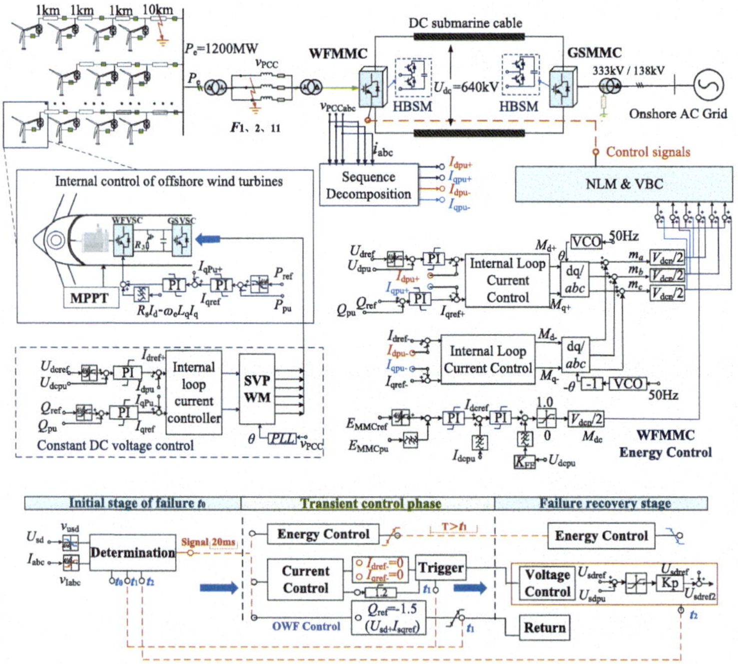

Fig. 7.1 The MSNEC and structure of MMC-HVDC integrated offshore wind farms

power compensation and current limiting control. Once the fault is cleared at t_1, the system transitions to the fault recovery phase. Here, the WFMMC initiates voltage compensation control, and the stored energy is gradually released back to the onshore AC grid. Simultaneously, the OWF returns to its initial steady-state control mode. The FRT process concludes at time t_2.

7.1.1 Equivalent Circuit Analysis of Offshore AC Asymmetric Faults

OWFs transmit offshore wind energy via submarine AC cables, which are collected at the WFMMC to convert from AC to DC and then transmitted to the onshore grid through submarine DC cables. Offshore faults are categorized into AC faults and DC faults [2, 3], with AC asymmetric faults being the most common. The fault current during offshore AC faults is supplied by power electronic converters on both sides, making the fault current characteristics complex. To analyze the fault characteristics during an offshore AC asymmetric fault, the WFMMC and the GSVSC of OWFs are considered power-equivalent. Unlike conventional synchronous machines, VSC and MMC can be modeled not only as positive sequence (PS) power sources but also as negative sequence (NS) power sources, contributing NS components to the NS loop. Additionally, the impedance of the line and converter is combined to represent the overall system impedance, leading to the equivalent circuit shown in Fig. 7.2. The combination of S_a, S_b, S_c and S_F represents offshore AC single-phase-to-ground (SPG) faults, double-phase-to-ground (DPG) faults, and phase-to-phase (PP) faults. Based on Fig. 7.2, the equivalent circuit relationship during the fault is derived as shown in Eq. (7.1).

When power electronic converters on both sides are equated, the WFMMC typically serves as a voltage support source for an offshore wind farm during normal operation. However, this equivalence to a voltage source is valid only if the inner-loop current controller operates within its control margins without reaching its limit. If the WFMMC reaches its current limit during a fault, it transitions from an equivalent voltage source to an equivalent current source, thereby losing its grid-supporting capability as a voltage source. On the other hand, the GSVSC of the OWFs is approximated as a current source under fault conditions, as the fault current amplitude reaches the inner loop control limit.

$$\begin{cases} E_{\mathrm{MMC}}^{abc} + Z_{\mathrm{MMC}} I_{\mathrm{MMC}}^{abc} = U_{\mathrm{F}}^{abc} \\ E_{\mathrm{OWF}}^{abc} + Z_{\mathrm{OWF}} I_{\mathrm{OWF}}^{abc} = U_{\mathrm{F}}^{abc} \end{cases} \tag{7.1}$$

where, E_{MMC}^{abc} represents the equivalent electromotive force at the MMC side, E_{OWF}^{abc} represents the equivalent electromotive force at the OWF side, Z_{MMC} represents the equivalent equivalent reactance at the MMC side, Z_{OWF} represents the equivalent reactance at the OWF side, I_{MMC}^{abc} represents the AC current at the MMC side, I_{OWF}^{abc} represents the AC current at the OWF side, and U_{F}^{abc} represents the fault voltage at the location.

Then the corresponding voltage and current relationship at the location of the fault is deduced as shown in (7.2).

$$\begin{cases} Z_{\mathrm{F}} I_{\mathrm{F}}^{abc} = U_{\mathrm{F}}^{abc} \\ I_{\mathrm{F}}^{abc} = I_{\mathrm{MMC}}^{abc} + I_{\mathrm{OWF}}^{abc} \end{cases} \tag{7.2}$$

The PMSG transmits power via a back-to-back full power converter equipped with an internal current limiter. Thus, the magnitude of fault current remains within predefined limits, making the short-circuit current provided by OWFs resemble a limited-magnitude current source. Upon the occurrence of a fault, the converter's short-circuit current is dictated by the fault ride-through control objectives. During the transient response of an offshore wind farm inverter, the dq-axis current response is predominantly governed by the current control loop, with the impact of the outer control loop being negligible [4].

$$\begin{gathered} \begin{cases} \frac{\mathrm{d}^2 i_d}{\mathrm{d}t^2} + \frac{k_{\mathrm{ip}}}{L}\frac{\mathrm{d}i_d}{\mathrm{d}t} + \frac{k_{\mathrm{ii}}}{L} i_d = \frac{k_{\mathrm{ii}}}{L} i_d^* \\ \frac{\mathrm{d}^2 i_q}{\mathrm{d}t^2} + \frac{k_{\mathrm{ip}}}{L}\frac{\mathrm{d}i_q}{\mathrm{d}t} + \frac{k_{\mathrm{ii}}}{L} i_q = \frac{k_{\mathrm{ii}}}{L} i_q^* \end{cases} \Rightarrow \begin{cases} i_{\mathrm{d}} = i_{\mathrm{d}}^* + (i_{\mathrm{d0}} - i_{\mathrm{d}}^*)\frac{\mathrm{e}^{-\xi\omega_{\mathrm{n}}t}\sin(\omega_{\mathrm{d}}t+\beta)}{\sqrt{1-\xi^2}} \\ i_{\mathrm{q}} = i_{\mathrm{q}}^* - \frac{i_{\mathrm{q}}^*\mathrm{e}^{-\xi\omega_{\mathrm{n}}t}\sin(\omega_{\mathrm{d}}t+\beta)}{\sqrt{1-\xi^2}} \end{cases} \\ I_{\mathrm{OWF}_i} = I_{\mathrm{dref+}}\cos(\omega t + \theta_{\varphi}) - I_{qref+}\sin(\omega t + \theta_{\varphi_i}) \\ \Rightarrow +(I_{d0} - I_{dref+})\frac{e^{-\xi\omega_{\mathrm{n}}t}\sin(\omega_{\mathrm{d}}t+\beta_i)}{\sqrt{1-\xi^2}}\cos(\omega t + \theta_{\varphi_i}) + I_{\mathrm{qref+}}\frac{e^{-\xi\omega_{\mathrm{n}}t}\sin(\omega_{\mathrm{d}}t+\beta_i)}{\sqrt{1-\xi^2}}\sin(\omega t + \theta_{\varphi_i}) \end{gathered} \tag{7.3}$$

where, ξ is the damping ratio of the second-order system; ω_{n} is the second-order is the natural oscillation angle frequency of the system; ω_{d} is the damped oscillation frequency; β_i is the damping angle; and ω is the fundamental frequency angular frequency. θ_{φ_i} is the initial phase angle after the occurrence of φ phase fault.

According to the low voltage FRT requirements of OWFs, OWFs are required to output reactive current to support the grid voltage recovery when the grid voltage is 0.2 pu-0.9 pu. Then, during the fault period, the reactive current is required to satisfy the following equation.

$$i_{\mathrm{sqref}} \geq 1.5 \times (0.9 - U_{\mathrm{sd}}) I_{\mathrm{s}} \tag{7.4}$$

where, U_{sd} represent the d-axis component of the grid-connected voltage.

The reference value is determined based on the d-axis component of the grid-connected voltage at the offshore wind farm. During single-phase, two-phase, and three-phase faults, this d-axis voltage component will decrease to varying extents, triggering the GSC to activate reactive current compensation control.

Given the limitations of the GSC, the maximum reference value for the inner loop current is capped at 1.2. The reference value for the d-axis current is then calculated as follows:

$$i_{\mathrm{sdref}} = \sqrt{i_{\mathrm{smax}}^2 - i_{\mathrm{sqref}}^2} = \sqrt{1.2^2 - i_{\mathrm{sqref}}^2} \tag{7.5}$$

where, i_{smax} represent inner loop current limit value of GSC. i_{sdref} represents the D-axis current reference value of the inner loop controller. i_{sqref} represents the reference value of the q axis current of the inner loop controller.

Thus, (7.6) can be substituted into (7.4) to obtain the equivalent reference value requirements of reactive power during the fault. The reference value of the transient reactive power output from the GSC during a voltage dip is obtained as the following equation. The reactive power reference satisfies the lower limit of the FRT requirement.

$$Q_{\text{ref}} = -\frac{3}{2}U_{\text{sd}}i_{\text{sqref}} \geq -1.5 \times U_{\text{sd}} \times 1.5 \times (0.9 - U_{\text{sd}})I_s \tag{7.6}$$

where, Q_{ref} represent transient reactive power reference output from the GSC.

During the fault, the fault current of the WFMMC contains both PS and NS components. And the WFMMC internal loop current controller also has a current limiting link, combined with (7.3) to obtain the expression of the fault current.

$$\begin{aligned} I_{F\,\text{max}} &= \tfrac{333}{230} I_{\text{MMC max}} + \sum_{i=1}^{n} I_{\text{OWF}_i\,\text{max}} \\ &= \tfrac{333}{230}\left[\tfrac{E_{\text{MMC}}\cos(\omega t+\theta_\varphi)}{Z_{\text{eq}}} + k(\tfrac{E_{\text{MMC}}}{Z_{\text{eq}}} + |i_{\text{OWF}}|)\cdot\cos(\omega t+\theta_\varphi)\right] + \sum_{i=1}^{n} I_{\text{OWF}_i\,\text{max}} \end{aligned} \tag{7.7}$$

where, Z_{eq} represent equivalent impedance between WFMMC and fault location.

From Eq. (7.7) above, it can be obtained that the fault current provided by the PMSG will also be reflected in the WFMMC fault current. This makes the WFMMC fault current amplitude may exceed the power electronics withstand capability, so the WFMMC needs to be equipped with over-current FRT capability. Since the analysis object is AC asymmetric fault, it is necessary to focus on PS and NS network diagram analysis. Firstly, the three-phase circuit parameters are converted into those of the circuit in the sequential network.

$$\begin{cases} P_{\text{Fault}}^{abc} = AP_{\text{Fault}}^{+-0},\ P_{\text{Fault}}^{+-0} = A^{-1}P_{\text{Fault}}^{abc} \\ A = \begin{bmatrix} 1 & 1 & 1 \\ 1 & \alpha & \alpha^2 \\ 1 & \alpha^2 & \alpha \end{bmatrix},\ A^{-1} = \frac{1}{3}\begin{bmatrix} 1 & \alpha & \alpha^2 \\ 1 & \alpha^2 & \alpha \\ 1 & 1 & 1 \end{bmatrix} \end{cases} \tag{7.8}$$

where, $\alpha = 1\angle 120°$, it rotates the unit vector counterclockwise by 120°, and the A matrix is the transformation matrix.

Therefore, this part analyzes the sequence network diagram of the offshore AC asymmetric fault. The focus is on explaining the constraints between the physical quantities in the fault process and elucidating the fault mechanism of the offshore AC asymmetric fault, so as to clarify the objects that need to be controlled during the transient process.

7.1.2 Sequence Network Analysis of SPG Faults

In Fig. 7.2, the closed S_a and S_F represent the occurrence of SPG fault in phase A. The main constraints to define the fault behavior are as follows: The fault current of the healthy phase is zero, so phase A is the source of the fault current and the fault voltage of phase A is determined by the fault impedance. The equivalent sequence network is shown in Fig. 7.3 below. SPG fault is asymmetrical fault and will form a zero sequence (ZS) path at the location of the fault. The transformers on both sides of the marine AC fault are wired in Y/Δ type, so the ZS current will not be transmitted to the converter side. The OWF is equivalent to the parallel connection of multiple equivalent current sources. WFMMC is equivalent to voltage source or current source according to the fault depth of grid-connected voltage, so there is an equivalent switching process. Because it is a SPG fault, the PS, NS and ZS networks are in series.

According to Fig. 7.3, the circuit relationship shown in Eqs. (7.9) and (7.10) below can be obtained. PS, NS and ZS currents together constitute the fault current, and all fault currents are only composed of phase A.

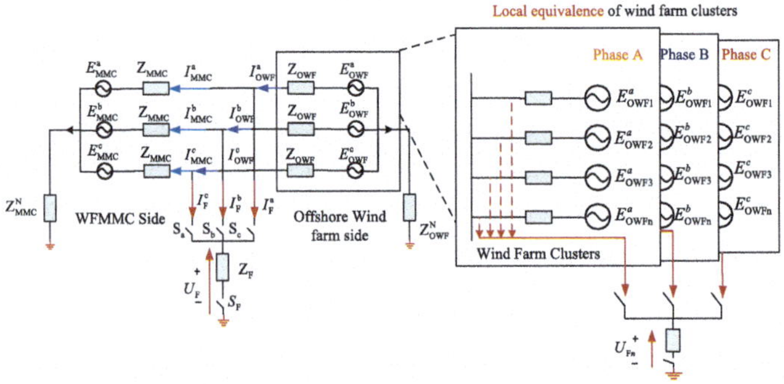

Fig. 7.2 Equivalent circuit for offshore AC asymmetric faults

Fig. 7.3 Sequence diagram of SPG faults

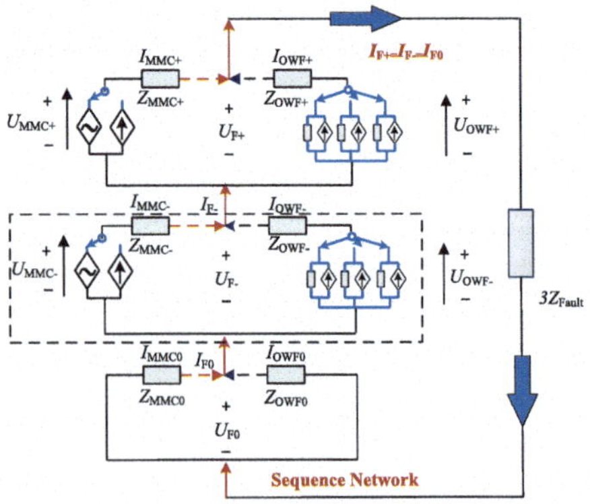

$$
\begin{aligned}
&U_{\mathrm{F}}^{\mathrm{a}} = Z_{\mathrm{F}}^{\mathrm{a}} I_{\mathrm{F}}^{\mathrm{a}} = U_{\mathrm{F}}^{0} + U_{\mathrm{F}}^{+} + U_{\mathrm{F}}^{-} \\
&\rightarrow U_{\mathrm{F}}^{0} + U_{\mathrm{F}}^{+} + U_{\mathrm{F}}^{-} = Z_{\mathrm{F}}^{a}(I_{\mathrm{F}}^{0} + I_{\mathrm{F}}^{+} + I_{\mathrm{F}}^{-})
\end{aligned}
\tag{7.9}
$$

$$
\begin{cases}
I_{\mathrm{F}} = I_{\mathrm{F}}^{a} + I_{\mathrm{F}}^{\mathrm{b}} + I_{\mathrm{F}}^{\mathrm{c}} = I_{\mathrm{F}}^{a} = I_{\mathrm{F}}^{0} + I_{\mathrm{F}}^{+} + I_{\mathrm{F}}^{-} \\
I_{\mathrm{F}}^{\mathrm{b}} = 0 \rightarrow \alpha^{2} I_{\mathrm{F}}^{+} + \alpha I_{\mathrm{F}}^{-} + I_{\mathrm{F}}^{0} = 0 \\
I_{\mathrm{F}}^{\mathrm{c}} = 0 \rightarrow \alpha I_{\mathrm{F}}^{+} + \alpha^{2} I_{\mathrm{F}}^{-} + I_{\mathrm{F}}^{0} = 0
\end{cases}
\tag{7.10}
$$

where, U_{F}^{+-0} and I_{F}^{+-0} represents the PS, NS and ZS components of the fault voltage and current formed at the fault location.

The transformers are wired in Y/Δ type that there is no zero equence voltage source in the system. The magnitude of the ZS voltage will be constrained by the PS and NS voltages, and its relationship with the ZS current always satisfies (7.11).

$$
\begin{cases}
U_{F}^{0} = \frac{Z_{\mathrm{MMC}}^{0} \cdot Z_{\mathrm{OWF}}^{0}}{Z_{\mathrm{MMC}}^{0} + Z_{\mathrm{OWF}}^{0}} I_{F}^{0} \\
I_{\mathrm{MMC}}^{0} = \frac{Z_{\mathrm{OWF}}^{0}}{Z_{\mathrm{MMC}}^{0} + Z_{\mathrm{OWF}}^{0}} I_{F}^{0} \\
I_{\mathrm{OWF}}^{0} = \frac{Z_{\mathrm{MMC}}^{0}}{Z_{\mathrm{MMC}}^{0} + Z_{\mathrm{OWF}}^{0}} I_{F}^{0}
\end{cases}
\tag{7.11}
$$

Thus, the current constraint relationship under SPG faults is obtained as Eq. (7.9). As shown in Fig. 7.3, the circuit constraint of the sequence network under SPG faults, its PS, NS and ZS paths are connected in series. The PS, NS and ZS fault currents are equal, so its fault current satisfies the current relationship of (7.12).

$$
\begin{aligned}
&\begin{cases}
I_{\mathrm{F}}^{0} = I_{\mathrm{OWF}}^{0} + I_{\mathrm{MMC}}^{0} \\
I_{\mathrm{F}}^{+} = I_{\mathrm{OWF}}^{+} + I_{\mathrm{MMC}}^{+} \\
I_{\mathrm{F}}^{-} = I_{\mathrm{OWF}}^{-} + I_{\mathrm{MMC}}^{-} \\
I_{\mathrm{F}}^{+} = I_{\mathrm{F}}^{-} = I_{\mathrm{F}}^{0}
\end{cases}
\Rightarrow I_{\mathrm{OWF}}^{+} + I_{\mathrm{MMC}}^{+} = I_{\mathrm{OWF}}^{0} + I_{\mathrm{MMC}}^{0} = I_{\mathrm{OWF}}^{-} + I_{\mathrm{MMC}}^{-} \\
&\Rightarrow I_{\mathrm{F}}^{\mathrm{a}} = 3(I_{\mathrm{OWF}}^{+} + I_{\mathrm{MMC}}^{+}) = 3(I_{\mathrm{OWF}}^{-} + I_{\mathrm{MMC}}^{-}) = 3(I_{\mathrm{OWF}}^{0} + I_{\mathrm{MMC}}^{0})
\end{aligned}
\tag{7.12}
$$

Given the sequence network and circuit relationships mentioned earlier, the following conclusions can be drawn: When the PS or NS currents on both sides are balanced, the resulting fault current is zero. By controlling the NS currents of both the WFMMC and GSC to zero, fault current suppression can be achieved. However, in the case of asymmetric faultsłespecially SPG and PP faultsłthe PS current control capability of WFMMC is constrained by the NS current magnitude supplied by the system. When both the MMC and GSC regulate NS currents to zero, the PS current of MMC becomes coupled with the GSC's current, making it challenging to simultaneously reduce NS currents to zero without altering the PS current. Therefore, it is necessary for WFMMC and GSC to control the NS current at one side of the fault location to zero while managing the current on the other side to prevent overcurrent in the converter on the opposite side. Typically, NSCC can be implemented on either the GSC or WFMMC. However, since large-scale OWFs usually consist of many turbines, installing NSCC and tuning parameters on all turbines simultaneously is complex. Thus, for simplicity in practical engineering applications, this chapter focuses on designing NSCC on the WFMMC, while treating the GSC as multiple current-limiting sources providing a pathway for NS current.

Because the ZS impedance of the system is very small when the transformer neutral is directly grounded, when the NS current is suppressed there is $I_{\mathrm{F}}^{0} \approx 0$ $U_{\mathrm{F}}^{0} \approx 0$.

Set the PS voltage in the transient state to $U^{+} = U_{\mathrm{Fm}}^{+} \sin(\omega t + \varphi^{+})$. The fault voltage expression is obtained as shown in the following Eq. (7.13).

$$\begin{cases} U_{F}^{+} = U_{Fm}^{+} \sin(\omega t + \varphi^{+}) \\ U_{F}^{-} = -U_{Fm}^{+} \sin(\omega t + \varphi^{+}) \end{cases} \tag{7.13}$$

Combined with the above analysis, the expression of the three-phase voltage during the faults can be obtained as shown below.

$$\begin{cases} U_{\mathrm{F}}^{a} = U_{\mathrm{F}}^{+} + U_{\mathrm{F}}^{-} + U_{\mathrm{F}}^{0} = 0 \\ U_{\mathrm{F}}^{b} = \alpha^{2} U_{\mathrm{F}}^{+} + \alpha U_{\mathrm{F}}^{-} + U_{\mathrm{F}}^{0} = \sqrt{3} U_{\mathrm{Fm}}^{+} \sin(\omega t + \varphi^{+} - \frac{\pi}{2}) \\ U_{\mathrm{F}}^{c} = \alpha U_{\mathrm{F}}^{+} + \alpha^{2} U_{\mathrm{F}}^{-} + U_{\mathrm{F}}^{0} = \sqrt{3} U_{\mathrm{Fm}}^{+} \sin(\omega t + \varphi^{+} + \frac{\pi}{2}) \end{cases} \tag{7.14}$$

The equation above indicates that during SPG faults, the voltage at the faulted phase drops to zero. Meanwhile, the non-faulted phase faces a risk of overvoltage during the fault event. Specifically, when the amplitude of the PS voltage at the fault location meets a certain condition, the non-faulted phase experiences an overvoltage phenomenon while remaining connected to the grid.

$$U_{Fm}^{+} \geq \frac{1}{\sqrt{3}} U_{\mathrm{N}} \tag{7.15}$$

7.1.3 Sequence Network Analysis of DPG Faults

In Fig. 7.2 , when S_{b}, S_{c} and S_{F} are closed, it means that DPG fault occurs in phase B and phase C. The constraints under the fault are as follows: Fault currents are composed of phase B and C of WFMMC and GSC. Phase A does not provide short-circuit fault current. The OWF consists of multiple equivalent network current sources in parallel. WFMMC needs to switch between equivalent voltage source and current source according to the degree of voltage drop. In this case, the PS and NS networks are in parallel. Its fault characteristics are shown in Eq. (7.16) below:

$$\begin{cases} I_{\mathrm{F}} = I_{\mathrm{F}}^{b} + I_{\mathrm{F}}^{c} = (\alpha^2 I_{\mathrm{F}}^{+} + \alpha I_{\mathrm{F}}^{-} + I_{\mathrm{F}}^{0}) + (\alpha I_{\mathrm{F}}^{+} + \alpha^2 I_{\mathrm{F}}^{-} + I_{\mathrm{F}}^{0}) \\ = 2I_{\mathrm{F}}^{0} + (\alpha^2 + \alpha)(I_{\mathrm{F}}^{+} + I_{\mathrm{F}}^{-}) = 2I_{\mathrm{F}}^{0} + (-1)(-I_{\mathrm{F}}^{0}) = 3I_{\mathrm{F}}^{0} \\ I_{\mathrm{F}}^{a} = I_{\mathrm{F}}^{0} + I_{\mathrm{F}}^{+} + I_{\mathrm{F}}^{-} = 0 \\ U_{\mathrm{F}}^{b} = U_{\mathrm{F}}^{c} = Z_{\mathrm{F}}(I_{\mathrm{F}}^{b} + I_{\mathrm{F}}^{c}) = Z_{\mathrm{F}}[(\alpha^2 I_{\mathrm{F}}^{+} + \alpha I_{\mathrm{F}}^{-} + I_{\mathrm{F}}^{0}) + (\alpha I_{\mathrm{F}}^{+} + \alpha^2 I_{\mathrm{F}}^{-} + I_{\mathrm{F}}^{0})] \\ = Z_{\mathrm{F}}[2I_{\mathrm{F}}^{0} + (\alpha^2 + \alpha)(I_{\mathrm{F}}^{+} + I_{\mathrm{F}}^{-})] = Z_{\mathrm{F}}[2I_{\mathrm{F}}^{0} + (-1)(-I_{\mathrm{F}}^{0})] = 3Z_{\mathrm{F}}I_{\mathrm{F}}^{0} \end{cases} \tag{7.16}$$

Since the transformers are wired in Y/Δ type, WFMMC and OWFs do not affect the ZS component. The ZS component is only related to the ZS impedance of the line and the secondary side of transformer.

$$\begin{cases} U_{\mathrm{F}}^{0} = I_{\mathrm{MMC}}^{0} Z_{\mathrm{MMC}}^{0} = I_{\mathrm{OWF}}^{0} Z_{\mathrm{OWF}}^{0} \\ I_{\mathrm{F}}^{0} = I_{\mathrm{MMC}}^{0} + I_{\mathrm{OWF}}^{0} \end{cases} \Rightarrow U_{\mathrm{F}}^{0} = \frac{Z_{\mathrm{MMC}}^{0} \cdot Z_{\mathrm{OWF}}^{0}}{Z_{\mathrm{MMC}}^{0} + Z_{\mathrm{OWF}}^{0}} I_{\mathrm{F}}^{0} \tag{7.17}$$

The residual PS voltage formed by the fault sequence network at the fault location in Fig. 7.4 can define the magnitude of the NS voltage and the ZS voltage, which are related as shown below.

$$\begin{aligned} U_{F}^{+} &= U_{F}^{-} = U_{F}^{0} - 3Z_{F}I_{F}^{0} \\ &= -\left(\frac{Z_{\mathrm{MMC}}^{0} \cdot Z_{\mathrm{OWF}}^{0}}{Z_{\mathrm{MMC}}^{0} + Z_{\mathrm{OWF}}^{0}} + 3Z_{F}\right)I_{F}^{0} = \left(\frac{Z_{\mathrm{MMC}}^{0} \cdot Z_{\mathrm{OWF}}^{0}}{Z_{\mathrm{MMC}}^{0} + Z_{\mathrm{OWF}}^{0}} + 3Z_{F}\right)I_{F}^{+} \end{aligned} \tag{7.18}$$

Set the PS voltage in the transient state to $U^{+} = U_{\mathrm{Fm}}^{+} \sin(\omega t + \varphi^{+})$ The fault voltage expression under DPG faults is obtained in Eq. (7.19).

$$\begin{cases} U_{\mathrm{F}}^{a} = U_{\mathrm{F}}^{+} + U_{\mathrm{F}}^{-} + U_{\mathrm{F}}^{0} = 2U_{\mathrm{Fm}}^{+} \sin(\omega t + \varphi^{+}) + U_{\mathrm{F}}^{0} \\ U_{\mathrm{F}}^{b} = \alpha^2 U_{\mathrm{F}}^{+} + \alpha U_{\mathrm{F}}^{-} + U_{\mathrm{F}}^{0} = -U_{\mathrm{Fm}}^{+} \sin(\omega t + \varphi^{+} -) + U_{\mathrm{F}}^{0} \\ U_{\mathrm{F}}^{c} = \alpha U_{\mathrm{F}}^{+} + \alpha^2 U_{\mathrm{F}}^{-} + U_{\mathrm{F}}^{0} = -U_{\mathrm{Fm}}^{+} \sin(\omega t + \varphi^{+} -) + U_{\mathrm{F}}^{0} \end{cases} \tag{7.19}$$

During DPG faults, the PS and NS networks are connected in parallel. While the NS current can be suppressed to zero, the WFMMC and GSC cannot actively disengage after fault recovery. To achieve NS current suppression, an NS voltage is present within the NS network during the fault. However, when the system switches the sequence network diagram at the moment of fault recovery, a residual NS voltage

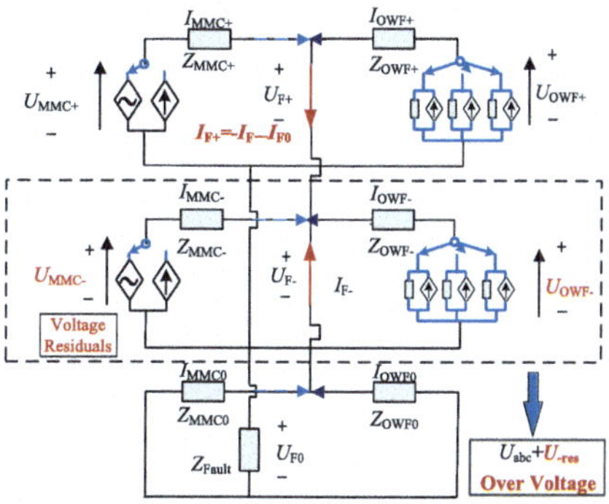

Fig. 7.4 Sequence diagram of DPG faults

remains in the normally operating NS loop. This residual voltage can lead to transient overvoltage during the recovery phase.

$$\begin{aligned} U_{\mathrm{R}}^{a} &= U^{+} + U^{-} + U^{0} = U_{m}^{+}\sin(\omega t + \varphi^{+}) + U_{\mathrm{s}}^{-} \\ &= U_{m}^{+}\sin(\omega t + \varphi^{+}) + U_{\mathrm{Fm}}^{+}\sin(\omega t_2 + \varphi_2)e^{-\delta t} \end{aligned} \tag{7.20}$$

where, δ represents the decay rate of NS voltage. φ_2 represents the angle of the fault clearing moment.

Given the presence of inductance in the transmission line, the voltage recovery process can result in a significant voltage increase. If the system has low damping, this recovery process may also lead to oscillations, making the issue of overvoltage during voltage recovery particularly pronounced. The overvoltage can peak at up to 1.3 times the rated voltage, posing a serious threat to the transmission line's insulation and the overall safety of the equipment in system.

7.1.4 Sequence Network Analysis of PP Faults

In Fig. 7.2, when S_{b} and S_{c} are closed, it means that PP fault occurs in phase B and phase C. The constraint condition is that the fault current is supplied by phases B and C. Since there is no ZS network in the sequence network, there is no ZS current. In this case, the PS and NS networks are in parallel, due to the PP fault, there is no ZS path in the sequential network diagram.

Based on Fig. 7.5 above, the following circuit relation expression of (7.21) can be obtained:

$$\begin{cases} I_{\mathrm{F}}^{0} = I_{\mathrm{OWF}}^{0} + I_{\mathrm{MMC}}^{0} = 0 \\ I_{\mathrm{F}}^{+} = I_{\mathrm{F}}^{-} = (I_{\mathrm{OWF}}^{+} + I_{\mathrm{MMC}}^{+}) = \left(I_{\mathrm{OWF}}^{-} + I_{\mathrm{MMC}}^{-}\right) \\ I_{\mathrm{F}} = I_{\mathrm{F}}^{b} + I_{\mathrm{F}}^{c} = (\alpha^2 I_{\mathrm{F}}^{+} + \alpha I_{\mathrm{F}}^{-} + I_{\mathrm{F}}^{0}) + (\alpha I_{\mathrm{F}}^{+} + \alpha^2 I_{\mathrm{F}}^{-} + I_{\mathrm{F}}^{0}) \\ = 2I_{\mathrm{F}}^{0} + (\alpha^2 + \alpha)(I_{\mathrm{F}}^{+} + I_{\mathrm{F}}^{-}) = 2I_{\mathrm{F}}^{0} + (-1)(-I_{\mathrm{F}}^{0}) = 3I_{\mathrm{F}}^{0} = 0 \\ I_{\mathrm{F}}^{a} = I_{\mathrm{F}}^{0} + I_{\mathrm{F}}^{+} + I_{\mathrm{F}}^{-} = 0 \\ U_{\mathrm{F}}^{c} = U_{\mathrm{F}}^{b} + Z_{\mathrm{F}} I_{\mathrm{F}}^{b} \\ U_{\mathrm{F}}^{+} = U_{\mathrm{F}}^{-} - I_{\mathrm{F}}^{-} Z_{\mathrm{F}} \end{cases} \tag{7.21}$$

Set the PS voltage in the transient process to $U^{+} = U_{\mathrm{Fm}}^{+}\sin(\omega t + \varphi^{+})$. Since there is no ZS network in the sequence network of PP faults, there is no ZS current and ZS voltage. Similarly the fault voltage expression under PP faults is obtained as shown in Eq. (7.22).

$$\begin{cases} U_{\mathrm{F}}^{a} = U_{\mathrm{F}}^{+} + U_{\mathrm{F}}^{-} + U_{\mathrm{F}}^{0} = 2U_{\mathrm{Fm}}^{+}\sin(\omega t + \varphi^{+}) \\ U_{\mathrm{F}}^{b} = \alpha^2 U_{\mathrm{F}}^{+} + \alpha U_{\mathrm{F}}^{-} + U_{\mathrm{F}}^{0} = -U_{\mathrm{Fm}}^{+}\sin(\omega t + \varphi^{+} -) \\ U_{\mathrm{F}}^{c} = \alpha U_{\mathrm{F}}^{+} + \alpha^2 U_{\mathrm{F}}^{-} + U_{\mathrm{F}}^{0} = -U_{\mathrm{Fm}}^{+}\sin(\omega t + \varphi^{+} -) \end{cases} \tag{7.22}$$

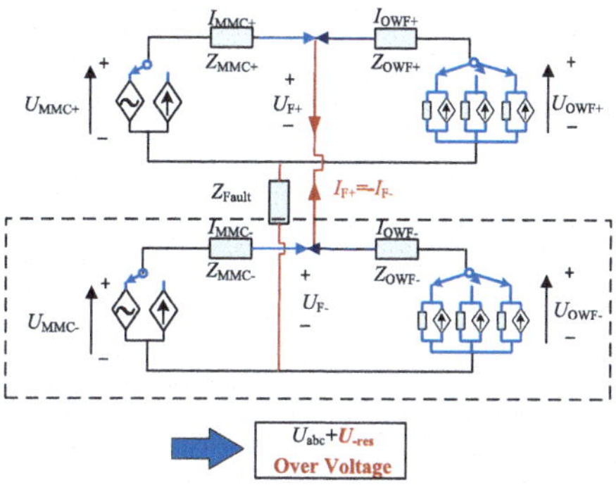

Fig. 7.5 Sequence diagram of during PP faults

The equation above indicates that when a PP fault occurs between phases B and C, the fault voltages in these two phases are equal. However, if the amplitude of the PS voltage at the fault location satisfies the condition described in Eq. (7.23), an overvoltage condition can arise in the healthy phase of the system.

$$U_{Fm}^{+} \geq \frac{1}{2} U_{\text{N}} \tag{7.23}$$

Similar to the DPG faults, there is a NS voltage residual in the normal sequence network after the fault is cleared. This is not beneficial to the AC voltage recovery after the fault. During fault recovery, the recovery phase overvoltage problem needs to be resolved.

7.2 MSNEC for Offshore AC Asymmetric Faults

7.2.1 Analysis of Transient Current Composition of Offshore AC Asymmetric Faults

When an offshore AC asymmetric fault occurs, the WFMMC as an equivalent power source feeds fault current to the fault point. The offshore wind energy output from the OWFs feeds fault current due to the low impedance at the fault point. The short-circuit current in transient state consists of the above two components.

7.2.2 Positive and Negative Sequential Network Control

The MMC-HVDC systems currently in operation typically utilize half-bridge submodules. The half-bridge MMC topology generally offers two degrees of control freedom, enabling separate control of PS and negative sequence (NS) components. Control is achieved by converting the three-phase electrical quantities into their corresponding PS and NS components, which are then further transformed into d-axis and q-axis components [5]. The derivation process for the PS and NS controller is outlined in the following equation.

$$I_{\text{sdref}}^{+} = K_{o\_p}(U_{\text{sdref}}^{+} - U_{\text{sd}}^{+}) + K_{o\_i}\int (U_{\text{sdref}}^{+} - U_{\text{sd}}^{+})\text{d}t \tag{7.24}$$

$$I_{\text{sqref}}^{+} = I_{\text{sdref}}^{-} = I_{\text{sqref}}^{-} = 0 \tag{7.25}$$

$$\begin{cases} m_{\text{d}}^{+-} = K_{i\_p}^{+-}(I_{\text{sdref}}^{+-} - I_{\text{sd}}^{+-}) + K_{i\_i}^{+-}\int (I_{\text{sdref}}^{+-} - I_{\text{sd}}^{+-})\text{d}t \\ m_{\text{q}}^{+-} = K_{i\_p}^{+-}(I_{\text{sqref}}^{+-} - I_{\text{sq}}^{+-}) + K_{i\_i}^{+-}\int (I_{\text{sqref}}^{+-} - I_{\text{sq}}^{+-})\text{d}t \end{cases} \tag{7.26}$$

where, I_{sdref}^{+} and U_{sdref}^{+} represent the reference values of current and voltage in axis d of positive sequence, K_{o_p} and K_{o_i} represent the control parameters of PI controller, m_{d}^{+} and m_{q}^{+}, m_{d}^{-} and m_{q}^{-} respectively represent the modulation ratio obtained by PS and NS control.

7.2.3 Transient Energy Support Strategy

The Modular Multilevel Converter (MMC) is fundamentally an energy conversion device, with power switches and energy storage components serving as the core units of this conversion system. By reconsidering the control characteristics of MMC from the perspective of energy conversion and transfer, it is possible to enhance the overall control performance of the equipment.

Reference [6] derives the principle of MMC energy control by decoupling the internal energy storage of the MMC from the DC voltage, so that the MMC does not affect the DC voltage after taking energy control. The energy control can be decoupled from the DC voltage U_{dc} by setting the corresponding AC/DC side input ratios $\alpha_{\text{ac}j}$ and $\alpha_{\text{dc}j}$ so that the DC side voltage U_{dc} is decoupled from the voltage $U\text{s}$ of the submodule (Fig. 7.6). In order to avoid affecting the calculation of the input ratio $\alpha_{\text{ac}j}$ and $\alpha_{\text{dc}j}$ of the upper and lower arm submodules, the nearest level modulation is changed to submodule capacitor voltage modulation. The average capacitance voltage of submodules is detected, which is taken as the base value of the capacitor voltage of the neutron module modulated at the nearest level at the moment, so as to realize the correct input of the number of submodules of the bridge arm.

$$2C_{\text{arm}}\frac{\text{d}\,u_{\bar{\text{C}}\text{s}j}}{\text{d}t} = \alpha_{\text{ac}j}i_{\text{v}j} + 2\alpha_{\text{dc}j}i_{\text{cir}j} \tag{7.27}$$

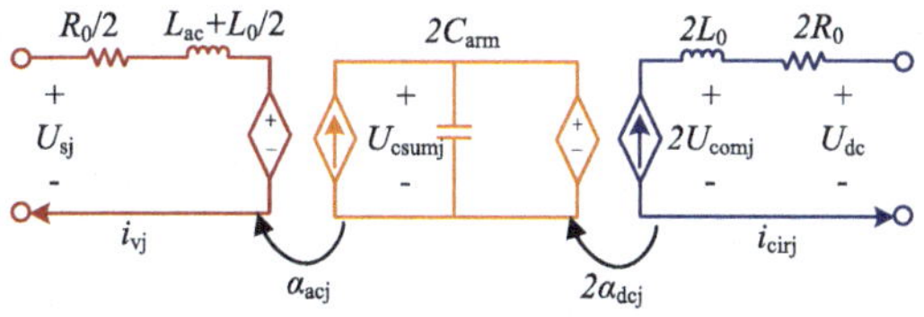

Fig. 7.6 The system equivalent dynamic circuit of phase j

The primary challenge of the transient energy support strategy lies in maximizing the MMC's inherent energy control capabilities to address the issues of overcurrent in the faulted phase and overvoltage in the non-faulted phase during offshore asymmetric AC faults. In the offshore wind power system, the MMC serves as a crucial energy storage component, with its structure illustrated in Fig. 7.7 below. The submodule capacitor acts as the vital energy storage unit within the MMC, and charging of the submodule occurs through the switching operations within the MMC submodule [7, 8].

Taking phase A as an example, the submodule's behavior during the fault process is depicted below. When energy control is activated, the WFMMC fully utilizes its redundant submodules, achieving the proper submodule engagement by controlling the conduction of IGBT and diodes. The rapid response of the redundant submodules during the fault increases the equivalent impedance between the fault location and the WFMMC. This process ultimately enables fault ride-through by utilizing the energy storage capabilities of the redundant submodules. Once the fault is cleared, the energy stored in the redundant submodule capacitors is released back to the onshore AC grid in the direction indicated by the arrows.

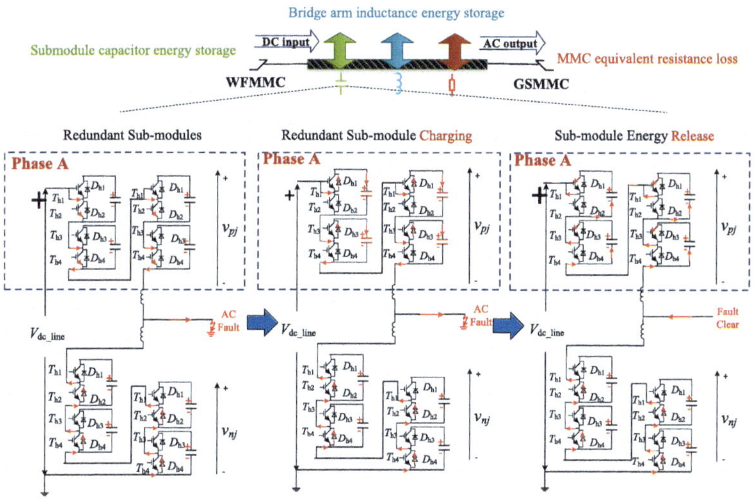

Fig. 7.7 Energy transmission of WFMMC

The rated energy storage capacity of converter stations is shown in Eq. (7.28):

$$W_{\mathrm{MMC0}} = \frac{1}{2} \times C_{\mathrm{SM}} \times V_{\mathrm{SM}}^2 \times 6N \tag{7.28}$$

The maximum margin of energy can reach 1.56 pu because the overvoltage protection of the sub-module has a fixed value of 1.6 pu. Considering a certain margin on the basis of the fixed value of the protection, the upper limit of the sub-module voltage is 1.5 pu [7].

$$\Delta W_{\mathrm{MMC}} = \frac{1}{2} \times 6 \times C_{\mathrm{SM}} \times (1.5^2 - 1^2) \times U_{\mathrm{SM}} \times N \tag{7.29}$$

The energy margin of WFMMC is calculated to be 1.25 pu.

With the presence of NS components in asymmetric faults, the system is prone to superposition periodic components with double frequency. Therefore, the change rate of grid-connected voltage also presents a periodic change, and the period is $\Delta T = 10$ ms. In order to reduce the control complexity, the average slope method is used to calculate the average rate of change of offshore grid-connected voltage in 2 ΔT. The change slope of the energy control curve is shown in the following equation.

$$k_1 = k_0 + \frac{U_{\mathrm{s}j2} - U_{\mathrm{s}j1}}{2\Delta T} \cdot \frac{W_{\mathrm{WFMMCmax}} - 1}{U_{\mathrm{s}j\max} - 1} \tag{7.30}$$

where, $U_{\mathrm{s}j1}$ and $U_{\mathrm{s}j2}$ represents before and after value of non-faulty phase voltage in $\Delta T = 10$ ms. $U_{\mathrm{s}j\max}$ represents maximum value of non-fault phase voltage.

Furthermore, the classification and discussion process of energy variation curve is obtained by considering the case of non-fault phase overvoltage.

$$k_1 = \begin{cases} P_{\mathrm{s}}(1.25 \times \frac{|P_{\mathrm{ac\,max}} - P_{\mathrm{dc0}}|}{\Delta W_{\mathrm{WFMMC}}} \times 10^{-3}), U_{\mathrm{s}j2} \le U_{\mathrm{s}j1} \\ P_{\mathrm{s}}(1.25 \times \frac{|P_{\mathrm{ac\,max}} - P_{\mathrm{dc0}}|}{\Delta W_{\mathrm{WFMMC}}} \times 10^{-3}) + \frac{U_{\mathrm{s}j2} - U_{\mathrm{s}j1}}{2\Delta T} \cdot \frac{W_{\mathrm{WFMMCmax}} - 1}{U_{\mathrm{s}j\max} - 1}, U_{\mathrm{s}j2} \ge U_{\mathrm{s}j1} \end{cases} \tag{7.31}$$

where, P_{s} represents active power output from wind farms. $P_{\mathrm{ac\,max}}$ represents maximum active power of WFMMC input. P_{dc0} represents DC power when fault occurs.

The maximum output power on the AC side of the onshore network is 1.2 per unit (pu). During the energy release phase of the WFMMC, the maximum allowable DC output power of the WFMMC is also set to 1.2 pu to prevent an excessive increase in energy within the GSMMC due to overloading the DC power. The energy release process during the WFMMCs fault recovery is structured as follows:

$$\begin{aligned} k_2 &= \frac{dW_{\mathrm{WFMMC}}}{dt} = \frac{S_{\mathrm{N}}}{W_{\mathrm{WFMMC0}}} \cdot (P_{\mathrm{dc}} - P_{\mathrm{ac}}) \\ &\ge \frac{S_{\mathrm{N}}}{W_{\mathrm{WFMMC0}}} \cdot (-1.2 + 1) \ge -0.2 \cdot \frac{S_{\mathrm{N}}}{W_{\mathrm{WFMMC0}}} \end{aligned} \tag{7.32}$$

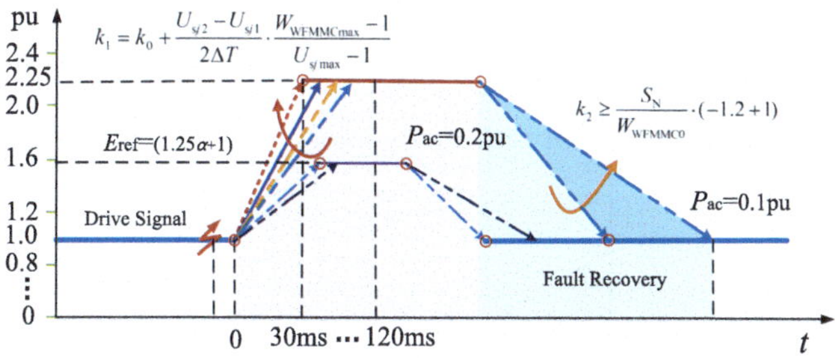

Fig. 7.8 Energy adaptive adjustment control process of WFMMC

The energy control process of the WFMMC is depicted in Fig. 7.8 below. The absorption slope is dynamically adjusted based on the presence or absence of overvoltage in the non-faulted phase. In situations where there is no overvoltage in the non-faulted phase, the slope adjustment is governed by the difference between the active power output of the OWFs and the power output on the DC side. However, when the faulted positive sequence PS voltage reaches a level that triggers fault overvoltage, the non-faulted phase may experience overvoltage. In such cases, it is challenging to mitigate the overvoltage by merely adjusting the reference in the outer voltage loop. Since this overvoltage is detected by the outer loop, additional control mechanisms can be employed for energy management.

To address this, adaptive regulation links are added to the original k0 control. An increase in the slope corresponds to the increase in the number of submodules engaged during transients. These additional submodules are inserted into the non-faulted phase, enabling series voltage division and actively preventing overvoltage in the non-faulted phase during fault conditions.

7.2.4 Fault Ride-Through Scheme Under Offshore AC Asymmetric Faults

When an offshore AC fault occurs, particularly in the region between the offshore wind farms (OWFs) and the WFMMC, it is crucial that the OWFs maintain stable operation without going off-grid during the fault. This stability is essential to prevent significant economic losses that could result from the widespread disconnection of the OWFs. The requirements for high-voltage FRT and low-voltage FRT for the wind farm are illustrated in Fig. 7.9 below. According to these requirements, OWFs can continue to operate stably even when the voltage drops to as low as 0.2 pu for a duration of 0.625 s, and they should recover to 0.9 pu within 2 s. For high-voltage FRT, OWFs can temporarily handle voltages of 1.3 pu and 1.25 pu for 0.5 s, after which they should be able to sustain 1.2 pu for up to 10 s. Additionally, the wind

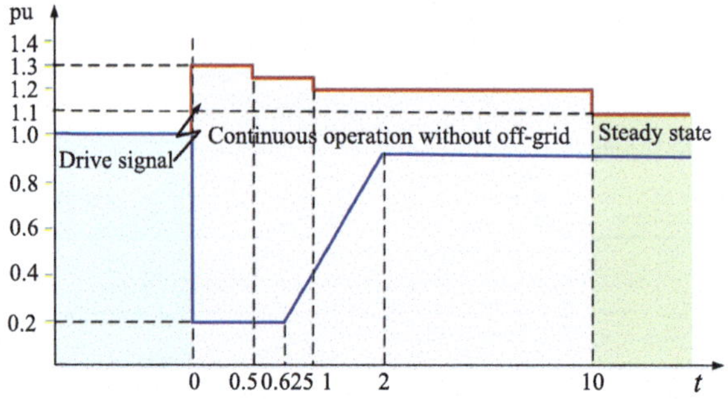

Fig. 7.9 High and low voltage FRT standards for OWFs

turbines are designed to ensure continuous operation without going off-grid at a voltage level of 1.1 pu.

Therefore, when an offshore AC asymmetric fault occurs, OWFs must meet the low-voltage FRT requirements during the initial voltage dip and ensure compliance with the high-voltage FRT requirements during fault recovery. This dual compliance is necessary for the OWFs to smoothly achieve FRT.

Based on the above discussion, the following explanation outlines the principles of the control strategy from different perspectives:

- Positive and Negative Sequence Control Perspective: As illustrated in Fig. 7.10a below, from the viewpoint of positive and negative sequence control, the WFMMC employs NSCC to suppress negative sequence components during faults. This approach helps maintain stable grid-connected voltage in the OWFs and minimizes fluctuations in the grid-connected voltage caused by negative sequence components.
- Power Electronics Equivalence Perspective: From the perspective of power electronic power equivalence, when the voltage at the point of common coupling

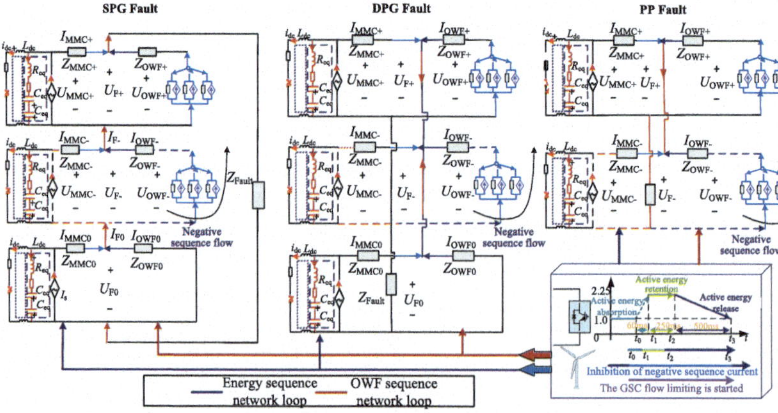

Fig. 7.10 Analysis process of energy sequential network

(U_{pcc}) drops but has not yet caused saturation in the loop current controller, the WFMMC behaves as a voltage source. In this scenario, the WFMMC's critical role as a supporting voltage source is preserved. However, if the U_{pcc} drop is substantial, leading to the saturation of the WFMMC's inner loop current controller, the WFMMC then acts as a current source with a controlled reduction in amplitude. This adjustment helps to mitigate the impact of fault currents and alleviates the overcurrent issues in the WFMMC. Additionally, the wind turbines on the OWF side, functioning as multiple parallel equivalent current sources, become the primary path for negative sequence current. To prevent overcurrent problems arising from the OWF side becoming the path for negative sequence current, the GSC employs a current inner loop limiting mechanism to restrict the magnitude of the fault current.

The MSNEC, illustrated in Fig. 7.10, is derived from the proposed concepts of Energy Control and Sequential Network Control. During an asymmetric fault, the converter station's energy control in the WFMMC progresses through three key phases: Energy Recovery, Energy Retention, and Energy Release. From t_0 to t_1, energy is actively managed, with sub-modules deployed in the fault phase of the WFMMC to reduce the amplitude of the fault current. Additionally, sub-modules are introduced in the non-faulted phase to achieve series voltage division and to prevent overvoltage, as depicted in Fig. 7.10.

The Energy Retention phase, which occurs between t_1 and t_2, serves as a transitional period, maintaining the energy within the WFMMC for roughly 250 ms. The Energy Release phase, from t_2 to t_3, is the final stage. During this phase, the equivalent impedance in the PS, NS, and ZS networks of the WFMMC increases, leading to a reduction in fault current (Fig. 7.11). This also stabilizes the grid-connected voltage and allows negative sequence control to suppress the NS current. Once the FRT process is complete, the energy stored in the WFMMC is released into the onshore AC grid, which minimizes the waste of offshore wind energy and promotes the efficient use of new energy. This process provides a buffer time of 80–100 ms for OWFs to perform FRT, thereby improving the engineering feasibility of existing solutions.

From a broader perspective, considering the multi-stage control process: when a fault is detected, the NSCC enters a normal start-up mode. The energy control process is divided into three stages: energy absorption, energy retention, and energy

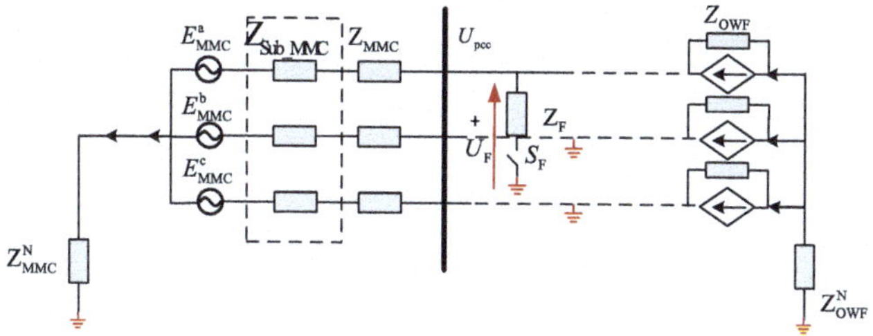

Fig. 7.11 Fault phase and non-fault phase equivalent circuits for energy control process

recovery. The GSC consistently maintains an inner loop current-limiting control to prevent overcurrent in offshore wind farms. Due to the presence of NS voltage residuals during the fault recovery process, a fault voltage control scheme, as shown in Fig. 7.1, is employed. When local fault detection confirms that the fault has been cleared, a signal triggers the overvoltage suppression scheme, adjusting the voltage control outer loop command during fault recovery. After the energy control process ends, the WFMMC issues a voltage control blocking signal to restore the steady-state voltage reference, thereby preventing overvoltage during fault recovery.

7.3 Simulation Verification

7.3.1 Simulation Analysis of SPG Faults

To verify the effectiveness of MSNEC, an OWFs grid-connected system is established as shown in Fig. 7.1, and its system parameters are shown in Table 7.1. Set at the moment of 6 s, a SPG fault occurs in phase A at sea, and the fault lasts 60ms, and obtain the simulation results shown in Fig. 7.12.

From an energy control perspective, the results in Fig. 7.12 demonstrate that when an offshore AC SPG fault occurs at 6 s, the WFMMC quickly detects the fault through local current sensing and triggers an activation signal to both the WFMMC and OWFs. As shown in Fig. 7.12e, after a 20 ms delay, the WFMMC engages redundant submodules to manage energy, effectively reducing fault current and maintaining voltage stability in the non-faulted phases. This adjustment causes the WFMMC voltage to rise to 2.25 pu, with a corresponding change in the reference value. After the fault is cleared, the stored energy is released back into the onshore AC grid, enabling the reuse of energy and promoting the efficient utilization of offshore wind power.

Table 7.1 The main parameters of simulation system

Parameter	Value
Rated voltage of parallel network U_{pcc}	230 kV
Rated DC voltage U_{dc0}	±320 kV
Number of SMs per arm N_{HBSM}	320
DC Cable Equivalent Capacitance C_{cable}	19.53 μF
Rated voltage of MMC sub-module U_{sm}	2 kV
Rated DC transmission power P_{dc0}	±600 MW
MMC sub-module capacitance C_{sm}	12500 μF
Rated power of wind farms P_{WF0}	1200 MW

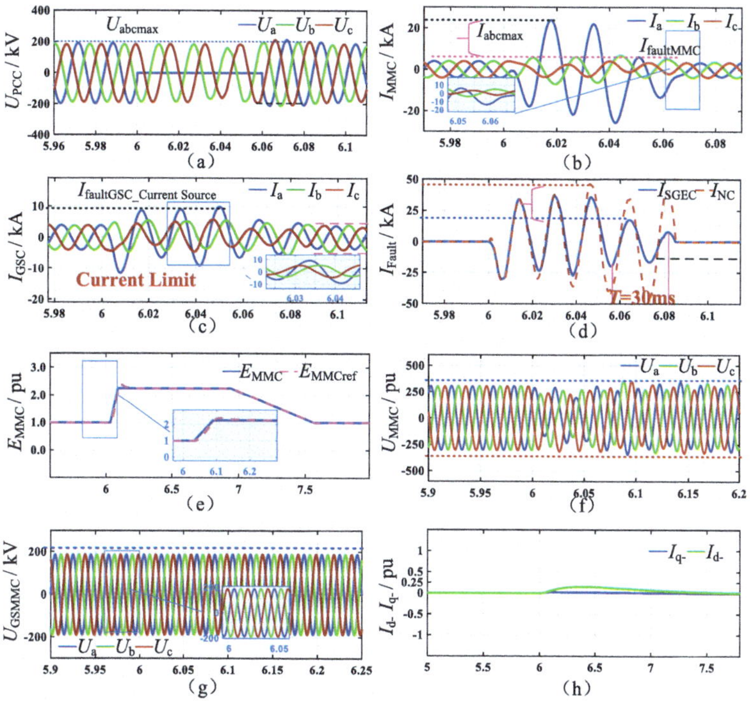

Fig. 7.12 SPG fault simulation analysis. **a** The Upcc of OWFs. **b** The fault current of MMC. **c** The fault current of GSC. **d** Comparison of fault current. **e** Energy variation of WFMMC. **f** Valve-side voltage of MMC. **g** Grid-connected voltage for onshore AC systems. **h** Each sequence component of the grid-connected voltage. **i** NS component of fault current

In terms of voltage FRT, Fig. 7.12a shows that during the fault, the voltage in phase A drops to zero due to grounding, while phases B and C continue to operate normally, with no signs of overvoltage. As depicted in Fig. 7.12f, the voltage in phase A of the WFMMC remains above zero due to the transformer impedance, and the overall WFMMC voltage stays stable during the fault, returning to balance after recovery. Figure 7.12g indicates that the onshore AC grid voltage maintains a stable three-phase balance during the fault, and there is no overvoltage during recovery, ensuring compliance with high-voltage FRT requirements for OWFs.

From a fault current control perspective, Fig. 7.12b, c show that the fault current supplied by the WFMMC decreases to nearly steady-state levels within 40 ms. Figure 7.12h reveals that the NS current component of the WFMMC during the fault remains below 0.1 pu, demonstrating that the fault current in the OWFs is effectively limited during the short circuit, thanks to the inner-loop current-limiting mechanism. After the fault is cleared, the current returns to steady-state levels. Additionally,

Fig. 7.12d shows that the fault current at the short-circuit point is significantly suppressed, dropping from 5 pu to around 2 pu reduction of 60%. This reduction in fault current helps to mitigate the impact and stress on power electronics.

7.3.2 DPG Faults Analysis

The applicability of MSNEC under SPG fault conditions is discussed in Sect. 7.3.1. To further verify its effectiveness, simulations were conducted for DPG and PP faults. The same system parameters were used as in the previous analysis. At the 6 s mark, a DPG fault and a PP fault were introduced in phases B and C, respectively, occurring offshore and lasting for 60 ms. The results of these fault simulations are presented in Figs. 7.13 and 7.14.

From the perspective of energy control, the simulation results show that when a DPG fault occurs, the WFMMC detects the fault through local current monitoring and initiates a start signal. After a 20 ms delay, the WFMMC adaptively engages redundant sub-modules to manage energy, causing the WFMMC voltage (W_{MMC}) to rise to 2.25 pu in line with the reference value. This reduces the fault current and stabilizes the voltage in the non-faulted phases. Following a transition phase, the energy absorbed during the fault is released into the onshore AC grid, enabling the efficient utilization of energy.

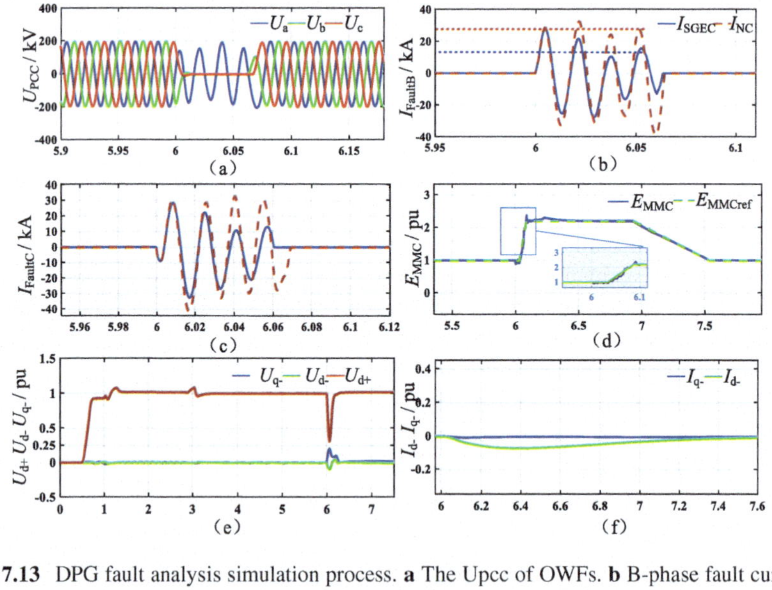

Fig. 7.13 DPG fault analysis simulation process. **a** The Upcc of OWFs. **b** B-phase fault current. **c** C-phase fault current. **d** Energy variation of WFMMC. **e** Each sequence component of the grid-connected voltage. **f** NS component of fault current

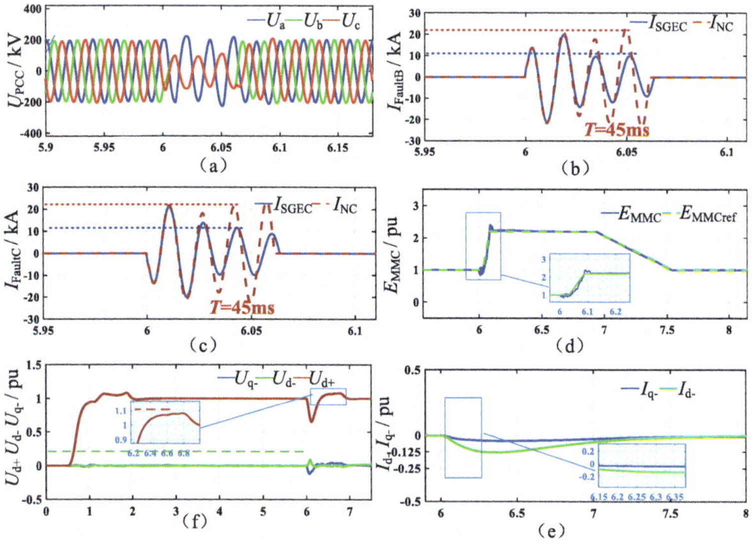

Fig. 7.14 **a** PP fault simulation analysis. **b** B-phase fault current. **c** C-phase fault current. **d** Energy variation of WFMMC. **e** Each sequence component of the grid-connected voltage. **f** NS component of fault current

In terms of voltage FRT, Fig. 7.13a shows that the voltages of phases B and C drop to zero due to fault grounding, while the voltage of phase A remains stable after the control is initiated. Figure7.13e shows that the PS voltage drops to 0.35 pu, meeting the low-voltage FRT requirements for OWFs. During fault recovery, the voltage smoothly returns to normal levels, without any signs of overvoltage, fulfilling the high-voltage FRT requirements.

Regarding fault current control, Fig. 7.13b, c illustrate that the fault currents in phases B and C begin to decrease in amplitude after 35 ms. The maximum reduction in fault current amplitude in these phases is approximately 66%, indicating significant suppression of the fault current. Additionally, Fig. 7.13f shows that the NS current of the WFMMC remains below 0.1 pu, suggesting effective suppression of the NS current. This reduction minimizes NS fluctuations in the WFMMC and ensures the quality of the grid-connected voltage.

7.3.3 PP Fault Analysis

In the context of offshore AC faults, a PP fault does not provide a pathway for ZS currents due to the constraints of the sequence network circuit topology. Only the PS and NS current paths exist, with the PS and NS networks connected in parallel, resulting in equal PS and NS voltages.

From an energy control perspective, once a fault is detected, start signals are issued, prompting both the WFMMC and OWFs to initiate control measures. As shown in Fig. 7.14d, after a 20 ms delay, redundant sub-modules are adaptively engaged to manage energy, thereby reducing the fault current and stabilizing the voltage in the non-faulted phase. The WFMMC voltage rises to 2.25 pu, according to the reference value. Following a transition phase, the absorbed energy is released into the onshore AC grid.

In terms of voltage FRT analysis, Fig. 7.14a, e show that the voltage in phase A is maintained at a steady-state level, while a fault voltage of 100 kV forms between phases B and C. The PS grid voltage briefly drops to 0.6 pu, which meets the low-voltage FRT requirements for OWFs. After fault recovery, the NS voltage at the fault site creates a residual NS voltage within the normal operation state sequence network due to the switching of the sequence network. This can impede the recovery of the grid-connected voltage. During fault recovery, the grid-connected voltage initially rises before stabilizing, with its amplitude remaining below 1.1 pu, thus fulfilling the high-voltage FRT requirements for OWFs.

Regarding fault current control, Fig. 7.14b, c demonstrate that the fault current magnitudes in phases B and C decrease under fault conditions. The maximum reductions in fault current magnitudes are 66% and 50% for phases B and C, respectively, which mitigates the impact on power electronics. Additionally, as shown in Fig. 7.14f, the maximum NS current magnitude of the WFMMC is constrained within 0.15 pu.

7.3.4 Simulation Analysis of Offshore Local AC Asymmetric Faults

The analysis in the previous Sects. 7.3.1 and 7.3.2 focused on AC asymmetric faults occurring after the centralized convergence of offshore wind farms (OWFs). Specifically, the scenario considered involves the convergence of a cluster of different OWFs located far offshore. The previous analysis addressed the most severe fault scenario, which has the largest impact on the entire system and presents the greatest challenge for FRT.

This section aims to demonstrate the effectiveness of MSNEC under more detailed and specific offshore AC local asymmetric faults. A comprehensive analysis of these specific offshore AC asymmetric fault scenarios is provided. In this context, the focus shifts to offshore AC local asymmetric faults, with the system topology shown in Fig. 7.15 serving as the analysis target. The OWFs consist of three clusters, each with a transmission power of 400 MW. These three clusters are connected via a 10 km cable to bus 1 for convergence. At the 6 s mark, SPG, PP, and DPG faults are introduced at the location of OWF cluster 1, with each fault lasting for 60 ms. The comprehensive simulation results of these fault scenarios are presented below.

Figure 7.16a–d present the simulation results for an SPG fault occurring at the outlet cable of wind farm cluster 1. From a voltage FRT perspective, Fig. 7.16a, b

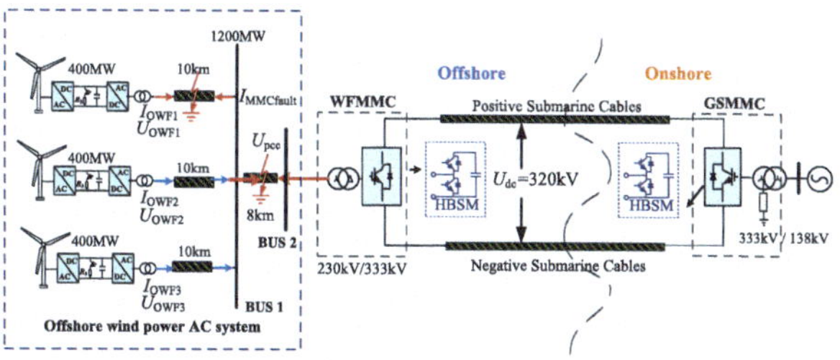

Fig. 7.15 Offshore AC local asymmetry fault location

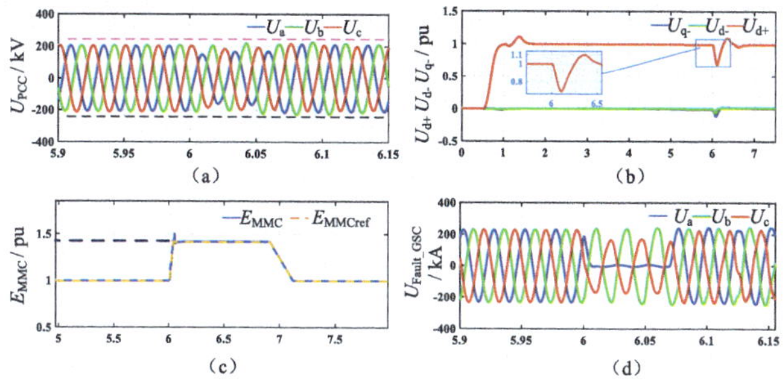

Fig. 7.16 Simulation results of local SPG fault. **a** The U_{pcc} of OWFs. **b** Each sequence component of the grid-connected voltage. **c** Energy variation of WFMMC. **d** The U_{owf} of GSC1

demonstrate that the Upcc is minimally affected by the local AC asymmetric fault. During this time, the inner-loop current controller of the WFMMC still has control margins, effectively functioning as a voltage source and providing voltage support to the OWFs. The PS voltage drops to 0.7 pu, and after the fault is cleared, the voltage amplitude remains below 1.1 pu. The grid-connected voltage quickly returns to its normal operating level following a brief voltage rise.

From an energy control standpoint, when the local detection system identifies the fault, control measures are initiated. As shown in Fig. 7.16c, the peak energy value is less than 2.56 pu, which does not trigger the sub-module capacitor voltage protection. Regarding fault current, Fig. 7.16d shows that the voltage at the Uowf1 of wind farm cluster 1 drops to zero following the offshore asymmetric fault, while the voltage in the non-faulted phases remains stable. After the fault is cleared, the voltage stabilizes smoothly.

Similarly, Fig. 7.17a–d presents the results for a DPG fault at the outlet cable of wind farm cluster 1. Figure 7.17a, d show that the impact of the local DPG fault at sea on Upcc is significant. This is accompanied by the energy control response depicted in Fig. 7.17b. The fault current in phase B decreases from 13 to 5 kA, while the fault current in phase C reduces from 15 to 7 kA. The envelope amplitude of the fault current is reduced to less than half of the original envelope amplitude through the use of NSCC. The fault management process successfully meets the high and low voltage FRT requirements for OWFs.

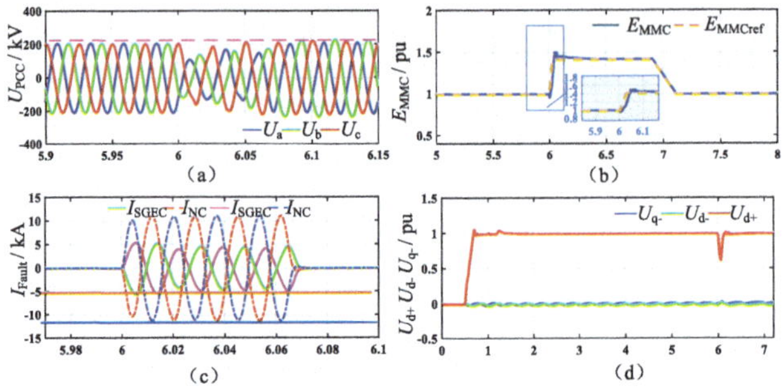

Fig. 7.17 Simulation results of local DPG fault. **a** The U_{pcc} of OWFs. **b** Energy variation of WFMMC. **c** Comparision of fault current. **d** Each sequence component of the grid-connected voltage

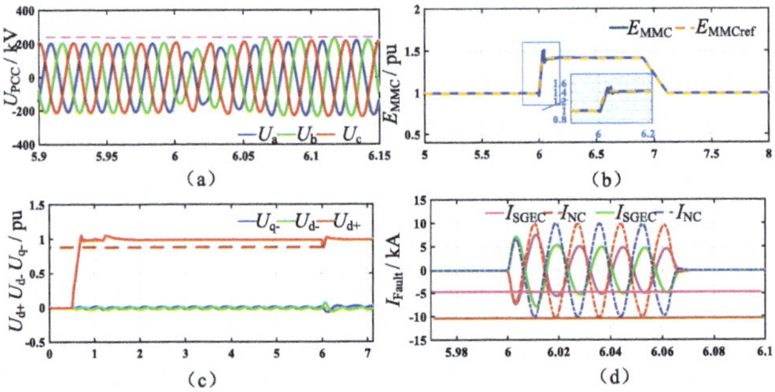

Fig. 7.18 Simulation Results of Local PP fault. **a** The U_{pcc} of OWFs. **b** Energy variation of WFMMC. **c** Each sequence component of the grid-connected voltage. **d** Comparision of fault current

Figure 7.18a–d presents the results for a DPG fault at the outlet cable of wind farm cluster 1. Figure 7.18a, b show that the PS voltage drops to 0.9 pu, indicating that the offshore wind cluster operates at 90% of the rated voltage under fault conditions. During this period, the inner-loop current controller of the WFMMC still retains control margins, effectively functioning as a voltage source and providing voltage support to the OWFs. In conjunction with the analysis of Fig. 7.18c, d, the fault currents in phases B and C decrease from 11 to 5 kA within the 20–50 ms time frame. The envelope amplitude of the fault current is reduced to less than half of its original value through the use of NSCC.

Further analysis considers a three-phase fault at the outlet cable of wind farm cluster 1. As shown in Fig. 7.19a, b, the Upcc experiences a more significant drop, but the system continues to operate stably without any overvoltage during the fault recovery phase. After implementing energy control, as depicted in Fig. 7.19c, the fault current amplitude is fully suppressed, with the envelope amplitude of the fault current reduced to about half of its original value using NSCC.

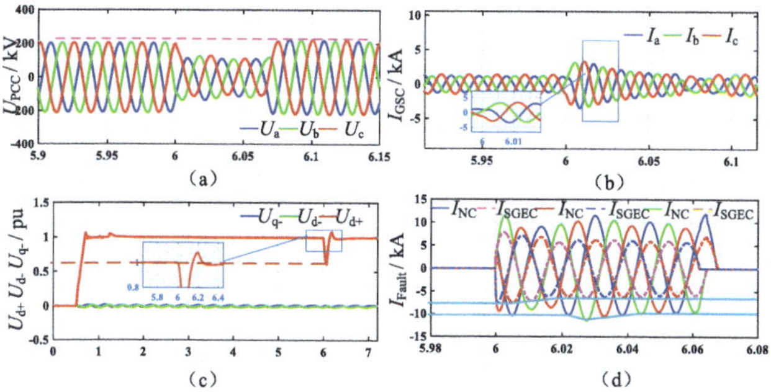

Fig. 7.19 Simulation Results of Local Three-phase Fault. **a** The U_{pcc} of OWFs. **b** Fault current of GSC. **c** Each sequence component of the grid-connected voltage. **d** Comparision of fault current

In summary, MSNEC can effectively reduce peak fault current under offshore AC asymmetric faults by at least 50%. Additionally, it can prevent fault-induced overvoltage issues and achieve multi-stage FRT. This strategy is applicable to scenarios involving AC asymmetric faults in different clusters of OWFs, significantly enhancing the FRT capability of OWFs.

7.3.5 Comparative Discussion Between Different Methods

This subsection compares various schemes for asymmetric fault ride-through (FRT) of offshore AC faults, with a focus on the MSNEC scheme in comparison to five other approaches. The comparison is summarized in the following Table 7.2.

One key indicator in analyzing overcurrent during asymmetric faults is the negative sequence (NS) current. The first scheme, commonly used in current engineering, aims to control the NS current to zero to achieve asymmetric FRT. However, this method overlooks the PS current of the WFMMC, which, when coupled with the GSC output current, can lead to an excessive NS voltage in the grid, potentially triggering NS protection. This approach also risks overvoltage during both the fault and recovery phases, as the NSCC may not disengage promptly after the fault is cleared.

The second scheme allows the WFMMC and GSC to inject a certain percentage of NS current during the fault, rather than strictly limiting it to zero. However, after the fault is cleared, the NS component may persist in the normal sequence network due to the sudden transition from fault to normal state, potentially causing overvoltage during the recovery phase.

Table 7.2 Comparison of effects between different solutions

Existing methods	Control effects	Economic analysis	Energy consumption
Control I^{-}_{MMC} and I^{-}_{OWF} to 0 [9]	Transient U^{-}_{pcc} Overvoltage	No promotion	No promotion
Inject a certain proportion of I^{-}_{MMC} and I^{-}_{OWF} [10]	U^{-}_{pcc} remain Overvoltage in recovery phase	No promotion	No promotion
Equivalent MMC and OWF to voltage source [10] [11]	The boundary of voltage source support is ignored	No promotion	No promotion
MMC and OWF provide transient I^{+}_{F} and U^{+}_{F} [12]	Neglect the characteristics of negative sequence power supply	No promotion	No promotion
Construction of energy storage in OWF [13]	Compensate reactive power to improve U_{pcc}	Increase construction cost	Additional capacitor for energy supply
Multi-stage Sequential Network Energy Control (MSNEC)	Mining voltage source properties to improve U_{pcc}	Improve economy	Fully utilize transient energy

The third and fourth schemes adopt different strategies for analyzing the sequential network equivalence. In the third approach, both the WFMMC and GSC are treated as voltage sources during the fault. However, this equivalence is not entirely accurate, as the GSC, which belongs to the OWFs converter, operates under grid-following control and exhibits current source characteristics during the fault. The equivalence of the WFMMC depends on the severity of the fault and the saturation level of the inner-loop current controller. If the inner-loop controller is not saturated, the WFMMC can act as a voltage source, providing adequate voltage support to the grid. Conversely, if the inner-loop controller is saturated, the WFMMC should be treated as a current source. The fourth scheme overlooks the unique properties of the power electronic equivalents of the GSC and WFMMC, which differ fundamentally from traditional synchronous machines that can only serve as PS power supplies. In contrast, power electronic equivalents can serve as NS power supplies, providing NS components to the sequence network.

The first four schemes do not alter the structure of OWFs connected to the onshore AC grid via MMC-HVDC, focusing instead on control strategies to achieve FRT. Consequently, these approaches offer limited benefits in terms of economic efficiency and the effective utilization of new energy.

The fifth scheme introduces supercapacitors on the offshore platform to enhance voltage support during faults. However, this approach incurs additional construction costs for the offshore platform and diminishes the overall economic efficiency of the scheme. Moreover, the lifespan and reliability of supercapacitors are crucial factors, and their implementation requires an additional power supply.

In contrast, the MSNEC scheme offers a more comprehensive approach by conducting an equivalent analysis of dual-side power electronic sources, fully leveraging the voltage support capabilities of the WFMMC. The inclusion of redundant submodules enhances the energy loop between the WFMMC and the fault location, thereby reducing the fault current in the affected phase. This scheme also mitigates the risk of overvoltage during fault conditions and ensures a stable recovery of the grid-connected voltage during the recovery phase. By maximizing the energy margin of the WFMMC, the MSNEC scheme provides energy storage support for OWFs without additional economic costs. It achieves similar benefits to the supercapacitor-based energy storage in Scheme 5 by efficiently re-releasing the energy absorbed during the fault back into the onshore AC system, thus minimizing energy waste. The WFMMC facilitates its own energy storage and discharge processes, thereby promoting the efficient and reliable utilization of new energy.

7.4 Summary

In this chapter, offshore AC asymmetric faults are analyzed from three perspectives: mechanism, fault characteristics, and control strategy. The MSNEC scheme is proposed as a solution for achieving FRT in offshore wind farms (OWFs) and WFMMC. MSNEC effectively coordinates the control of energy, NS current, and voltage in both OWFs and WFMMC during offshore AC asymmetric faults. The following conclusions have been drawn:

The fault current in an offshore AC asymmetric fault consists of contributions from a double-ended power electronic supply, which differs from the characteristics of onshore AC faults. Analysis of the sequence network reveals the mechanisms of three types of asymmetric faults, identifying the NS component as the core target for fault control. The fault current characteristics and the equivalence principles of WFMMC and GSVSC are applied to the sequential network power supply, clarifying the control objectives for asymmetric faults.

This method approaches WFMMC from an energy conversion perspective, fully utilizing its energy potential. Energy control reduces the amplitude of the fault phase current by activating submodules in the faulted phase. Additional submodules are also engaged in the non-faulted phase to achieve a series voltage dividing effect, actively preventing overvoltage in the non-faulted phase. This approach enhances the economic efficiency of MMC-HVDC while providing energy storage capacity for the system. The fault process is flexibly managed to enable the charging and discharging processes of WFMMC.

MSNEC operates through three stages: fault transient, fault transition, and fault recovery. The control capability of WFMMC is fully utilized during faults, meeting the FRT requirements across multiple stages. Simulations demonstrate that MSNEC can initiate control actions rapidly after a fault occurs, verifying its effectiveness. This approach can reduce peak fault current during offshore AC asymmetric faults by at least 50% and mitigate overvoltage issues.

This strategy is applicable to scenarios involving AC asymmetric faults in different OWF clusters, significantly improving the FRT capability of OWFs. Future work will focus on fully leveraging the energy storage capacity of MMC-HVDC, with plans to apply this methodology to other offshore wind grid-connected scenarios to achieve comprehensive offshore and onshore AC/DC FRT.

References

1. Xiang W, Lin W, Xu L, Wen J (2018) Enhanced independent pole control of hybrid mmc-hvdc system. IEEE Trans Power Del 33(2):861–872. https://doi.org/10.1109/TPWRD.2017.2715040
2. Jing Y, Li R, Xu L, Wang Y (2018) Enhanced AC voltage and frequency control of offshore MMC station for wind farm connection. IET Renew Power Gener 12(15):1771–1777. https://doi.org/10.1049/iet-rpg.2018.5391
3. Hu X, Liang J, Rogers DJ, Li Y (2013) Power flow and power reduction control using variable frequency of offshore AC grids. IEEE Trans Power Syst 28(4):3897–3905. https://doi.org/10.1109/TPWRS.2013.2257884
4. Xiang W, Zhang H, Yang S, Zhou M, Lin W, Wen J (2021) A differential pilot protection scheme for mmc-based dc grid resilient to communication failure. IEEE J Emerg Sel Top Power Electron **9**(5), 5631–5645 (2021). https://doi.org/10.1109/JESTPE.2020.3035672
5. Wang W, Yin X, Jiang L, Cao Y, Li Y (2022) Perturbation observer-based nonlinear control of VSC-MTDC systems. Int J Electr Power Energy Syst **134** (2022). https://doi.org/10.1016/j.ijepes.2021.107387
6. Moon JW, Park JW, Kang DW, Kim JM (2015) A control method of hvdc-modular multilevel converter based on arm current under the unbalanced voltage condition. IEEE Trans Power Del 30(2):529–536. https://doi.org/10.1109/TPWRD.2014.2342229
7. Lu Z, Ye Y, Qiao Y (2019) An adaptive frequency regulation method with grid-friendly restoration for VSC-HVDC integrated offshore wind farms. IEEE Trans Power Syst 34(5):3582–3593. https://doi.org/10.1109/TPWRS.2019.2901986
8. Meng Y, Zou Y, Wang H, Kong Y, Du Z, Wang X (2021) Novel submodule topology with large current operation and dc-fault blocking capability for MMC-HVDC. IEEE Trans Power Del 36(3):1542–1551. https://doi.org/10.1109/TPWRD.2020.3011228
9. Seifi K, Moallem M (2019) An adaptive pr controller for synchronizing grid-connected inverters. IEEE Trans Ind Electron 66(3):2034–2043. https://doi.org/10.1109/TIE.2018.2838098
10. Li R, Yu L, Xu L (2019) Offshore ac fault protection of diode rectifier unit-based hvdc system for wind energy transmission. IEEE Trans Ind Electron 66(7):5289–5299. https://doi.org/10.1109/TIE.2018.2869357
11. Shi X, Yang R, Cai X, Fang Z, Dong P, Rao F (2023) Improved comprehensive energy-based control for mmc-hvdc system. Int J Electr Power Energy Syst **145**, 108,593 (2023). https://doi.org/10.1016/j.ijepes.2022.108593

12. Zhou H, Yao W, Sun K, Ai X, Wen J, Cheng S (2024) Characteristic investigation and overvoltage suppression of mmc-hvdc integrated offshore wind farms under onshore valve-side spg fault. IEEE Trans Power Syst 39(1):1346–1359. https://doi.org/10.1109/TPWRS.2023.3236095
13. Tong N, Tang Z, Wang Y, Lai CS, Lai LL (2022) Semi ai-based protection element for mmc-mtdc using local-measurements. Int J Electr Power Energy Syst **142**, 108,310 (2022). https://doi.org/10.1016/j.ijepes.2022.108310

Part III
Active Frequency Support in Large-Scale Offshore Wind Farms Connected Via VSC-HVDC

As offshore wind energy offers stable conditions, high utilization rates, proximity to load centers, and a minimal footprint, it has become a vital component of global wind energy development. For small, near-shore offshore wind farms, alternating current (AC) integration is typically used. However, with the development of large-scale offshore wind farms with long-distance transmission, voltage source converter-based multi-terminal high voltage direct current (VSC-MTDC) systems have gained prominence as a solution for clustering, long-distance, and multi-regional wind power grid connections. The VSC-MTDC system decouples offshore and onshore AC systems, posing challenges for frequency support from offshore wind farms during onshore system frequency fluctuations. Therefore, researching active frequency support control strategies for offshore wind farms connected via VSC-MTDC systems is crucial for ensuring the safe and stable grid integration of these large-scale installations.

Large-scale offshore wind farms (OWFs), often located far from onshore power systems, are economically integrated using VSC-HVDC systems. Various frequency regulation strategies for VSC-HVDC systems have been extensively studied [1–10]. Some approaches utilize the electrostatic energy stored in converter capacitors or the kinetic energy from wind turbines (WTs) for frequency support [1–3]. However, these methods often require significant investments in DC capacitors or can cause fluctuations in DC voltage, potentially limiting their effectiveness. Other communication-free strategies convert frequency deviations into changes in DC voltage [8–10], but they may not fully reflect frequency deviations, impacting their overall performance.

In VSC-MTDC systems, WTs can provide kinetic energy for frequency regulation during disturbances. Enhancing frequency support can be achieved by optimally distributing frequency support power among onshore VSC stations. Similar power distribution control strategies have been proposed to optimize power sharing among onshore VSC stations and enhance frequency regulation performance [11]. However, methods that require WTs to reserve power through de-loading strategies can reduce the economic efficiency of OWFs [12].

This part explores active frequency support control strategies for VSC-MTDC integrated offshore wind farms, considering scenarios with both minimal and significant communication time delays. It investigates coordination between offshore and onshore systems, interaction among various onshore systems, communication time delays, different operational states of wind turbines, and smooth rotor speed recovery. The details can be presented as follows.

Chapter 8 proposes a communication-based coordinated control strategy for offshore wind farms with minimal communication delays, focusing on step start-up control at the wind farm level and adaptive droop control at the converter station level to distribute frequency support power effectively among onshore stations.

Chapter 9 introduces an improved communication-free coordinated control strategy to address communication delays. Adaptive droop control at the onshore converter station converts onshore AC frequency changes into DC voltage changes, with offshore wind farms responding to these estimated changes. The strategy also includes an asymptotic recovery control to reduce secondary frequency drops during rotor speed recovery.

Chapter 10 develops a distributed coordinated control scheme based on a consistency algorithm, accounting for the varying operational states of offshore wind turbines due to geographical distribution and wake effects. Faster turbines release more frequency modulation power, optimizing the wind farm's frequency support capabilities while ensuring safe operation.

Chapter 11 proposes a communication-free coordinated frequency support control strategy for VSC-MTDC integrated offshore wind farms connected to different asynchronous AC areas. The adaptive dual droop control quickly identifies disturbed and undisturbed AC areas, enabling undisturbed systems to provide frequency support. The frequency support power is limited based on estimated system inertia for safe operation.

References

1. Zhu J, Booth CD, Adam GP, Roscoe AJ, Bright CG (2013) Inertia emulation control strategy for VSC-HVDC transmission systems. IEEE Trans. Power Syst. 28(2):1277–1287. https://doi.org/10.1109/TPWRS.2012.2213101
2. Zhu J, Guerrero JM, Hung W, Booth CD, Adam GP (2014) Generic inertia emulation controller for multi-terminal voltage-source-converter high voltage direct current systems. IET Renew. Power Gener. 8(7):740–748. https://doi.org/10.1049/iet-rpg.2014.0109
3. Liu H, Chen Z (2015) Contribution of VSC-HVDC to frequency regulation of power systems with offshore wind generation. IEEE Trans. Energy Convers. 30(3):918–926. https://doi.org/10.1109/TEC.2015.2417130

4. Leon AE (2018) Short-term frequency regulation and inertia emulation using an MMC-based MTDC system. IEEE Trans. Power Syst. 33(3):2854–2863. https://doi.org/10.1109/TPWRS.2017.2757258
5. Junyent-Ferre A, Pipelzadeh Y, Green TC (2015) Blending HVDC-link energy storage and offshore wind turbine inertia for fast frequency response. IEEE Trans. Sustain. Energy 6(3):1059–1066. https://doi.org/10.1109/TSTE.2014.2360147
6. Li Y, Xu Z, Wong KP (2017) Advanced control strategies of PMSG-based wind turbines for system inertia support. IEEE Trans. Power Syst. 32(4):3027–3037. https://doi.org/10.1109/TPWRS.2016.2616171
7. Wang W, Li Y, Cao Y, Häger U, Rehtanz C (2018) Adaptive droop control of VSC-MTDC system for frequency support and power sharing. IEEE Trans. Power Syst. 33(2):1264–1274. https://doi.org/10.1109/TPWRS.2017.2719002
8. Phulpin Y (2012) Communication-free inertia and frequency control for wind generators connected by an HVDC-link. IEEE Trans. Power Syst. 27(2):1136–1137. https://doi.org/10.1109/TPWRS.2011.2175817
9. Kabsha MM, Rather ZH (2020) A new control scheme for fast frequency support from HVDC connected offshore wind farm in low inertia system. IEEE Trans. Sustain. Energy. 11(3):1829–1837. https://doi.org/10.1109/TSTE.2019.2942541
10. Kou P, Liang D, Wu Z, Ze Q, Gao L (2018) Frequency support from a DC-grid offshore wind farm connected through an HVDC link: a communication-free approach. IEEE Trans. Energy Convers. 33(3):1297–1310. https://doi.org/10.1109/TEC.2018.2814604
11. Davari M, Mohamed YAI (2018) Robust droop and DC-bus voltage control for effective stabilization and power sharing in VSC multiterminal DC grids. IEEE Trans. Power Electron. 33(5):4373–4395. https://doi.org/10.1109/TPEL.2017.2715039
12. Lu Z, Ye Y, Qiao Y (2019) An adaptive frequency regulation method with grid-friendly restoration for VSC-HVDC integrated offshore wind farms. IEEE Trans. Power Syst. 34(5):3582–3593. https://doi.org/10.1109/TPWRS.2019.2901986

Chapter 8
Two-Level Combined Control Scheme of VSC-MTDC Integrated Offshore Wind Farms for Onshore System Frequency Support

Abstract This chapter presents a two-level combined control (TLCC) approach for voltage source converter-based multi-terminal high-voltage direct current (VSC-MTDC) networks that integrate offshore wind farms, aiming to enhance frequency support for the mainland grid. The TLCC operates across two hierarchical levels: the offshore wind turbine level and the onshore VSC station level. At the turbine level, each wind turbine employs adaptive inertial and droop control while maintaining maximum power point tracking (MPPT) without energy reserves. To mitigate secondary frequency drops (SFD), wind turbines are grouped into clusters based on rotor speed, and a step start-up control scheme is employed to sequentially activate these clusters, providing coordinated frequency support. At the system level, a communication-free allocation control strategy is used, leveraging local frequency signals from onshore VSC stations to efficiently distribute active power among them. This TLCC approach effectively enhances the frequency stability of the onshore grid and reduces SFD, while keeping the turbines in MPPT mode. Case studies conducted on a 3-area, 4-terminal VSC-MTDC system with integrated offshore wind farms validate the TLCC strategy's effectiveness and adaptability across diverse scenarios.

8.1 Structure of the Proposed TLCC Scheme

8.1.1 Overview of the Proposed TLCC Scheme

The proposed Two-Level Combined Control (TLCC) scheme integrates a step start-up and adaptive inertial droop control at the wind turbine (WT) level, along with a communication-free allocation control strategy at the onshore VSC station level. The basic structure of this TLCC scheme is illustrated in Fig. 8.1. As depicted, the droop coefficients of the onshore VSC stations revert to their initial values after a delay period, denoted as Δt_m.

W. Yao et al., *Active Control of Large-Scale Offshore Wind Farms Connected Via VSC-HVDC*, Power Systems, https://doi.org/10.1007/978-981-97-9346-4_8

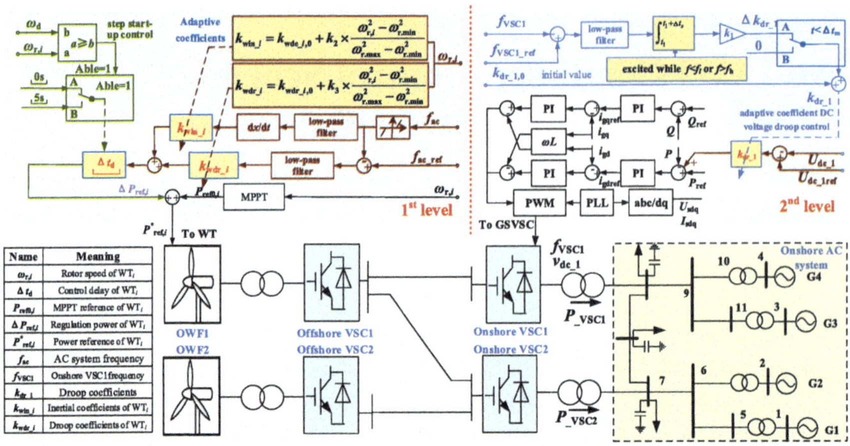

Fig. 8.1 Control block diagram of the proposed TLCC scheme for VSC-MTDC based OWFs

At the wind turbine level, adaptive inertial droop control is employed while all WTs operate in maximum power point tracking (MPPT) mode. The WTs are grouped into clusters based on their rotor speeds, with the number of clusters tailored to the specific scenario. For illustrative purposes, this chapter divides the WTs in an offshore wind farm (OWF) into two clusters. To minimize both the frequency nadir and the secondary frequency drop (SFD), WTs are sorted by rotor speed. The top 60% of WTs by rotor speed form the first cluster, while the remaining 40% constitute the second cluster during load increase frequency events, with the critical rotor speed denoted as ω_d. For example, in an OWF with 25 WTs, the top 15 WTs are placed in the first cluster, and the remaining 10 in the second cluster, as shown in Fig. 8.2.

In this TLCC scheme, WTs in different clusters participate in frequency regulation with distinct time delays. According to [1], inertial frequency support typically operates within 0–5 s following a frequency event. Therefore, once the rotor speeds of the WTs in the first cluster recover, the WTs in the second cluster are activated to provide frequency support with a 5 s delay, aiming to reduce the SFD. The primary goal for the WTs in Cluster 1 is to reduce the frequency nadir. After the WTs in Cluster 1 return to their initial conditions, the WTs in Cluster 2 engage in frequency regulation after a 5 s delay to mitigate the SFD.

At the VSC station level, onshore VSC stations implement power-voltage droop control. The droop coefficient of each onshore VSC station is adaptively adjusted based on local frequency signals, which vary due to the different electrical distances from the disturbance bus node, as defined in [2, 3]. Consequently, the frequency support power provided by the OWFs is appropriately allocated among the various VSC stations. The closer a station is to the disturbance bus node, the more power it transmits, thereby keeping power flow changes in the onshore AC system within a small range, which is beneficial for frequency recovery. The start-up sequence of

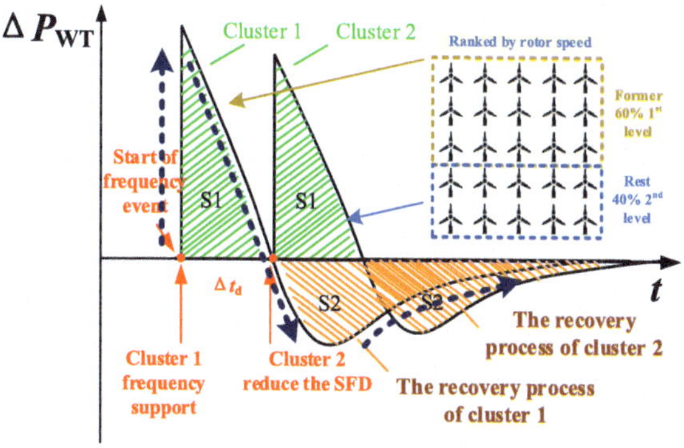

Fig. 8.2 Start-up sequence of the proposed control scheme

the TLCC scheme is illustrated in Fig. 8.2. As shown, the ΔP_{WT} of WTs in different clusters returns to its initial state after frequency support.

8.1.2 The WT Level: Step Start-Up Adaptive Inertial and Droop Control Strategy

WTs can support frequency stability by releasing kinetic energy. However, if there is no additional power injection to offset the recovery time in the onshore AC system, a secondary frequency drop (SFD) may occur. Improperly configured recovery times and control strategies can worsen the SFD, potentially making it more severe than the initial frequency drop.

To address this issue, this chapter introduces a step start-up adaptive inertial and droop control strategy at the wind turbine level. A schematic overview of this strategy is provided in Fig. 8.1. In this approach, WTs operating in maximum power point tracking (MPPT) mode are organized into two clusters based on their rotor speeds. This study considers an offshore wind farm (OWF) with 25 equivalent WTs. The critical rotor speed, labeled as ω_d, is set at 0.9 pu. WTs with rotor speeds above 0.9 pu are placed in the first cluster, while the remaining WTs form the second cluster.

Under this control strategy, WTs in Cluster 1 immediately begin providing frequency support when the frequency crosses a predefined threshold. WTs in Cluster 2 are delayed by 5 s before engaging in frequency regulation, which helps to further mitigate the SFD. The sequence for activating the WTs is illustrated in Fig. 8.2.

The step start-up adaptive inertial and droop control strategy consists of the following steps:

- WTs are divided into two clusters based on rotor speed. Those with rotor speeds higher than the critical value ω_d (set at 0.9 pu in this study) form the first cluster, while the rest belong to the second cluster.

- When the onshore AC system frequency f_{ac} crosses the threshold ($f_{ac} \geq 50.1$, Hz or $f_{ac} \leq 49.9$, Hz in this study), the WTs in Cluster 1 immediately participate in frequency support by releasing their kinetic energy.
- The inertial and droop coefficients of WTs in Cluster 1 are adaptively adjusted as their rotor speeds change, ensuring the speeds remain within a safe operating range (0.9–1.2 pu).
- After Δt_d (5 s in this chapter), once the rotor speeds in Cluster 1 have recovered, WTs in Cluster 2 begin to provide frequency support, releasing additional kinetic energy. This power contribution from Cluster 2 aids in reducing the SFD.

The calculation formulas for $\Delta P_{\mathrm{win},i}$, $\Delta P_{\mathrm{wdr},i}$, $k_{\mathrm{win},i}$, and $k_{\mathrm{wdr},i}$ are presented below:

$$\Delta P_{\mathrm{win},i} = -k_{\mathrm{win},i} \times \frac{\mathrm{d}f}{\mathrm{d}t} \tag{8.1}$$

$$\Delta P_{\mathrm{wdr},i} = -k_{\mathrm{wdr},i} \times (f_{\mathrm{ac,ref}} - f_{\mathrm{ac}}) \tag{8.2}$$

$$k_{\mathrm{win},i} = k_{\mathrm{win,i,0}} + k_2 \times \frac{\omega_{\mathrm{r},i}^2 - \omega_{\mathrm{r,min}}^2}{\omega_{\mathrm{r,max}}^2 - \omega_{\mathrm{r,min}}^2} \tag{8.3}$$

$$k_{\mathrm{wdr},i} = k_{\mathrm{wdr,i,0}} + k_3 \times \frac{\omega_{\mathrm{r},i}^2 - \omega_{\mathrm{r,min}}^2}{\omega_{\mathrm{r,max}}^2 - \omega_{\mathrm{r,min}}^2} \tag{8.4}$$

In Eqs. (8.1)–(8.4), $\Delta P_{\mathrm{win},i}$ and $\Delta P_{\mathrm{wdr},i}$ represent the frequency regulation power provided by the adaptive inertial control and adaptive droop control of wind turbine WTi, respectively. The variables f_{ac} and $f_{\mathrm{ac,ref}}$ denote the measured onshore AC system frequency and its reference value, which are communicated to the offshore wind farms (OWFs). The coefficients $k_{\mathrm{win},i}$ and $k_{\mathrm{wdr},i}$ correspond to the adaptive inertial and droop coefficients, respectively, and are calculated according to Eqs. (8.3)–(8.4) to ensure that the rotor speed remains within a safe range [4]. If the rotor speed of the wind turbines exceeds this limit under certain operating conditions, it will be constrained to maximum or minimum values, thereby limiting the frequency support power they can provide.

The parameters $k_{\mathrm{win},i,0}$ and $k_{\mathrm{wdr},i,0}$ denote the initial values of the inertial and droop coefficients, while the constants k_2 and k_3 represent the slopes of the inertia and droop coefficients, respectively. The rotor speed of WTi, denoted as $\omega_{\mathrm{r},i}$, is maintained within a safe range (0.9-1.2 pu) by the adaptive coefficients while the WTs participate in frequency regulation. The detailed controller design will be elaborated in Sect. 8.3. After the adaptive inertial and droop control, the WTs return to their initial operational state. Further details on the recovery control strategy can be found in [5].

8.1.3 The VSC Station Level: Communication-Free Allocation Control Strategy of Onshore VSC Stations

Large-scale offshore wind farms (OWFs) integrated via VSC-MTDC systems often connect to the onshore grid through multiple VSC stations due to the substantial transmission power required. As illustrated in Fig. 8.1, OWF1 is connected to the AC system through two onshore VSC stations. When a load increase occurs on an AC bus, the frequency deviations detected by different onshore VSC stations (e.g., VSC_1 and VSC_2 in Fig. 8.1) will vary.

For example, consider a scenario where a 600 MW load increase occurs at Bus 7 in Area 1. The frequency deviations measured by VSC_1 and VSC_2 will differ, as shown in Fig. 8.3. The station closest to the disturbance bus will experience a larger rate of change of frequency (RoCoF) and a greater frequency deviation. Therefore, during frequency events, the frequency support power provided by OWFs should be appropriately distributed between onshore VSC stations to prevent significant power flow changes in the onshore AC system.

To optimize frequency support performance, a larger portion of the frequency support power should be injected into the onshore AC system through the VSC station nearest to the disturbance bus. This approach effectively manages power flow changes and enhances the stability of the onshore grid.

In this section, we propose a communication-free control strategy for allocating frequency support among onshore VSC stations. The block diagram of this strategy is shown in Fig. 8.1. In this figure, $f_{\mathrm{VSC}i}$ and $f_{\mathrm{VSC}i,\mathrm{ref}}$ represent the measured frequency and its reference value at onshore station VSC_i. The initial value of the power-voltage droop coefficients is represented by $k_{\mathrm{dr},0}$. The variable $\Delta k_{\mathrm{dr},i}$ denotes the change in $k_{\mathrm{dr},0}$ during a frequency event, calculated by multiplying k_1 with the integral of the difference between $f_{\mathrm{VSC}i}$ and $f_{\mathrm{VSC}i,\mathrm{ref}}$ over 0.01 s (refer to Eq. (8.5)). The adaptive droop coefficient $k_{\mathrm{dr},i}$ is the sum of $k_{\mathrm{dr},0}$ and $\Delta k_{\mathrm{dr},i}$. For VSC1 and VSC2, both k_1 and $k_{\mathrm{dr},0}$ are set to the same values (see Table 8.2), but $\Delta k_{\mathrm{dr},i}$ differs due to the varying electrical distances from the disturbance bus. Consequently, frequency support power is more effectively distributed among the different onshore VSC stations compared to conventional constant droop control. The process for calculating the adaptive droop coefficient $k_{\mathrm{dr},i}$ and the power injected by onshore station VSC_i is detailed below.

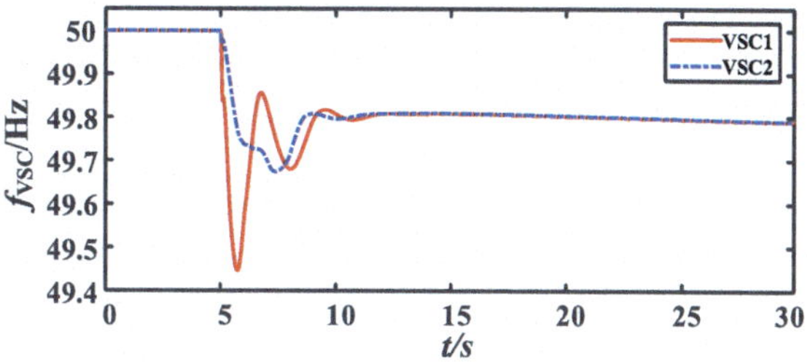

Fig. 8.3 Measured frequency of different VSC stations (with load decrease 600 MW)

Table 8.1 Other parameters of the whole system

Description	Value (unit)
Rated AC grid rms voltage	220 kV
Nominal AC frequency	50 Hz
Frequency dead band	0.1 Hz
Rated VSC AC rms voltage	110 kV
Nominal DC voltage	200 kV
DC capacitor	0.048 mF
DC voltage range	0.9–1.1 pu
Rated wind speed	12 m/s

$$k_{\mathrm{dr},i} = k_{\mathrm{dr},0} + \int_{t_1}^{t_1+\Delta t_s} k_1 \times (f_{\mathrm{VSC}i,\mathrm{ref}} - f_{\mathrm{VSC}i}(t)), dt \tag{8.5}$$

$$P^*_{i,\mathrm{ref}} = P_{i,\mathrm{ref}} + k_{\mathrm{dr},i} \times (U_{\mathrm{dc}i,\mathrm{ref}} - U_{\mathrm{dc}i}) \tag{8.6}$$

In Eqs. (8.5) and (8.6), t_1 represents the moment when the change in onshore AC system frequency exceeds the deadband threshold. Δt_s is the integration time, set to 0.01 s in this study. This duration reflects the electrical distance between various onshore VSC stations and the disturbed bus, enabling prompt adjustment of droop coefficients following a frequency event. It is important to note that the integration period is limited to 0.01 s, and achieving zero steady-state error in frequency is not required within the primary frequency response timescale.

The variables $k_{\mathrm{dr},i}$ and $k_{\mathrm{dr},0}$ represent the adaptive and initial values of the power-voltage droop coefficients for different onshore VSC stations. The frequencies $f_{\mathrm{VSC}i}$ and $f_{\mathrm{VSC}i,\mathrm{ref}}$ indicate the measured and reference frequencies at onshore station VSCi, respectively. Similarly, $P^*_{i,\mathrm{ref}}$ and $P_{i,\mathrm{ref}}$ denote the reference and initial values of injected active power for VSC_i. Lastly, $U_{\mathrm{dc}i,\mathrm{ref}}$ and $U_{\mathrm{dc}i}$ represent the reference and measured DC voltages, respectively.

As shown in Fig. 8.1, under normal conditions, the droop coefficients of each onshore VSC station remain unchanged. However, during frequency events, the droop coefficients $k_{\mathrm{dr},1}$ and $k_{\mathrm{dr},2}$ for onshore stations VSC_1 and VSC_2 adjust adaptively in response to changes in the measured frequency. This adaptive adjustment ensures that the frequency regulation power from the offshore wind farms (OWFs) is optimally distributed, with more power directed through the VSC stations closest to the disturbance bus. Throughout this control process, only local frequency signals are used, making the proposed strategy entirely communication-free. The comprehensive flowchart of the TLCC scheme is illustrated in Fig. 8.4.

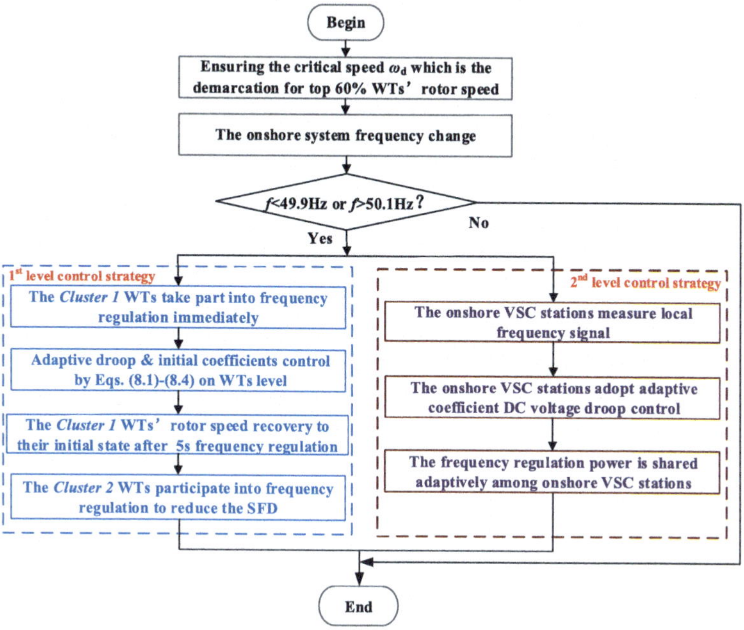

Fig. 8.4 The flow chart of the proposed two-level combined control scheme

8.2 Controller Design

8.2.1 Test System

The test VSC-MTDC-based offshore wind farm and onshore AC system model are shown in Fig. 8.5, which consists of three areas and four terminals.

In Fig. 8.5, Area 1 represents the onshore AC system, which consists of four synchronized generators (G1-G4) connected across 11 bus nodes and 10 lines. The onshore AC system operates at a rated voltage of 220 kV and a frequency of 50 Hz. Areas 2 and 3 depict offshore wind farms, each containing five equivalent wind turbine models (G5-G9 in Area 2 and G10-G14 in Area 3). Each equivalent wind turbine model represents a group of 20 wind turbines, each rated at 2 MW and operating in maximum power point tracking (MPPT) mode. Only the wind turbines in Area 2 (G5-G9) participate in frequency support.

The offshore wind farms are connected to the onshore AC system via a DC grid, which includes two offshore VSC stations and two onshore VSC stations. Stations VSC_1 and VSC_2 implement a voltage-active power droop control strategy, while stations VSC_3 and VSC_4 employ an AC voltage magnitude control strategy. Only OWF1 contributes to onshore frequency regulation. Due to varying wind conditions, the initial steady-state rotor speeds of G5-G9 in OWF1 are 0.9, 0.95, 0.85, 1.0,

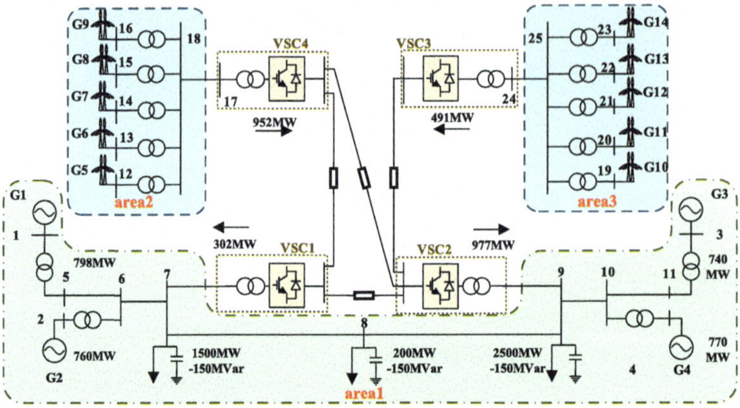

Fig. 8.5 The 3-area 4-terminal of VSC-MTDC based offshore wind farm integration system

Table 8.2 Parameters of offshore wind turbine controller

Parameters	Description	Value (unit)
$\omega_{r,i}$	The rotor speed of WT_i	0.7–1.2 pu
ω_d	The critical rotor speed of WT_i	0.9 pu
$\omega_{r,max}$	The maximum rotor speed limit of WTs	1.2 pu
$\omega_{r,min}$	The minimum rotor speed limit of WTs	0.7 pu
$k_{win,i,0}$	The initial value of inertial coefficient	–100 to –135 (MW · s/Hz)
$k_{wdr,i,0}$	The initial value of droop coefficient	–90 to –160 (MW/Hz)
$k_{win,i}$	The value of inertial coefficient	–20 to –180 (MW · s/kV)
$k_{wdr,i}$	The value of droop coefficient	–40 to –180 (MW/kV)
Δt_d	The time delay for cluster 2	5 s
f_{VSCi}	The reference value of VSC_i	50 Hz
$k_{dr,0}$	The initial value of P-U droop coefficients	20 (MW/kV)
k_I	The integral coefficient	10000
$\Delta k_{dr,i}$	The change value of $k_{dr,0}$ for VSC_i	–20 to 100 MW/kV)
$k_{dr,i}$	The adaptive droop coefficient	0–120 (MW/kV)

and 1.05 pu, respectively. Consequently, G6, G8, and G9 form Cluster 1, while G5 and G7 constitute Cluster 2 during steady-state operation. Additional parameters are provided in Table 8.1.

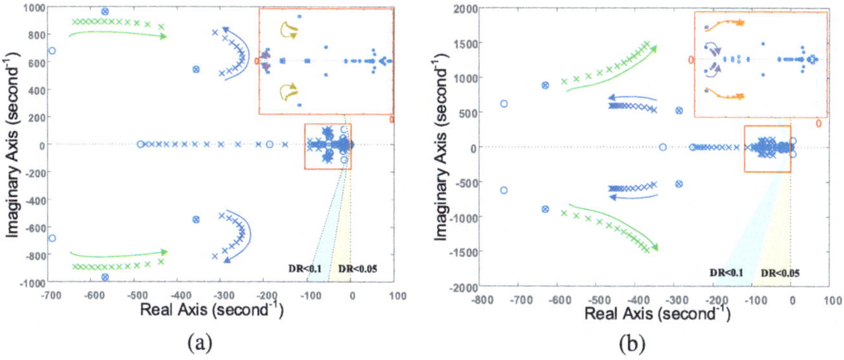

Fig. 8.6 The zero-pole map for: **a** $k_{dr,1}$ from 0 to 120; **b** $k_{dr,2}$ from 0 to 120

8.2.2 Controller Parameter Design

The controller parameter design follows the methodology outlined in [5]. Offshore wind turbines utilize step start-up adaptive inertial and droop control to provide frequency regulation power during onshore AC system frequency events. The parameter design for this control scheme is detailed in Eqs. (8.1)–(8.4). Onshore VSC stations employ a communication-free coordinated control strategy to efficiently allocate frequency regulation power among the stations, with droop coefficients calculated using Eq. (8.5). A detailed description of the controller parameters is provided in Table 8.2.

The adaptive droop coefficients $k_{\mathrm{dr},i}$ across all onshore VSC stations influence the power transmission of each station VSCi. To illustrate the impact of $k_{\mathrm{dr},i}$, a small-signal analysis was conducted with varying droop coefficient values $k_{\mathrm{dr},1}$ and $k_{\mathrm{dr},2}$ at VSC1 and VSC2, respectively, under nominal system operation. The trajectory of eigenvalue changes with different values of $k_{\mathrm{dr},1}$ and $k_{\mathrm{dr},2}$ is depicted in Fig. 8.6. The results show that eigenvalues vary with changes in droop coefficients. Notably, the damping ratios (DR) of critical eigenvalues remain above 0.1 across the range as $k_{\mathrm{dr},1}$ and $k_{\mathrm{dr},2}$ vary from 0 to 120, demonstrating the system's stability and adaptability within this range.

8.3 Simulation Verification

To demonstrate the effectiveness and applicability of the proposed TLCC scheme, a 3-area, 4-terminal VSC-MTDC-based offshore wind farm and onshore AC system model, referenced from [6, 7], was developed and simulated using the MATLAB/Simulink platform. The model incorporates electromechanical transient models for two-level VSC-based converter stations and Type 3 Doubly Fed Induction Generator (DFIG) models for wind turbines (WTs). Standard vector control strategies with inner current control loops are assumed for the converters, and π-section equivalent circuits represent the AC and DC transmission lines. Synchronous generators are modeled with sixth-order sub-transient models equipped with Type AC-4A excitation systems, while loads are represented by constant impedance models. More detailed information on system modeling and parameters can be found in the Appendix and in [6, 7].

Various frequency events were simulated using different control schemes within the 3-area, 4-terminal VSC-MTDC based offshore wind farm system, as shown in Fig. 8.5. These schemes include:

- Control scheme 1 (NO FSC): WTs do not participate in frequency regulation.
- Control scheme 2 (CIDC): WTs use inertial and droop control with constant parameters for frequency support.
- Control scheme 3 (CIDC&DL): WTs employ conventional inertial and droop control with a 10% de-loading strategy for frequency support, as described in [5].
- Control scheme 4 (TLCC): WTs utilize the proposed TLCC scheme for frequency support.

All frequency events are initiated at Bus 7 at $t = 5$ s, with a 200 ms time delay for frequency signal transmission to the offshore wind farms in all scenarios. The simulations are conducted using the MATLAB/Simulink package to evaluate the performance of each control scheme.

8.3.1 Load Increase

In this scenario, a 200 MW load increase at Bus 7 occurs at precisely $t = 5s$. The initial steady-state rotor speeds of G5-G9 in OWF1 are 0.9, 0.95, 0.85, 1.0, and 1.05 pu, respectively. Consequently, WTs G6, G8, and G9 form Cluster 1, which actively participates in frequency regulation immediately following the frequency event. In contrast, WTs G5 and G7 constitute Cluster 2, which release their kinetic energy once the WTs in Cluster 1 return to their initial states, thereby mitigating secondary frequency drops (SFD). This scenario tests four distinct control schemes.

The frequency measured at VSC_1, along with the transmitted power from VSC_1 and VSC_2, is depicted in Fig. 8.7. As shown in Fig. 8.7a, during the 200 MW load increase at Bus 7 at $t = 5s$, the frequency nadirs at VSC_1 are 49.78, 49.78, 49.83, and 49.85 Hz for the NO FSC scheme, CIDC scheme, CIDC&DL scheme, and the proposed TLCC scheme, respectively. Moreover, SFD occurs when WTs participate in frequency regulation under the CIDC, CIDC&DL, and TLCC schemes. However, the proposed TLCC scheme effectively mitigates SFD. The nadir frequencies for Control Schemes 2–4 are 49.81, 49.83, and 49.85 Hz, respectively. Since WTs do not participate in frequency regulation under the NO FSC control strategy, SFD does not occur due to the absence of WTs in the recovery process.

Additionally, as shown in Fig. 8.7b, c, VSC_1 injects more frequency support power into the onshore AC system under the TLCC scheme during the frequency event. The

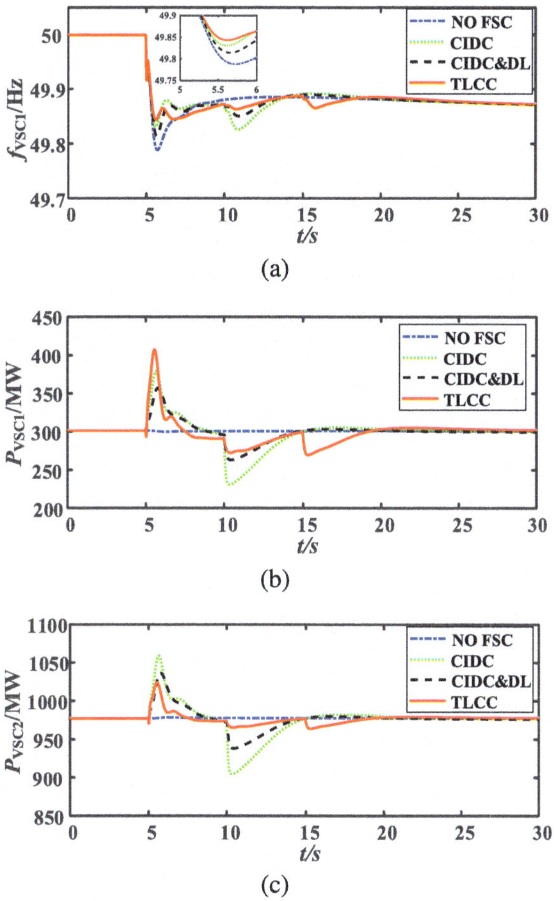

Fig. 8.7 The simulation results of different control schemes (with load increase 200 MW): **a** The measured frequency in VSC1 station; **b** The transmitted active power of VSC1 station; **c** The transmitted active power of VSC2 station

peak transmission power of VSC_1 reaches 410 MW with the proposed TLCC scheme, compared to 380 MW with the CIDC scheme and 355 MW with the CIDC&DL scheme. During the recovery period, the power drop of VSC_1 is 82 MW with the proposed TLCC scheme, 45 MW with the CIDC scheme, and 32 MW with the CIDC&DL scheme. The proposed TLCC scheme integrates more frequency support power via VSC_1, thereby avoiding significant power flow changes and reducing SFD. The third frequency drop, caused by the recovery of WTs in Cluster 2, is quickly mitigated.

Compared to the other three control schemes, the proposed TLCC scheme demonstrates superior performance by adaptively adjusting the inertial and droop coefficients of WTs to optimize kinetic energy utilization. This scheme also effectively allocates frequency support power among onshore VSC stations, minimizing power

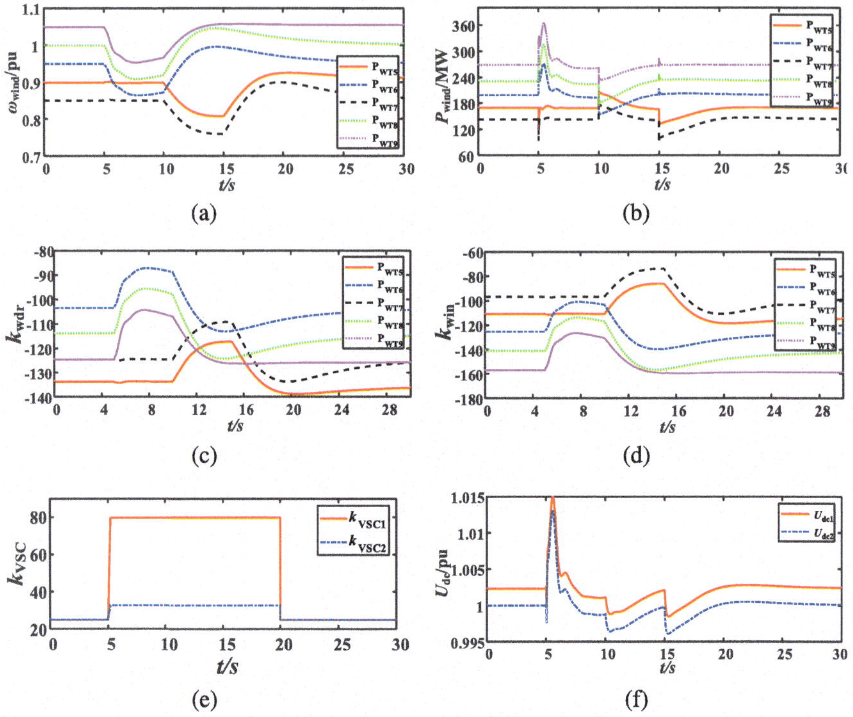

Fig. 8.8 The changes of coefficients with the proposed two level control strategies(with load increase 200 MW): **a** The rotor speed of WTs in WF1; **b** The active power of WTs in WF1; **c** The adaptive droop coefficients of WTs in WF1; **d** The adaptive inertial coefficients of WTs in WF1; **e** The adaptive voltage droop coefficients of VSC1 and VSC2 stations; **f** The DC voltage of VSC1 and VSC2 stations

flow fluctuations within the onshore system. Despite a higher number of WTs participating in frequency regulation under the CIDC scheme, the TLCC scheme surpasses it in control effectiveness.

In contrast, the CIDC&DL scheme, with its fixed droop coefficients at onshore stations, fails to optimally distribute frequency regulation power among VSC stations, compromising frequency regulation performance. Additionally, as observed in Fig. 8.7, the third frequency droop resulting from the recovery of WTs in Cluster 2 is minor and swiftly mitigated under the proposed TLCC scheme. If this third frequency droop were substantial, further segmentation of clusters could effectively reduce its impact.

To explore the detailed control process of the proposed TLCC scheme, Fig. 8.8a–d illustrate the rotor speed, active power, adaptive droop coefficients, and adaptive inertial coefficients of WTs in Area 1.

As shown in Fig. 8.8a, the rotor speeds of G6, G8, and G9 from Cluster 1 decrease between $t = 5s$ and $t = 10s$ to release kinetic energy. After $t = 10s$, the rotor speeds of Cluster 1 WTs begin recovering, while Cluster 2 WTs (G5, G7) commence releasing their kinetic energy. Subsequently, the rotor speeds of all WTs return to their initial states, as depicted in Fig. 8.8a. This is achieved through a step start-up control strategy that enables OWF1's WTs to participate in frequency regulation and mitigate SFD without requiring de-loading reserves.

Furthermore, Fig. 8.8c, d highlight the adaptive nature of the droop and inertial coefficients of WTs in Area 1. This adaptability ensures that the rotor speeds of WTs are constrained within safe limits, thereby enhancing operational stability. Notably, Fig. 8.8c demonstrates how the power-voltage droop coefficients of VSC_1 and VSC_2 adjust dynamically. This adaptability facilitates optimal allocation of frequency support power among onshore VSC stations during frequency events, thereby minimizing fluctuations in the onshore AC system's power flow.

8.3.2 Load Decrease

To further illustrate the effectiveness of the proposed control scheme during a load decrease scenario, consider a 200 MW load reduction at Bus 7 occurring at $t = 5s$. During this event, the WTs are required to reduce their output power accordingly. Unlike in the load increase event, the top 60% of WTs, ranked by rotor speed from smallest to largest, form Cluster 1. Therefore, WTs G5, G6, and G7 constitute Cluster 1, while G8 and G9 form Cluster 2.

Figure 8.9a shows the frequency measured at VSC_1. It can be observed that with the four different control schemes, the frequency of the onshore system increases to 50.35, 50.25, 50.31, and 50.23 Hz, respectively. Notably, the proposed TLCC scheme limits the secondary frequency increase to a lower value of 50.24 Hz compared to 50.29 Hz with the CIDC scheme, as depicted in Fig. 8.9a.

Figures 8.9b, c display the transmitted power by VSC_1 and VSC_2, respectively. It can be observed that the transmission power valley of VSC_1 is 202 MW with the proposed TLCC scheme, compared to 242 MW with the CIDC scheme and 271 MW with the CIDC&DL scheme. This reduction in transmitted power output helps to limit the frequency increase during the load decrease event, highlighting the effectiveness of the proposed TLCC scheme in managing system dynamics.

However, due to the presence of only two clusters in this scenario, the third frequency increase with the proposed TLCC scheme after $t = 15s$ reaches 50.24 Hz, which is promptly restrained. Future research could explore employing more clusters in real-world applications to further mitigate such frequency increases, although this aspect is not addressed in the current study.

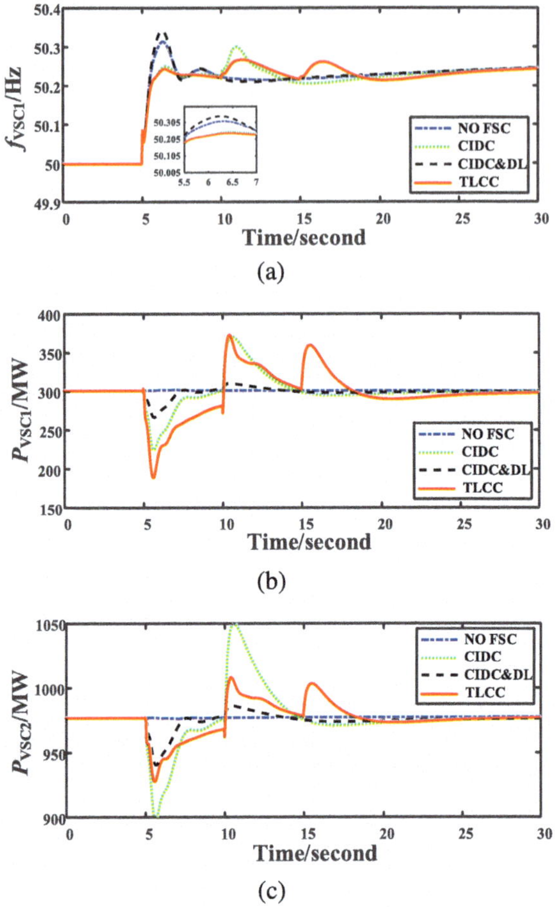

Fig. 8.9 The simulation results of different control schemes (with load decrease 200 MW): **a** The measured frequency in VSC1 station; **b** The transmitted active power of VSC1 station; **c** The transmitted active power of VSC2 station

8.3.3 Partial Outage of the Generator

In this scenario, a 250 MW partial outage at G2 in Area 1 occurs at $t = 5s$. The initial steady-state rotor speeds of G5-G9 in OWF1 are 0.9, 0.95, 0.85, 1.0, and 1.05 pu, respectively. Consequently, WTs G6, G8, and G9 are grouped into Cluster 1 to participate in frequency regulation immediately after the frequency event, while G5 and G7 are assigned to Cluster 2 to release their kinetic energy and reduce the secondary frequency drop (SFD) after the Cluster 1 WTs recover to their initial states.

Four different control schemes are tested in this scenario. The frequency measured at the VSC_1 station, along with the transmitted power by VSC_1 and VSC_2, is shown in Fig. 8.10.

As shown in Fig. 8.10a, during the 250 MW partial outage of G2 at $t = 5s$, the frequency nadirs at VSC_1 are 49.68, 49.74, 49.76, and 49.78 Hz with the NO FSC

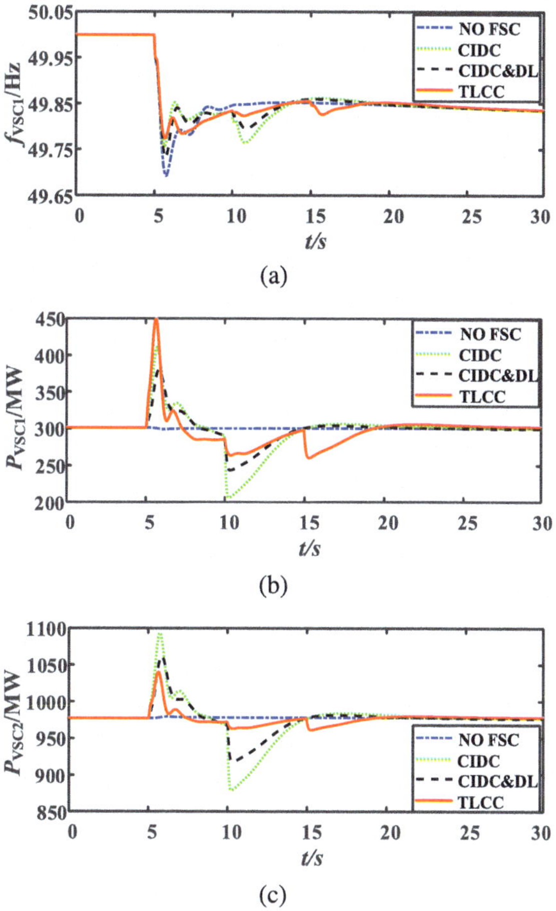

Fig. 8.10 The simulation results of different control schemes (with 250 MW partial outage of the G2): **a** The measured frequency in VSC1 station; **b** The transmitted active power of VSC1 station; **c** The transmitted active power of VSC2 station

scheme, CIDC scheme, CIDC&DL scheme, and the proposed TLCC scheme, respectively. Although an SFD occurs when WTs participate in frequency regulation with the CIDC, CIDC&DL, and TLCC schemes, the proposed TLCC scheme effectively reduces the SFD. The nadirs of the SFD with Control Schemes 2–4 are 49.76, 49.78, and 49.80 Hz, respectively.

Additionally, Fig. 8.10b, c illustrate that more frequency support power is injected into the onshore AC system by VSC_1 with the TLCC scheme during the frequency event. The peak transmission power of VSC_1 reaches 450 MW with the proposed TLCC scheme, compared to 420 MW with the CIDC scheme and 380 MW with the CIDC&DL scheme. Consequently, large-scale power flow changes are avoided, and the SFD is reduced, demonstrating the effectiveness of the TLCC scheme in maintaining system stability during partial outages.

Table 8.3 The results of different case studies

The load increase/MW	NO FSC		CIDC		CIDC&DL [5]		URC [8]		TLCC	
	FFD	SFD	FFD	SFD	FFD	SFD	FFD	SFD	FFD	SFD
600 MW	0.55 Hz	/	0.49 Hz	0.26 Hz	0.45 Hz	0.22 Hz	0.43 Hz	0.31 Hz	**0.42 Hz**	**0.21 Hz**
500 MW	0.46 Hz	/	0.41 Hz	0.30 Hz	0.38 Hz	**0.22 Hz**	0.37 Hz	0.31 Hz	**0.36 Hz**	**0.22 Hz**
400 MW	0.37 Hz	/	0.35 Hz	0.28 Hz	0.31 Hz	**0.21 Hz**	0.31 Hz	0.27 Hz	**0.29 Hz**	**0.21 Hz**
300 MW	0.30 Hz	/	0.26 Hz	0.21 Hz	0.24 Hz	0.23 Hz	0.24 Hz	0.24 Hz	**0.23 Hz**	**0.18 Hz**
200 MW	0.21 Hz	/	0.17 Hz	0.17 Hz	0.18 Hz	0.15 Hz	0.17 Hz	0.18 Hz	**0.16 Hz**	**0.14 Hz**

8.3.4 Performance Comparison of Different Schemes

To further demonstrate the universality of the proposed TLCC scheme, various load-increase scenarios were conducted, comparing different control schemes. In Control Scheme 3, WTs utilize conventional inertial and droop control with 10% de-loading for frequency support, as referenced in [5] (referred to as CIDC&DL). In Control Scheme 4, WTs adopt the methods proposed in [8] (referred to as URC). The scenarios include load increases of 600, 500, 400, 300, and 200 MW at Bus 7. The system responses to these frequency events under different control schemes are summarized in Table 8.3.

The results show that both the first frequency drop (FFD) and the secondary frequency drop (SFD) are smallest with the proposed TLCC scheme, indicating superior frequency support performance compared to the other three schemes. In the 500 and 400 MW load increase scenarios, while the SFD of the TLCC scheme matches that of the CIDC&DL scheme, the FFD is smaller with the TLCC scheme.

Moreover, the CIDC&DL scheme requires WTs to reserve a significant amount of power, leading to revenue loss for OWFs due to de-loading control, as noted in [9]. In contrast, the proposed TLCC scheme effectively reduces the SFD without the need for energy reserves, making it more economical for OWFs. The CIDC scheme, with its constant inertial and droop coefficients, fails to optimally utilize the kinetic energy of different WTs, resulting in less effective control compared to the TLCC scheme.

The comparison of control performance across different frequency events highlights the effectiveness and universality of the proposed TLCC scheme. The URC scheme, which lacks the ability to allocate frequency support power effectively among different VSC stations, also exhibits inferior frequency regulation performance compared to the TLCC scheme.

8.4 Summary

This chapter proposes a two-level combined control (TLCC) scheme to coordinate OWFs with the VSC-MTDC system for onshore frequency support, aiming to reduce SFD while all WTs operate MPPT mode. The proposed TLCC scheme involves step start-up and adaptive inertial droop control at the offshore wind turbine level, as well as communication-free allocation control at the onshore VSC station level.

Compared to other control schemes, the proposed TLCC scheme effectively reduces frequency deviation in both load increase and load decrease scenarios. By utilizing adaptive inertial and droop control, WTs with higher rotor speeds release more kinetic energy, enhancing frequency support performance beyond that of schemes with constant inertial and droop coefficients. Additionally, the step start-up control allows WTs in Cluster 2 to provide power support while WTs in Cluster 1 recover their rotor speed. Moreover, the communication-free allocation control strategy ensures the reasonable distribution of frequency support power among different onshore VSC stations during frequency events, reducing power flow changes in the onshore AC system and decreasing the RoCoF. Various case studies confirm the effectiveness and universality of the proposed TLCC scheme.

It is important to note that the TLCC scheme is most effective when wind speed and WT rotor speed are sufficient to provide the necessary kinetic energy for frequency support. The scheme's limitations include the division of WTs into clusters, which, while reducing SFD, may slightly impact the effectiveness of frequency nadir reduction. More clusters can be employed based on the specific situation. Additionally, the time delay for Cluster 2 to participate in frequency regulation is fixed at 5 s. Since this chapter focuses on the proposed two-level control scheme, optimizing the number of clusters and the time delay will be subjects of future research. Furthermore, the consideration of asynchronous interconnection in the onshore system will also be explored in future work.

References

1. Dreidy M, Mokhlis H, Mekhilef S (2017) Inertia response and frequency control techniques for renewable energy sources: a review. Renew Sustain Energy Rev 69:144–155. https://doi.org/10.1016/j.rser.2016.11.170
2. Cotilla-Sanchez E, Hines PDH, Barrows C, Blumsack S, Patel M (2013) Multi-attribute partitioning of power networks based on electrical distance. IEEE Trans Power Syst 28(4):4979–4987. https://doi.org/10.1109/TPWRS.2013.2263886
3. Lagonotte P, Sabonnadiere JC, Leost JY, Paul JP (1989) Structural analysis of the electrical system: application to secondary voltage control in france. IEEE Trans Power Syst 4(2):479–486. https://doi.org/10.1109/59.193819
4. Wang Z, Wu W (2018) Coordinated control method for dfig-based wind farm to provide primary frequency regulation service. IEEE Trans Power Syst 33(3):2644–2659. https://doi.org/10.1109/TPWRS.2017.2755685

5. Lu Z, Ye Y, Qiao Y (2019) An adaptive frequency regulation method with grid-friendly restoration for VSC-HVDC integrated offshore wind farms. IEEE Trans Power Syst 34(5):3582–3593. https://doi.org/10.1109/TPWRS.2019.2901986
6. Vennelaganti SG, Chaudhuri NR (2018) Selective power routing in mtdc grids for inertial and primary frequency support. IEEE Trans Power Syst 33(6):7020–7030. https://doi.org/10.1109/TPWRS.2018.2854647
7. Chaudhuri NR, Chaudhuri B, Majumder R, Yazdani A (2014) Multi-terminal direct-current grids: modeling, analysis, and control. Wiley, Hoboken, NJ, USA
8. Rouzbehi K, Zhang W, Candela JI, Luna A, Rodriguez P (2017) Unified reference controller for flexible primary control and inertia sharing in multi-terminal voltage source converter-HVDC grids. IET Gener Transm Distrib 11(3):750–758. https://doi.org/10.1049/iet-gtd.2016.0665
9. Vidyanandan KV, Senroy N (2013) Primary frequency regulation by deloaded wind turbines using variable droop. IEEE Trans. Power Syst. 28(2):837–864 May. https://doi.org/10.1109/TPWRS.2012.2208233

Chapter 9
Improved Communication-Free Coordinated Control of VSC-MTDC Integrated Offshore Wind Farms for Onshore System Frequency Support

Abstract To mitigate the adverse impact of communication latency, the direct current (DC) voltage of offshore converters is typically employed as input signals in existing communication-free frequency support schemes for offshore wind farms (OWFs). Nevertheless, the offshore DC voltage fluctuates with the output power of OWFs, and this distortion degrades the frequency support performance. To resolve this problem, an improved communication-free coordinated control (ICFCC) scheme for VSC-MTDC integrated OWFs is proposed. The ICFCC scheme incorporates the onshore frequency variation into the DC voltage deviation when frequency events occur. Additionally, an estimator is proposed to determine the onshore DC voltage at offshore VSC stations, enabling rapid frequency support and mitigating the distortion effect. Meanwhile, to enhance support performance, the frequency support power is distributed logically by adaptively adjusting droop coefficients, allowing more power to be injected via the station nearest to the disturbance point. Following frequency support, the offshore wind turbines will restore their rotor speed using asymptotic control to minimize a second frequency drop. Case studies are performed on the 4-terminal and 5-terminal test systems, respectively, considering parameter uncertainties and noise disturbance. Various control methods are compared to demonstrate the effectiveness of the ICFCC scheme.

9.1 Formation of The ICFCC Scheme

9.1.1 Conspectus of the ICFCC Scheme

The ICFCC scheme illustrated in Fig. 9.1 comprises three components: the adaptive allocation control scheme with DC voltage limitation, the communication-free onshore DC voltage estimator, and the asymptotic rotor speed recovery control scheme. As shown in Fig. 9.1, the DC voltage is measured on the DC side, while the reactive power is measured at the PCC-AC bus between the transformer and the grid.

W. Yao et al., *Active Control of Large-Scale Offshore Wind Farms Connected Via VSC-HVDC*, Power Systems, https://doi.org/10.1007/978-981-97-9346-4_9

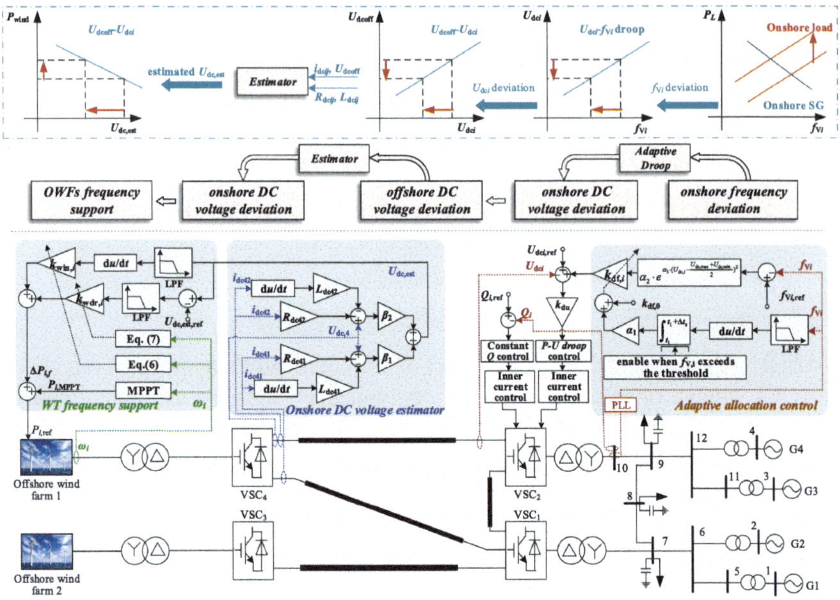

Fig. 9.1 The block diagram of the proposed improved communication-free coordinated control scheme

- **Adaptive allocation control scheme with DC voltage limitation**: Using adaptive droop control, changes in onshore frequency can affect DC voltage variations, facilitating the logical distribution of frequency support power among onshore systems while maintaining the safe operation of DC voltage.
- **Communication-free onshore DC voltage estimator**: The onshore DC voltage can be approximated using Kirchhoff's law, allowing the identification of onshore frequency deviations at offshore stations without needing extra remote communication between onshore and offshore systems.
- **Asymptotic rotor speed recovery control scheme**: To mitigate the SFD, an asymptotic restoration method is recommended to recover rotor speed using a particular P–ω curve, which will be gradually modified to align with the MPPT curve, ensuring a smooth recovery.

The primary steps of the ICFCC strategy are outlined below. First, onshore frequency changes are integrated into DC voltage variations. To avoid remote communication between onshore and offshore units, an estimator is employed to calculate the onshore DC voltage from local readings at the offshore VSC terminals for rapid frequency support. Concurrently, to improve frequency support efficiency, the regulation power is logically distributed among onshore VSC units by adjusting the droop coefficients, ensuring more power is channeled through the station closest to the affected bus. After providing frequency support, offshore wind turbines will restore their rotor speed using asymptotic control to minimize SFD.

9.1.2 Adaptive Allocation Control Scheme with DC Voltage Limitation

The onshore stations typically manage the distribution of active power among AC systems [1], directing more energy to the mainland grids through the station nearer to the disturbed node for optimal frequency support [8]. To achieve efficient power distribution, an innovative adaptive droop control, illustrated in Fig. 9.1, is suggested for all onshore stations. This method adjusts droop coefficients based on the impulse integral of the RoCoF, meaning that the closer a station is to the disturbed bus, the larger its droop coefficient will be, resulting in more power being channeled into the mainland grid through that station.

The computation of droop coefficients and power reference across various stations is expressed in Eqs. (9.1)–(9.2). In Eq. (9.1), k_{du} denotes the P-U droop coefficients. The symbols $f_{\mathrm{V}i}$ and $f_{\mathrm{V}i,\mathrm{ref}}$ represent the frequency measurement and reference frequency at VSCi, respectively. The term $k_{\mathrm{df},0}$ stands for the initial value of the U-f droop coefficients. To adaptively adjust $k_{\mathrm{df},i}$, the integration of the RoCoF over Δt_s seconds is scaled by a constant α_1 (refer to (9.1)). Simultaneously, to prevent the DC voltage from surpassing its limits, the voltage margin index $\alpha_2 \cdot e^{\alpha_3 \cdot (U_{dc,i} - \frac{U_{dc,max}+U_{dc,min}}{2})^2}$ is applied in (9.1). This allows the U-f droop coefficient $k_{\mathrm{df},i}$ to adjust adaptively during frequency events.

For all onshore stations, constants α_1, α_2, α_3, $k_{\mathrm{df},0}$, and k_{du} are set to identical values, as detailed in Table 9.2. However, due to the integration process, the U-f droop coefficient $k_{\mathrm{df},i}$ varies among stations. Consequently, compared to traditional constant droop control, the ICFCC scheme enables more suitable distribution of frequency regulation power among different VSC converters. The procedures for calculating $k_{\mathrm{df},i}$ and the power reference $P_{\mathrm{V}i,\mathrm{ref}}$ are delineated in equation (9.2).

$$\begin{aligned} k_{\mathrm{df},i} &= \alpha_2 \cdot e^{\alpha_3 \cdot (U_{dc,i} - \frac{U_{dc,max}+U_{dc,min}}{2})^2} \cdot \left[k_{\mathrm{df},0} + \int_{t_1}^{t_1+\Delta t_s} (\alpha_1 \cdot \frac{d f_{\mathrm{V}i}(t)}{dt}) dt \right] \\ &= \alpha_2 \cdot e^{\alpha_3 \cdot (U_{dc,i} - \frac{U_{dc,max}+U_{dc,min}}{2})^2} \cdot [k_{\mathrm{df},0} + \alpha_1 \cdot (f_{\mathrm{V}i}(t_1 + \Delta t_s) - f_{\mathrm{V}i}(t_1))] \end{aligned} \tag{9.1}$$

$$P_{\mathrm{V}i,\mathrm{ref}} = P_{\mathrm{V}i,0} - k_{\mathrm{du}} \cdot [U_{\mathrm{dc}i,\mathrm{ref}} - U_{\mathrm{dc}i} + k_{\mathrm{df},i} \cdot (f_{\mathrm{V}i,\mathrm{ref}} - f_{\mathrm{V}i})] \tag{9.2}$$

In Eq. (9.1), t_1 marks the start time when the frequency deviation exceeds the deadband threshold (0.02 Hz in this context). The parameter Δt_{s} represents the integration duration, set at 0.01 s. The derivative of $f_{\mathrm{V}i}(t)$ corresponds to the RoCoF, while $k_{\mathrm{df},i}$ and $k_{\mathrm{df},0}$ denote the U–f droop coefficient for VSC$_{\mathrm{i}}$ and its initial value, respectively. The expression $\alpha_2 \cdot e^{\alpha_3 \cdot (U_{dc,i} - \frac{U_{dc,max}+U_{dc,min}}{2})^2}$ serves as the DC voltage margin index, illustrated in Fig. 9.2. If the DC voltage approaches its upper and lower limits (1.05 and 0.95 pu respectively in this chapter), this index will gradually reduce to zero to ensure secure operation.

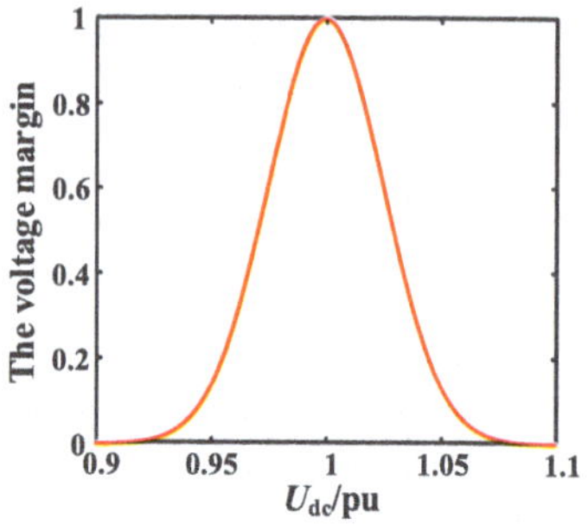

Fig. 9.2 The voltage margin index $\alpha_2 \cdot e^{\alpha_3 \cdot (U_{dc,i} - \frac{U_{dc,max}+U_{dc,min}}{2})^2}$

Additionally, the calculation of the adaptive droop control coefficient is streamlined by measuring the frequency at two distinct points within the short interval Δt_s, thus simplifying the integration in (9.1) to the deviation of $f_{Vi}(t)$ over these time points. An 8.8 kHz cut-off frequency one-order low-pass filter is employed to eliminate high-frequency noise in the measured frequency signal $f_{Vi}(t)$. Consequently, the droop coefficients adaptively adjust with the RoCoF while maintaining the DC voltage within a safe operating range.

The variables $P_{Vi,0}$ and $P_{Vi,\text{ref}}$ represent the initial and adjusted power references. Similarly, $U_{\text{dc}i,\text{ref}}$ and $U_{\text{dc}i}$ denote the reference and measured DC voltages, respectively. With a negative $U{-}f$ droop slope and a positive $U{-}P$ droop slope as per (9.2), the DC voltage will decrease while the transmission power reference will rise during under-frequency scenarios. The proposed adaptive allocation control scheme stabilizes power flow fluctuations and ensures the safe operation of DC voltage while maintaining effective frequency support. The detailed control procedures are outlined below.

- Measure the frequency f_{Vi} of VSC$_i$ and calculate the DC voltage margin index $e^{\alpha_3 \cdot (U_{dc,i} - \frac{U_{dc,max}+U_{dc,min}}{2})^2}$;
- When the measured frequency f_{Vi} exceeds threshold ($f_{Vi} \geq 50.02$ Hz or $f_{Vi} \leq 49.98$ Hz), calculate the adaptive droop coefficients $k_{\text{df},i}$ by (9.1);
- Calculate the injected power reference value $P_{Vi,\text{ref}}$ by (9.2);
- The more power could be transmitted to the mainland power grids through the onshore VSC converter nearer to the disturbance.

9.1.3 Communication-Free Onshore DC Voltage Estimator

The main role of offshore stations is to collect active power generated by offshore wind turbines (WTs) [1]. When offshore WTs support frequency by increasing output power, the offshore DC voltage rises. Thus, the offshore DC voltage cannot precisely reflect onshore frequency deviations during frequency support with existing communication-free methods as noted in [3]. Moreover, improper configuration of droop coefficients for onshore and offshore systems can lead to oscillations. To tackle this, a communication-free offshore DC voltage estimator is proposed. This estimator uses offshore DC voltage to estimate onshore DC voltage for frequency support, mitigating distortion effects.

In the proposed Improved Communication-Free Coordinated Control (ICFCC) scheme, onshore frequency deviation is factored into DC voltage variations as shown in (9.2). Offshore DC voltage and currents are then used to estimate onshore DC voltage for WTs to provide frequency support, as detailed in (9.3). The frequency set point for offshore Voltage Source Converter (VSC) stations is kept at a constant value (1 pu), and the communication-free estimator does not impact offshore AC frequency. The block diagram of the proposed offshore communication-free estimator for Multi-Terminal Direct Current (MTDC) based Offshore Wind Farms (OWFs) is depicted in Fig. 9.1.

In Fig. 9.1, there are $j-1$ onshore converters (VSC_1 to VSC_{j-1}) and $m-j+1$ offshore converters (VSC_j to VSC_m). Offshore DC voltage at VSC_j and DC line currents from VSC_j to other onshore converters can be measured at VSC_j. Thus, onshore DC voltage can be estimated using Kirchhoff's law, considering known DC line parameters. The detailed calculation process is presented in (9.3).

$$U_{\mathrm{dc,est}} = \sum_{i=1}^{j-1} \beta_i \cdot \left(U_{\mathrm{dc},j} - I_{\mathrm{dc},ij} \cdot R_{\mathrm{dc},ij} - \frac{\mathrm{d}I_{\mathrm{dc},ij}}{\mathrm{d}t} \cdot L_{\mathrm{dc},ij} \right) \tag{9.3}$$

$$\beta_i = \frac{I_{\mathrm{dc},ij}}{\sum_{i=1}^{j-1} I_{\mathrm{dc},ij}} \quad \left(\sum \beta_i = 1 \right) \tag{9.4}$$

where $L_{\mathrm{dc},ij}$ and $R_{\mathrm{dc},ij}$ denote the inductance and resistance of the DC Line $i-j$, respectively. These parameters can be acquired for a specific practical project. Hence, the onshore DC voltage can be estimated using only locally measured electrical quantities at the offshore station. Furthermore, the estimated onshore DC voltage $U_{\mathrm{dc,est}}$ is computed as an average value with a weighted factor β_i. The factor β_i is determined by the proportion of the DC line current, which represents the average of all onshore station DC voltages and can be calculated as shown in (9.4), with the sum of β_i equal to 1.

It is important to note that the ICFCC scheme is founded on Kirchhoff's law, making it adaptable to other DC grid topologies as well. The detailed control processes are illustrated below.

- Measure the local DC voltage and DC current of VSC_4;
- Calculate the weighted factor β_i by (11.16);
- Calculate the estimated onshore DC voltage $U_{\mathrm{dc,est}}$ by (11.15).

9.1.4 Asymptotic Rotor Speed Recovery Control Scheme

To enable OWFs to contribute to frequency support, supplementary adaptive droop and inertia control must be integrated with the MPPT control strategy and harness kinetic energy. The formulas for computing $\Delta P_{\text{win},i}$, $\Delta P_{\text{wdr},i}$, $k_{\text{win},i}$, and $k_{\text{wdr},i}$ are presented below.

$$\Delta P_{\text{win},i} = k_{\text{win},i} \times \frac{\text{d}U_{\text{on,est}}}{\text{d}t} \tag{9.5}$$

$$\Delta P_{\text{wdr},i} = -k_{\text{wdr},i} \times (U_{\text{dcest,ref}} - U_{\text{dc,est}}) \tag{9.6}$$

$$k_{\text{win},i} = k_{\text{win,i,0}} + k_1 \frac{\omega_{\text{r},i}^2 - \omega_{\text{r,min}}^2}{\omega_{\text{r,max}}^2 - \omega_{\text{r,min}}^2} \tag{9.7}$$

$$k_{\text{wdr},i} = k_{\text{wdr,i,0}} + k_2 \frac{\omega_{\text{r},i}^2 - \omega_{\text{r,min}}^2}{\omega_{\text{r,max}}^2 - \omega_{\text{r,min}}^2} \tag{9.8}$$

where, $\Delta P_{\text{wdr},i}$ and $\Delta P_{\text{win},i}$ denote the additional power contributed by the adaptive droop and inertia control of WTi, respectively. $U_{\text{dc,est}}$ and $U_{\text{dc,est,ref}}$ indicate the estimated onshore DC voltage and its reference value. $k_{\text{win},i}$ and $k_{\text{wdr},i}$ represent the control coefficients for adaptive inertial and droop control, respectively. $k_{\text{win},i,0}$ and $k_{\text{wdr,i,0}}$ are the initial values of inertial control and droop control parameters. k_1 and k_2 are constants that determine the rate of change of the control coefficients. $\omega_{\text{r},i}$ denotes the rotor speed of WTi, which is maintained within a safe range (0.7–1.2pu) using adaptive coefficients during frequency support operations. With negative values of inertial and droop coefficients $k_{\text{win},i}$ and $k_{\text{wdr},i}$ in (9.5) and (9.6), $\Delta P_{\text{wdr},i}$ and $\Delta P_{\text{win},i}$ are positive, providing the necessary frequency support power.

Following the provision of frequency support, OWFs must restore their rotor speeds. However, an inappropriate recovery strategy may lead to Second Frequency Drop (SFD) during this period [4]. The cause of SFD can be illustrated as depicted in Fig. 9.3a. Under normal operating conditions, the MPPT curve is represented by $P = k\omega^3$, with offshore WTs operating at the MPPT point. During a frequency event, WTs release kinetic energy for frequency support, shifting their operating point from MPPT to P_{op1}. Upon completion of frequency support, WTs need to return to their initial operational state, transitions from P_{op1} to P_{op2}. This abrupt change results in a significant disturbance in active power (see ΔP in Fig. 9.3a), which triggers the occurrence of SFD. Improperly configured control schemes can exacerbate SFD, potentially surpassing the magnitude of the primary frequency drop and causing severe repercussions.

To mitigate the Second Frequency Drop (SFD), a novel asymptotic recovery control scheme for WTs is introduced. When the WT starts recovering its rotor speed from point P_{op1}, a new P-ω curve is formulated as $P = \alpha k\omega^3$ (see (9.9)). The parameter α varies over time and is described in (9.10). Throughout the recovery duration

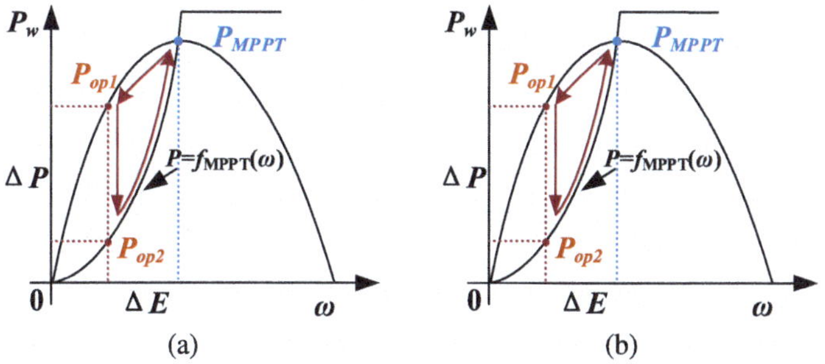

Fig. 9.3 The output power curve for OWFs: **a** conventional control scheme; **b** proposed improved ICFCC scheme

$(t_b - t_e)$, α gradually transitions from its initial value to 1, thereby minimizing significant active power disturbances. This approach effectively reduces the likelihood of SFD through the proposed control strategy.

$$P = \alpha f_{\mathrm{MPPT}}(\omega) \tag{9.9}$$

$$\alpha = \begin{cases} 1 & t < t_b \\ \frac{1-\alpha_0}{\Delta t_r} \cdot (t - t_b) + \alpha_0 \ (\alpha_0 = \frac{P_{op1}}{\omega_{op1}}) & t_b \le t \le t_e \\ 1 & t_e < t \end{cases} \tag{9.10}$$

where α_0 denotes the initial value at the onset of rotor speed recovery, determined by the real-time operational point of various wind turbines at the start of wind speed restoration, P_{op1} and ω_{op1} denote the measured power output and rotor velocity, respectively. t_{s} indicates the start time of frequency events. t_{b} and t_{e} represent the start and end times of rotor speed re-storage, $\Delta t_{\mathrm{r}} = t_{\mathrm{e}} - t_{\mathrm{b}}$ denotes the duration of re-storage ($\Delta t_r = 7s$ in this study). The evolving curve of α can be seen in Fig. 9.4.

Fig. 9.4 The piecewise curve of α

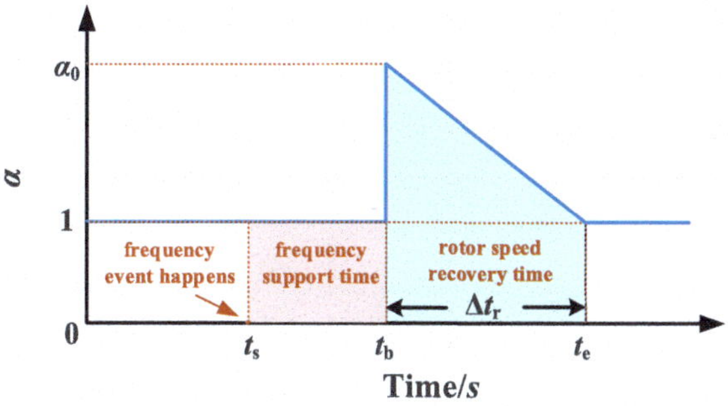

9.2 Case Study I: 3-Area 4-Terminal System

9.2.1 Benchmark Description

To demonstrate the effectiveness of the ICFCC method, the 4-terminal test system depicted in Fig. 9.5 is implemented in the Matlab/Simulink simulation platform. The basic model of the toolbox is available for download from [5]. Detailed parameters are listed in Tables 9.1 and 9.2, referenced from [2, 6, 7]. Various control schemes for comparison and their abbreviations are summarized in Table 9.3.

The 4-terminal test system, described in [6], comprises three main components: onshore AC systems, the MTDC network, and OWFs. This model includes two OWFs; OWF1 connects to the onshore AC system via VSC1 and VSC2, while OWF2 is linked to the onshore AC system solely through VSC2. The power generation of OWFs and onshore generators is illustrated in Fig. 9.4, and the crucial parameters of the MTDC system are detailed in [2].

9.2.2 Controller Parameters

The parameters for the adaptive droop and inertia controllers in the proposed ICFCC scheme are derived from [8]. The parameters for OWFs are presented in Table 9.1. As indicated by Eqs. (9.7) and (9.8), the droop and inertia coefficients can be dynamically adjusted.

For the 3-area 4-terminal system, the controller parameters are sourced from [2, 6], with their values and explanations detailed in Table 9.2. To examine the small-signal stability of the system under adaptive changes in the droop coefficients according

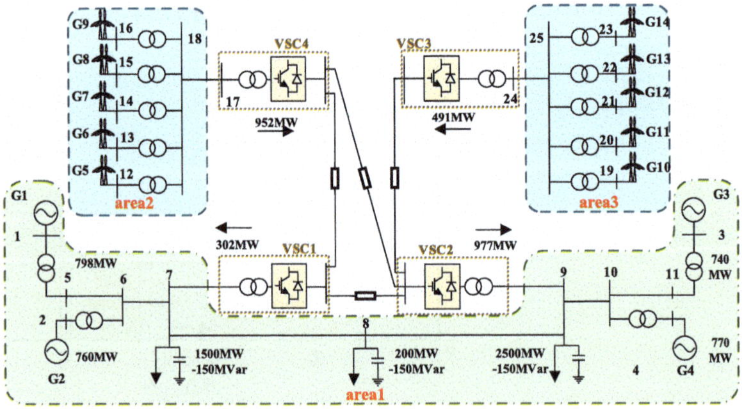

Fig. 9.5 The 4-terminal test systems

Table 9.1 Parameters of OWFs

Parameters	Meaning	Value (unit)
$\omega_{r,i}$	The WT_i's rotor speed	0.7–1.2 pu
$\omega_{r,max}$	The maximum limitation of the rotor speed	1.2 pu
$\omega_{r,min}$	The minimum limitation of the rotor speed	0.7 pu
$k_{win,i,0}$	The original parameter of inertial control	–5
$k_{wdr,i,0}$	The original parameter of droop control	–40
$k_{win,i}$	The adaptive inertial control coefficient	–10 to –1
$k_{wdr,i}$	The adaptive droop control coefficient	–60 to –10
k_1	The slope of the inertial coefficient	–10
k_2	The slope of the droop coefficient	–130

Table 9.2 Parameters of the 4-Terminal Test System

Parameters	Meaning	Value (unit)
$k_{df,0}$	The initial $U-f$ droop coefficients	–6.5
$k_{df,i}$	The $U-f$ droop coefficient	–30 to 5
k_{du}	The $P–U$ droop coefficient	20
α_1	The integral gain	2000
α_2	The DC voltage margin gain	1
α_3	The gain of DC voltage index	–800
Δt_s	The RoCoF integration time	0.01 s
$f_{Vi,ref}$	The reference value of VSC_i	50 Hz
P_{rate}	Converter rated capacity	1000 MVA
$V_{dc,rate}$	Rated DC voltage	400 kV
L_{dc}	DC reactor	100 mH
R_0	DC line resistance	0.0114 Ω/km
L_0	DC line inductance	0.9356 mH/km
C_0	DC line capacitance	0.0123 μF/km
L_{dc41}	The inductance of DC line 4–1	140.34 mH
L_{dc42}	The inductance of DC line 4–2	187.12 mH
R_{dc41}	The resistance of DC line 4–1	1.71 Ω
R_{dc42}	The resistance of DC line 4–2	2.28 Ω
β_1	The weight of VSC_1 DC voltage	0–1
β_2	The weight of VSC_2 DC voltage	0–1

Table 9.3 The compared control strategies and their abbreviations

Abbreviation	Description
ICFCC	The proposed communication-free coordinated control
TLCC	The two-level combined control strategy in [8]
CCR	Communication-free control in [9] with PI recovery
CC	Communication-based control
NOWT	Without offshore WTs participation

to the ICFCC scheme, a small-signal analysis is conducted with the U–f droop coefficients varying from −150 to −5, the P–U droop coefficients ranging from 5 to 150, the WTs droop coefficient k_{wdr} shifting from −150 to −10, and the inertia control coefficient k_{win} altering from −20 to −1. The results of this analysis are depicted in Fig. 9.6. As seen in Fig. 9.6, the damping ratios of the most critical eigenvalues remain above 0.05 when the droop control parameters of onshore converters are adjusted within the specified ranges in Table 9.2, indicating robust small-signal stability with parameter variations. However, the system becomes unstable when k_{wdr} drops below -100 and k_{win} falls below -16. To ensure safe operation, the default values and limits for all adaptive coefficients are provided in Table 9.4.

9.2.3 *Defects of Existing Schemes*

To address the limitations of current communication-free schemes, a 300 MW load is added to Bus 7 at $t = 10s$ in this scenario. The performance of the proposed offshore estimator and the existing control method from [3] are examined. As shown in Fig. 9.7, when OWFs provide frequency support, the offshore DC voltage increases due to the rising wind power, compared to $U_{dc,off}$ without OWF involvement (see Fig. 9.7a). Consequently, the DC voltage does not accurately reflect the onshore frequency deviation during the frequency support period. However, the proposed ICFCC can estimate the onshore DC voltage for frequency support (see Fig. 9.7b), mitigating the negative impact of offshore DC voltage variation. Additionally, the DC voltage trend in the existing scheme from [3] could lead to oscillations if the control parameters are not properly configured.

9.2.4 *Load Increase*

At Bus 7 of Area 1, shown in Fig. 9.5, a 300 MW load surge occurs at $t = 10s$, and diverse control strategies are contrasted with the ICFCC approach. The communication-based control is denoted as CC, primarily differing from ICFCC

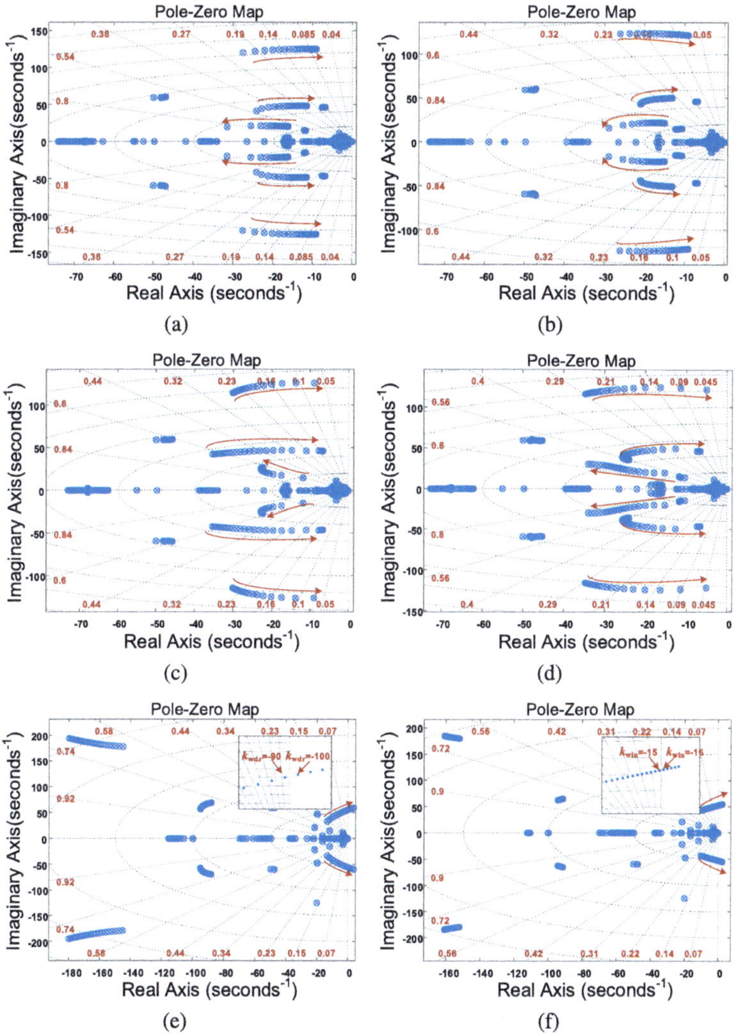

Fig. 9.6 The small-signal analysis with droop coefficients change of onshore stations: **a** $k_{df,1}$ from –150 to –5; **b** $k_{df,2}$ from –150 to –5; **c** $k_{du,1}$ from 5 to 150; **d** $k_{du,2}$ from 5 to 150; **e** k_{wdr} from –150 to –10; **f** k_{win} from –20 to –1

Table 9.4 The default values and limitations of the adaptive coefficients

Parameters	Default value	Limitation
The $U - f$ droop coefficient $k_{df,i}$	–6.5	[–30, –5]
The $P - U$ droop coefficient $k_{du,i}$	20	[5, 40]
The droop coefficients of WTs k_{wdr}	–40	[–60, –10]
The inertia coefficients of WTs k_{win}	–5	[–10, –1]

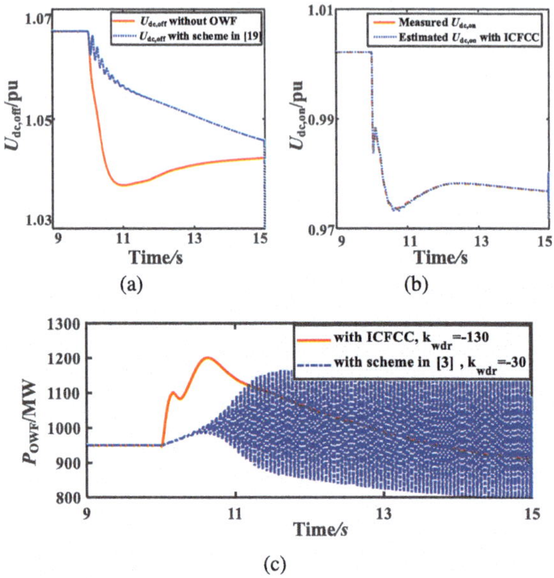

Fig. 9.7 The comparison results between the ICFCC scheme and the existing scheme: **a** $U_{dc,off}$, **b** $U_{dc,on}$, **c** P_{OWF}

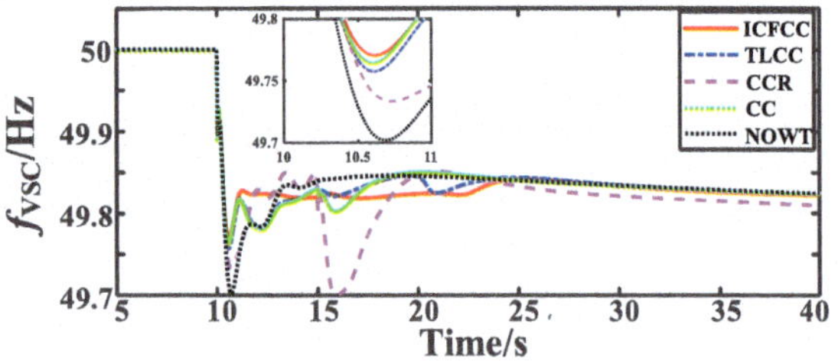

Fig. 9.8 The frequency response with different control schemes (with load increase 300 MW)

by the inclusion or lack of remote communication. The TLCC method is sourced from [8]. The CCR denotes the communication-less control method from [10] with PI control recovery. The method where OWFs abstain from frequency support is termed NOWT. The frequency response of VSC_1 is displayed in Fig. 9.8. The proposed ICFCC method offers frequency support performance akin to the CC approach. Compared to the NOWT scheme, all ICFCC, CC, TLCC, and CCR strategies enhance the frequency nadir, with ICFCC demonstrating superior performance over other methods. Moreover, compared to CCR and CC strategies, the asymptotic rotor speed recovery control of WTs allows smooth rotor speed recovery, notably minimizing the SFD.

To exhibit the frequency support performance of the ICFCC method, frequency nadirs at distinct buses in Area 1 are illustrated in Fig. 9.9. The ICFCC scheme effectively raises frequency nadirs across all buses in comparison to TLCC, CCR, and NOWT methods, with frequency deviations across different buses being nearly identical between ICFCC and CC strategies. Additionally, the ICFCC approach allocates

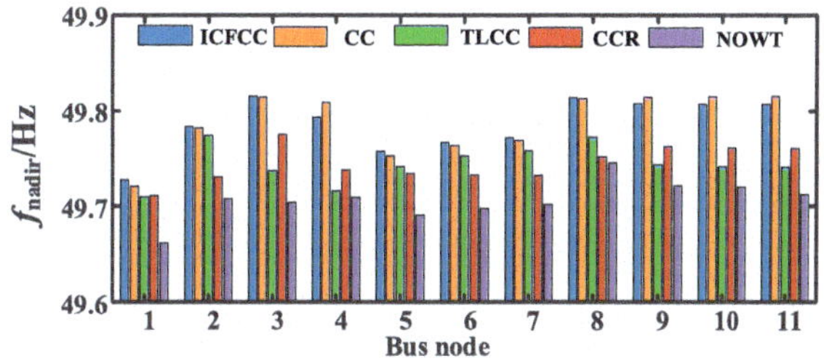

Fig. 9.9 The frequency nadirs of different buses in Area 1 (with load increase 300 MW)

frequency regulation power more appropriately among different converters, reducing power flow variations in Area 1 and improving frequency support performance. To further explore the frequency support processes of the ICFCC method, the output active power, rotor speed, and adaptive droop coefficients are presented in Fig. 9.9. According to Fig. 9.10a, b, only WTs in OWF1 (WT1-5) engage in frequency support by releasing kinetic energy in this scenario. However, with additional control in OWF2, WT6-10 also contribute to frequency support. All WTs in OWF1 operate within a safe range with rotor speeds between 0.7 and 1.2 pu as depicted in Fig. 9.10c. More frequency regulation power is supplied by WTs with higher rotor speeds and greater kinetic energy owing to adaptive control coefficients. Moreover, the $U-f$ droop coefficients of onshore stations $k_{df,1}$ and $k_{df,2}$ are adaptively modified during the frequency support process, distributing the frequency regulation power suitably and maintaining the DC voltage within the safe range of 0.95–1.05 pu. Figure 9.10e demonstrates the real and estimated values of onshore DC voltages U_{dc1} and U_{dc2}, along with the average estimated onshore DC voltage $U_{on,est}$ computed using (9.3). Despite slight discrepancies between the measured and estimated values of $U_{on,est}$ during the dynamic process, $U_{on,est}$ effectively reflects onshore frequency deviations. Additionally, all these estimated values are calculated at the offshore VSC stations using Kirchhoff's law, removing the necessity for remote communication between onshore and offshore systems. As illustrated in Fig. 9.10f, g, the frequency support power is judiciously allocated between onshore VSC stations, with more power injected into the onshore AC system via VSC1, which is nearer to the disturbed Bus 7 compared to VSC2.

9.2.5 Performance Comparison with Different Schemes

To further assess the efficacy of the proposed ICFCC method, various control approaches were evaluated, with their abbreviations listed in Table 10.3. The frequency responses under different control strategies are depicted in Fig. 9.11. When the onshore load was increased by 200, 300, and 400 MW, the frequency nadirs with the ICFCC scheme were 49.85, 49.78, and 49.70 Hz, respectively. These nadirs were higher than those achieved with the other four strategies. A notable SFD was observed with the CCR approach. However, the proposed asymptotic recovery control strategy

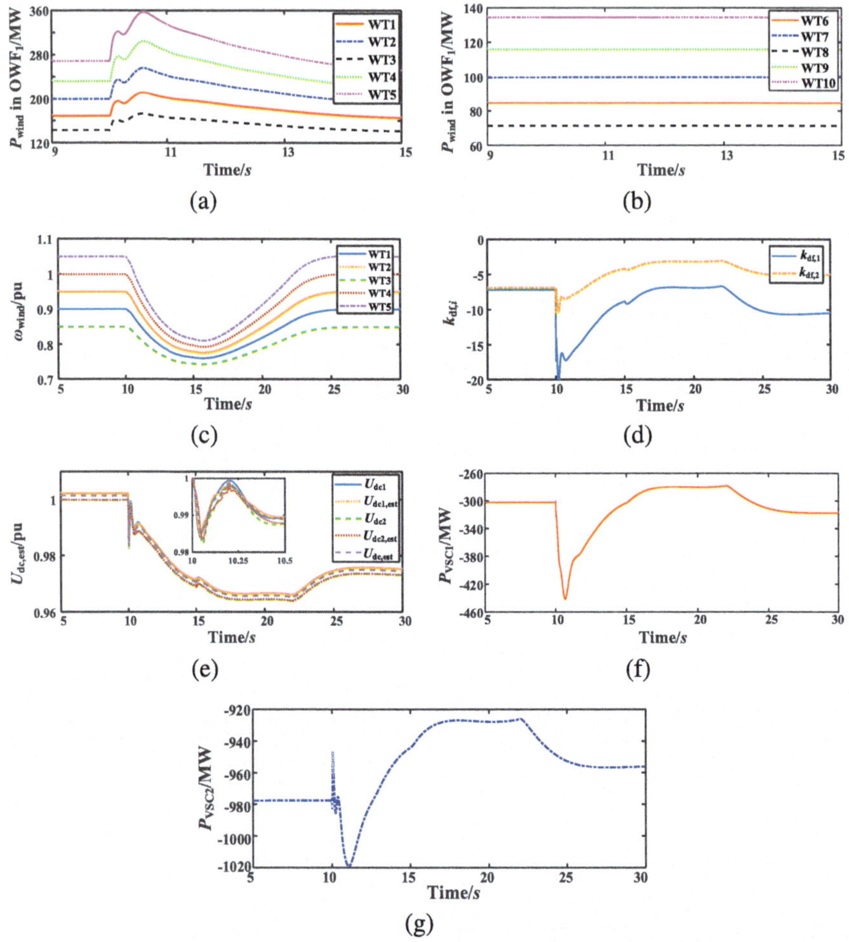

Fig. 9.10 The simulation results with the proposed ICFCC scheme: **a** the output power of WTs in OWF1; **b** the output power of WTs in OWF2; **c** the onshore droop control coefficients; **d** onshore DC voltage measurement and the estimated DC voltage; **e** the transmission power of VSC_1; **f** the transmission power of VSC_2

for offshore WTs helped to mitigate the SFD. Furthermore, in Fig. 9.11d–f, where the onshore load was decreased by 50, 100, and 200 MW, the ICFCC method effectively reduced the frequency peaks to 50.05, 50.1, and 50.2 Hz, respectively, and suppressed frequency fluctuations compared to the other strategies. It's worth noting that due to different steady-state transmission power of VSC stations, the steady-state frequency is not equivalent to the frequency without frequency support.

The detailed values of Δf and RoCoF under different conditions are illustrated in Fig. 9.12. The results indicate that, compared to the other strategies, the proposed ICFCC method performs better and reduces the SFD under most conditions.

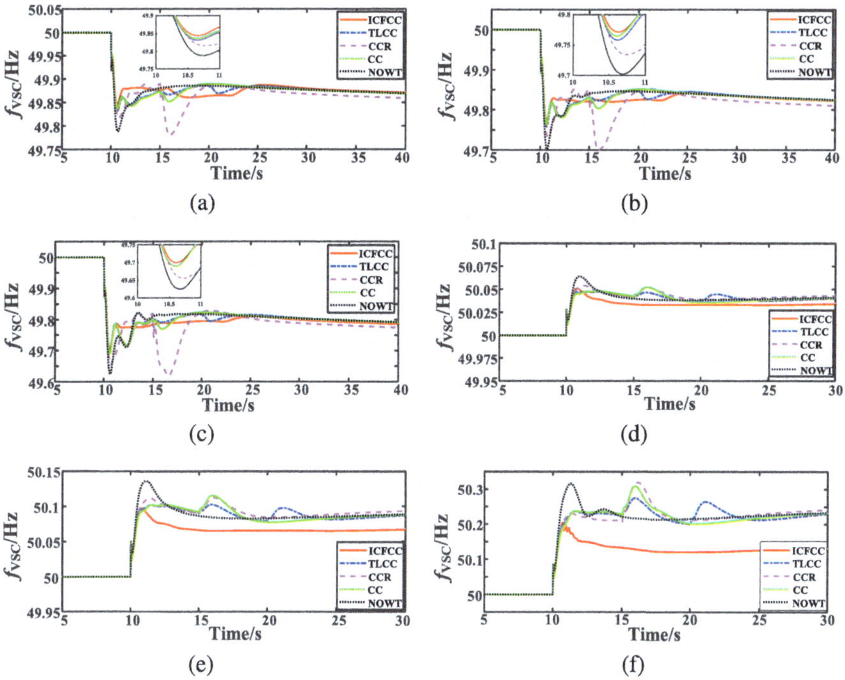

Fig. 9.11 The frequency response with diverse control methods and different disturbance: **a** with load increase 200 MW; **b** with load increase 300 MW; **c** with load increase 400 MW **d** with load decrease 50 MW; **e** with load decrease 100 MW; **f** with load decrease 200 MW

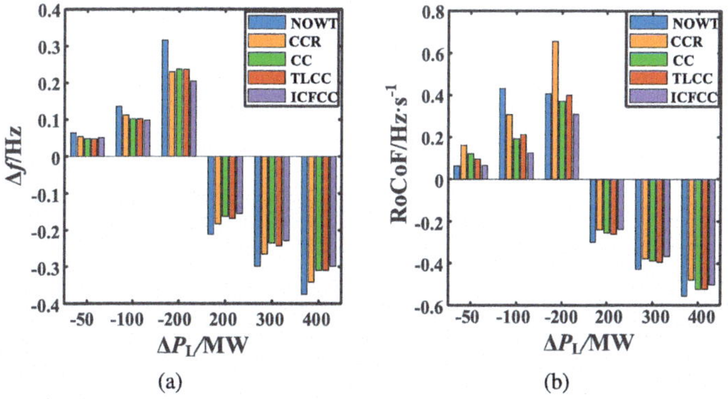

Fig. 9.12 The simulation results with different control schemes: **a** The Δf; **b** The RoCoF

9.2.6 *Parameter Uncertainty*

In a practical power system, parameter uncertainties can affect control performance [11]. To explore the impact of parameter uncertainties, both positive and negative uncertainties were applied to the DC line parameters L_{dc41}, L_{dc42}, R_{dc41}, and R_{dc42}. A 300 MW load increase at Bus 7 occurred at $t = 10s$, and the frequency response curves under different parameter uncertainties are depicted in Fig. 9.13.

As illustrated in Fig. 9.13a, the frequency nadir with a positive parameter uncertainty ranging from +5% to +25% remains nearly the same as that without parameter uncertainty (49.76 Hz). However, an oscillation occurs when the parameter uncertainty increases to +27%. Additionally, the frequency support performance slightly deteriorates with negative parameter uncertainties ranging from –5% to –25%. As shown in Fig. 9.13c, the oscillation is mainly caused by the positive parameter uncer-

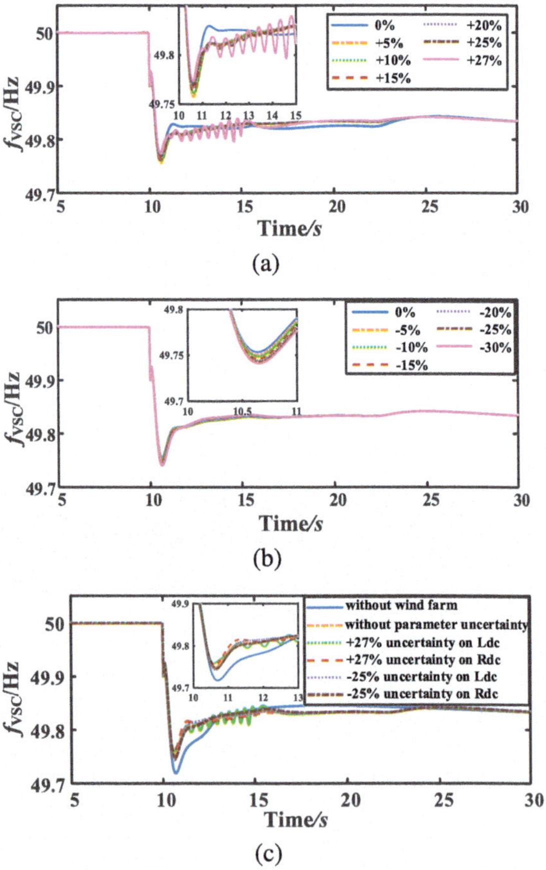

Fig. 9.13 The frequency response with different parameter uncertainty: **a** with parameter increase; **b** with parameter decrease; **c** with different components

tainty in L_{dc}. Consequently, the proposed ICFCC scheme maintains effective performance with parameter uncertainties ranging from –25% to +25%.

9.2.7 Gaussian Noise

In a practical power system, noise can also impact control performance. To evaluate this influence on the performance of the proposed ICFCC scheme, different levels of Gaussian noise are introduced in the scenario: (i) 1% Gaussian noise (signal-to-noise ratio (SNR) of 40 dB), and (ii) 5% Gaussian noise (SNR of 26 dB). A 300 MW load increase event occurs at $t = 10s$. The frequency response curves under different noise disturbances are displayed in Fig. 9.14.

The frequency response curve with 40 dB noise is almost identical to the curve without noise. However, when the SNR decreases to 26 dB, the frequency response exhibits slight fluctuations. According to Fig. 9.14, the proposed ICFCC scheme demonstrates tolerance to noise levels down to 26 dB.

9.2.8 DC Voltage Limitation

The proposed ICFCC scheme reduces the DC voltage when the onshore frequency drops. To ensure the DC voltage operates within a safe range, a DC voltage margin index is incorporated into the onshore droop control scheme as shown in Fig. 9.1. To study the impact of DC voltage limitation, a 300 MW load increase at Bus 7 was simulated, and the dynamic responses with and without DC voltage limitation were tested.

The frequency response curve and onshore DC voltage are shown in Fig. 9.15. According to the figure, the frequency nadir without DC voltage limitation (49.78 Hz) is slightly higher than that with voltage limitation (49.775 Hz). However, without the DC voltage limitation, the DC voltage drops to 0.92 pu. By including the DC voltage margin index and adaptively adjusting the droop coefficient during the frequency

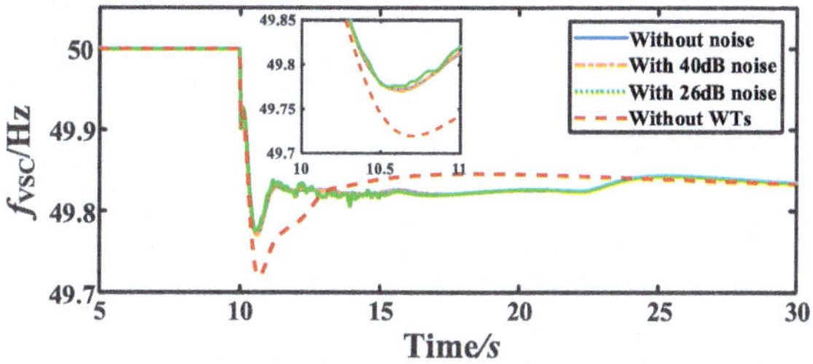

Fig. 9.14 The frequency response with different SNR

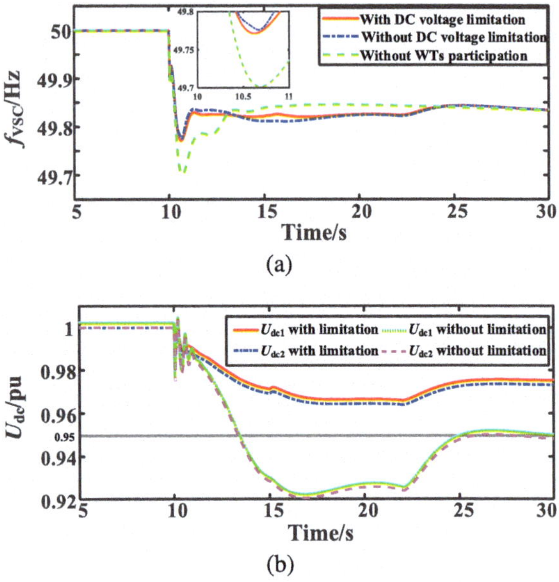

Fig. 9.15 The simulation results with and without DC voltage limitation: **a** The frequency response curve; **b** The onshore DC voltage

event, the onshore DC voltage U_{dc1} and U_{dc2} remain above the minimum value of 0.95 pu. Consequently, the DC voltage is maintained within the safe range, ensuring secure operation with the DC voltage margin index.

9.3 Case Study II: 4-Area 5-Terminal System

9.3.1 Benchmark Description

To further explore the effectiveness of the proposed ICFCC scheme under different network configurations, a more generalized 5-terminal system, as referenced in [12], is examined in this section. The parameters of onshore stations in this 4-area 5-terminal setup are similar to those in the 4-terminal system, as detailed in Table 9.2. The parameters of the MTDC network are sourced from [2, 12], with their values and meanings provided in Table 9.5. Additionally, to assess the impact of onshore droop coefficients on system stability, a stability analysis is conducted with the U–f droop coefficients ranging from −20 to 0, and the P–U droop coefficients varying from 15 to 35.

According to Fig. 9.17, most of the damping ratios remain above 0.05 when the droop coefficients of onshore stations are adjusted within the ranges specified in Table 9.5. This demonstrates robust small-signal stability even with parameter variations (Fig. 9.16).

Table 9.5 Parameters of the 4-Area 5-Terminal System

Parameters	Description	Value (unit)
L_{dc51}	The inductance of DC line 5–1	240.34 mH
L_{dc52}	The inductance of DC line 5–2	240.34 mH
L_{dc43}	The inductance of DC line 4–3	240.34 mH
L_{dc12}	The inductance of DC line 1–2	120.17 mH
L_{dc23}	The inductance of DC line 2–3	120.17 mH
R_{dc51}	The resistance of DC line 5–1	1.71 Ω
R_{dc52}	The resistance of DC line 5–2	1.71 Ω
R_{dc43}	The resistance of DC line 4–3	1.71 Ω
R_{dc12}	The resistance of DC line 1–2	0.855 Ω
R_{dc23}	The resistance of DC line 2–3	0.855 Ω

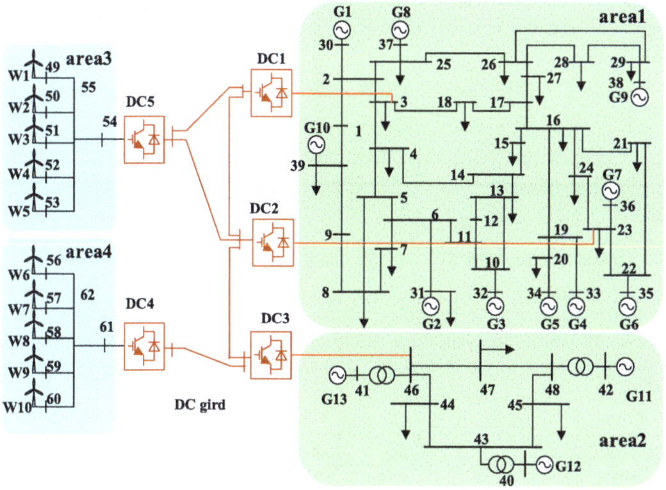

Fig. 9.16 The 4-area 5-terminal test systems

9.3.2 Simulation Verification

In this scenario, various load increase events are initiated at Bus 23 of Area 1 at $t = 10s$, resulting in different frequency response outcomes depicted in Fig. 9.18. According to Fig. 9.18, the ICFCC scheme demonstrates superior frequency support performance compared to other control strategies. The frequency nadir is elevated to 49.429 Hz, whereas it is 49.423 Hz with the CC scheme, 49.402 Hz with the TLCC scheme, 49.396 Hz with the CCR scheme, and 49.35 Hz with the NOWT scheme.

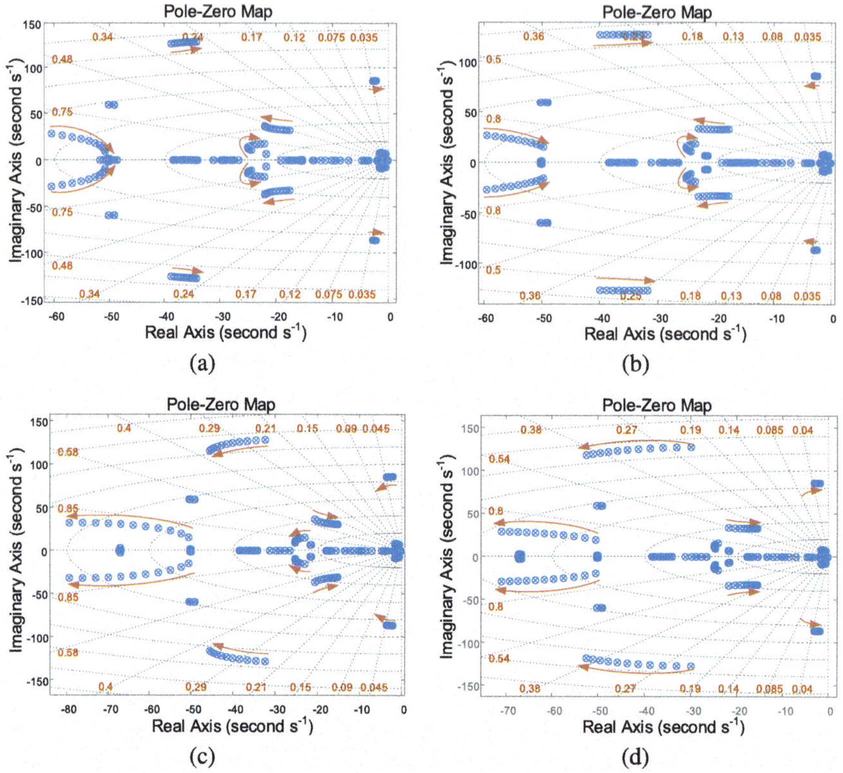

Fig. 9.17 The small-signal analysis with droop coefficients change of onshore stations for 5-terminal test system: **a** k_{df1} from 0 to –20; **b** k_{df2} from 0 to –20; **c** k_{du1} from 15 to 35; **d** k_{du2} from 15 to 35

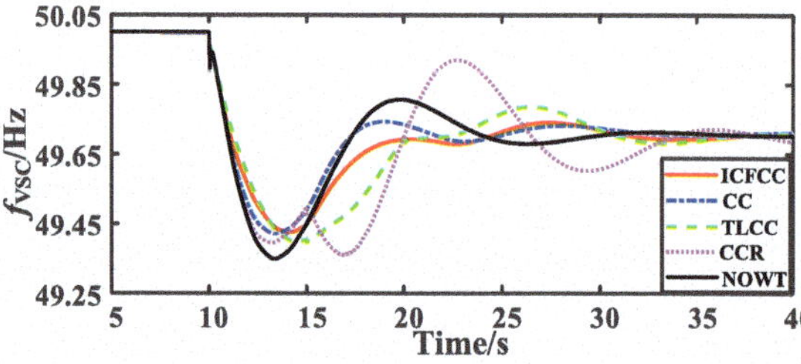

Fig. 9.18 The simulation results for 5-terminal test system with different control schemes (with load increase 500 MW)

Furthermore, to highlight the advantages of the ICFCC scheme under varying load increase conditions, the frequency deviation and RoCoF are illustrated in Fig. 9.19. It is evident that the proposed ICFCC scheme outperforms other schemes as the load increases from 100 to 500 MW, underscoring its adaptability across different network topologies.

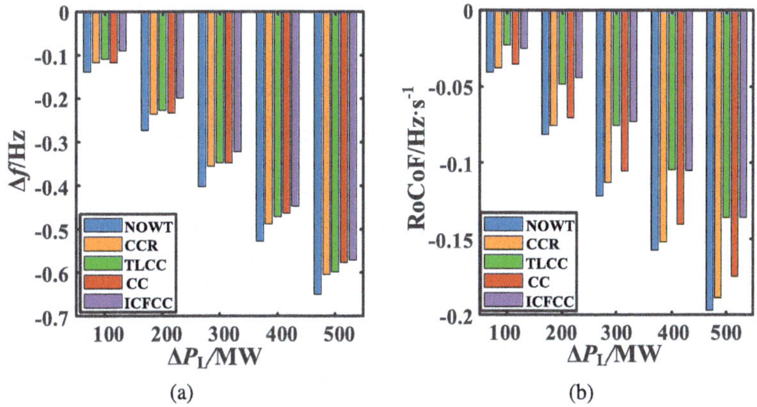

Fig. 9.19 The simulation results with different control schemes: **a** The Δf; **b** The RoCoF

Fig. 9.20 The simulation results with frequency support from the OWF in Area 4

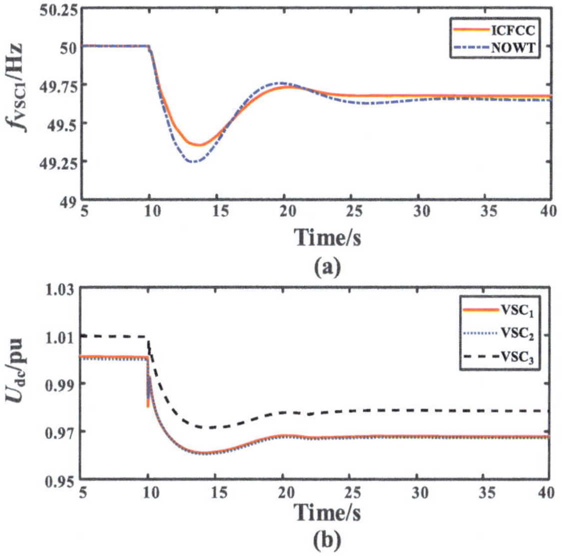

Furthermore, to validate the effectiveness of the proposed ICFCC scheme across different topologies, a 600MW load increase event occurs at Bus 23 of Area 1 at $t = 10s$. Only the offshore wind farm in Area 4 provides frequency support, as shown in the simulation results in Fig. 9.20. The DC voltage of onshore stations VSC_1, VSC_2, and VSC_3 changes consistently, while the DC voltage estimator in offshore station VSC_4 estimates the DC voltage deviation in VSC_3, enabling WTs in Area 4 to provide frequency support.

With adaptive droop control, the distribution of frequency support power among VSC_1 and VSC_2 is managed reasonably. Despite the absence of a direct connection between offshore station VSC_4 and onshore VSC stations VSC_1 and VSC_2, WTs in Area 4 still deliver effective frequency support performance to Area 1.

9.4 Summary

This chapter introduces an improved communication-free coordinated control scheme aimed at enabling onshore systems to receive frequency support from VSC-MTDC integrated offshore wind farms. Compared to various control methods, the proposed ICFCC scheme demonstrates superior performance under both load increase and decrease scenarios, while eliminating the need for remote communication between onshore and offshore systems.

Moreover, with the ICFCC scheme, frequency regulation power is appropriately distributed among onshore converters, ensuring the DC voltage remains within safe limits. Additionally, the ICFCC scheme mitigates frequency deviations caused by WT recovery control. Comparative case studies across 4-terminal and 5-terminal systems highlight the flexibility of the ICFCC scheme in different DC grid topologies, showcasing its effectiveness in raising frequency nadirs and reducing SFD.

The robustness of the ICFCC scheme against parameter uncertainties ranging from -25% to +25% and noise disturbances below 5% is also demonstrated. Future research will explore cluster and distributed control strategies for OWFs, as well as effective coordination with conventional generators to enhance frequency support capabilities.

References

1. Kou P, Liang D, Wu Z, Ze Q, Gao L (2018) Frequency support from a DC-grid offshore wind farm connected through an HVDC link: a communication-free approach. IEEE Trans Energy Convers 33(3):1297–1310. https://doi.org/10.1109/TEC.2018.2814604
2. Xiao L, Xu Z, Xiao H, Zhang Z, Wang G, Xu Y (2019) Electro-mechanical transient modeling of MMC-based multi-terminal HVDC system with dc faults considered. Int J Electr Power Energy Syst 113:1002–1013. https://doi.org/10.1016/j.ijepes.2019.06.003
3. Silva B, Moreira CL, Seca L, Phulpin Y, Lopes JAP (2012) Provision of inertial and primary frequency control services using offshore multiterminal HVDC networks. IEEE Trans Sustain Energy 3(4):800–808. https://doi.org/10.1109/TSTE.2012.2199774
4. Lu Z, Ye Y, Qiao Y (2019) An adaptive frequency regulation method with grid-friendly restoration for VSC-HVDC integrated offshore wind farms. IEEE Trans Power Syst 34(5):3582–3593. https://doi.org/10.1109/TPWRS.2019.2901986
5. Lin S, Yao W, Xiong Y, Zhao Y, Shi Z, Ai X, Wen J (2023) MatPSST: A Matlab/Simulink-based power system simulation toolbox for research and education, IET Gener Transm Distrib 17(10):2272–2288. https://doi.org/10.1049/gtd2.12805

6. Vennelaganti SG, Chaudhuri NR (2018) Ratio-based selective inertial and primary frequency support through MTDC grids with offshore wind farms. IEEE Trans Power Syst 33(6):7277–7287. https://doi.org/10.1109/TPWRS.2018.2850145
7. Cole S, Beerten J, Belmans R (2010) Generalized dynamic VSC MTDC model for power system stability studies. IEEE Trans Power Syst 25(3):1655–1662. https://doi.org/10.1109/TPWRS.2010.2040846
8. Xiong Y, Yao W, Wen J, Lin S, Ai X, Fang J, Wen J, Cheng S (2021) Autonomous synchronizing and frequency response control of multi-terminal DC systems with wind farm integration. IEEE Trans Power Syst 36(1):781–792. https://doi.org/10.1109/TSTE.2020.2964145
9. Akkari S, Dai J, Petit M, Guillaud X (2016) Interaction between the voltage-droop and the frequency droop control for multi-terminal hvdc systems. IET Gener Transm Distrib 10(6):1345–1352. https://doi.org/10.1049/iet-gtd.2015.0814
10. Sakamuri JN, Altin M, Hansen AD, Cutululis NA (2017) Coordinated frequency control from offshore wind power plants connected to multi terminal DC system considering wind speed variation. IET Renew Power Gener 11(8):1226–1236. https://doi.org/10.1049/iet-rpg.2016.0433
11. Chen J, Yao W, Zhang CK, Ren Y, Jiang L (2019) Design of robust mppt controller for grid-connected PMSG-based wind turbine via perturbation observation based nonlinear adaptive control. Renew Energy 134:478–495. https://doi.org/10.1016/j.renene.2018.11.048
12. Wang Z, Wu W (2018) Coordinated control method for dfig-based wind farm to provide primary frequency regulation service. IEEE Trans Power Syst 33(3):2644–2659. https://doi.org/10.1109/TPWRS.2017.2755685

Chapter 10
Adaptive Dual Droop Control of MTDC Integrated Offshore Wind Farms for Fast Frequency Support

Abstract This chapter introduces an adaptive dual droop control (ADDC) method, which delivers rapid frequency support to a disturbed onshore grid from both undisturbed onshore systems and VSC-MTDC (voltage source converter-based multi-terminal direct current) integrated offshore wind farms (OWFs). With standard droop control at onshore converters, any frequency disturbances cause changes in DC voltage and power flow, leading to frequency fluctuations in unaffected systems. The ADDC approach initially identifies disturbed and undisturbed systems, prompting the latter to supply additional frequency support while maintaining safety by setting support power limits and adjusting droop coefficients. Furthermore, offshore stations use onshore DC voltage as a control signal for swift frequency support. Subsequently, OWFs will recover rotor speed through an asymptotic control method to mitigate the secondary frequency drop. Case studies on 3-terminal and 5-terminal test benchmarks, utilizing the Opal-RT real-time simulation platform, compare various control methods, accounting for parameter uncertainties and noise disturbances, showcasing the ADDC scheme's efficiency and effectiveness.

10.1 Existing Schemes Formula

10.1.1 Existing Conventional Droop Control Scheme

The traditional power-voltage-frequency droop control strategy introduced in [1] is analyzed. Assuming there are i onshore stations, the power reference for the ith onshore converter is determined as:

$$P^*_{\text{on},i} = P_{\text{on0},i} - k_{\text{vp},i}(U_{\text{on0},i} - U_{\text{on},i}) + k_{\text{fp},i}(f_{\text{on0},i} - f_{\text{on},i}) \tag{10.1}$$

where $k_{\text{vp},i}$ and $k_{\text{fp},i}$ denote the voltage and frequency droop coefficients of converter i. The DC voltage references and measurements are $U_{\text{on0},i}$ and $U_{\text{on},i}$. The frequency

W. Yao et al., *Active Control of Large-Scale Offshore Wind Farms Connected Via VSC-HVDC*, Power Systems, https://doi.org/10.1007/978-981-97-9346-4_10

references and measurements are $f_{\text{on0},i}$ and $f_{\text{on},i}$. The initial and revised power references are $P_{\text{on0},i}$ and $P^*_{\text{on},i}$. When frequency events occur, the deviations in power reference can be expressed as:

$$\begin{cases} \Delta P^*_{\text{on},m} = -k_{\text{vp},m}\Delta U_{\text{on},m} + k_{\text{fp},m}\Delta f_{\text{on},m} & m \in N_d \\ \Delta P^*_{\text{on},n} = -k_{\text{vp},n}\Delta U_{\text{on},n} + k_{\text{fp},n}\Delta f_{\text{on},n} & n \in N_n \end{cases} \tag{10.2}$$

where N_d is the number of onshore stations connected to the disturbed systems, and N_n is the number of onshore stations connected to the undisturbed systems. If the frequency surpasses its threshold, the DC voltage will fluctuate, and the injection power of the undisturbed systems will also vary [1]. The offshore converters use the DC voltage-frequency droop to integrate the offshore DC voltage changes into the offshore frequency deviations, as described in [1]:

$$f^*_{\text{off}} = f_{\text{off0}} + k_{\text{vf}} \cdot (U_{\text{dcoff0}} - U_{\text{dcoff}}) \tag{10.3}$$

where k_{vf} represents the voltage-frequency droop coefficient for the offshore system. U_{dcoff0} and U_{dcoff} denote the reference and measured offshore DC voltage values. f^*_{off} and f_{off} signify the revised and initial frequency reference values, respectively.

10.1.2 Limitations of the Existing Droop Control Scheme

There are two primary drawbacks of the existing scheme in [1]: (1) The droop control intensifies the interconnection among various onshore systems and spreads the frequency disturbance to other unaffected systems. (2) The offshore DC voltage changes with the OWFs' output power, impacting frequency support performance.

Assuming no power loss in the DC network, no converter failures, and that OWFs do not offer inertial response or frequency support, the DC voltage deviations at all onshore stations are nearly identical. Thus, (10.1) can be reformulated as:

$$\Delta f_{\text{on},n} = \frac{1}{k_{\text{fp},n}}\left[(k_{\text{vp},m} + k_{\text{vp},m}) \cdot \Delta U_{\text{on}} - k_{\text{fp},m} \cdot \Delta f_{\text{on},m}\right] \tag{10.4}$$

where $\Delta f_{\text{on},n}$ and $\Delta f_{\text{on},m}$ denote the frequencies of the undisturbed and disturbed converters. Due to the linkage among onshore stations with droop control, the frequency of converter n connected to undisturbed AC grids, $\Delta f_{\text{on},n}$, will also vary during frequency events. This linkage is unfavorable, as the frequency deviation $\Delta f_{\text{on},n}$ might become significant and potentially cause more severe consequences.

Regarding the other limitation, the droop control in [1] incorporates the onshore frequency deviation into the DC voltage and uses offshore DC voltage variations as the frequency control signal. Assuming there are N onshore converters connected to a single offshore converter, the transmission power of the offshore converter can be expressed as follows:

$$P_{\mathrm{off}} = U_{\mathrm{dcoff}} \times \begin{bmatrix} \frac{U_{\mathrm{dcoff}} - U_{\mathrm{on},1}}{R_{\mathrm{dc},1}} & 0 & 0 \\ \vdots & \ddots & \vdots \\ 0 & 0 & \frac{U_{\mathrm{dcoff}} - U_{\mathrm{on},N}}{R_{\mathrm{dc},N}} \end{bmatrix} \tag{10.5}$$

where P_{off} and U_{dcoff} denote the transmission power and DC voltage of the offshore converter. $R_{\mathrm{dc},i}$, $(i \in N)$, represents the connected line resistance between the onshore converter i and the offshore converter. The expression for P_{off} can be rewritten as follows:

$$P_{\mathrm{off}} = \sum_{i=1}^{N} \left(\frac{U_{\mathrm{dcoff}} - U_{\mathrm{on},i}}{R_{\mathrm{dc},i}} \right) \cdot U_{\mathrm{dcoff}} \tag{10.6}$$

Assuming that the DC grid is a point-to-point path and $N = 1$, (10.6) can be simplified as:

$$U_{\mathrm{dcoff}} = \frac{U_{\mathrm{on},1} + \sqrt{U_{\mathrm{on},1}^2 + 4P_{\mathrm{off}}U_{\mathrm{on},1}}}{2} \tag{10.7}$$

It can be observed from (10.7) that U_{dcoff} is influenced not only by the onshore DC voltage but also by the transmission power of the offshore converter, P_{off}. Thus, during frequency support moments, the offshore transmission power P_{off} will increase, causing the DC voltage to inaccurately reflect the onshore frequency deviation during frequency events. According to (10.3), the offshore frequency reference signal f_{off}^*, which serves as the frequency control signal for OWFs, will also be affected. Consequently, the frequency support performance will be impacted with existing droop control schemes.

10.2 Structure of The Proposed ADDC Scheme

10.2.1 Overview of the Proposed ADDC Scheme

To overcome the limitations of the traditional power-voltage-frequency droop control in [1], the ADDC scheme for VSC-MTDC integrated OWFs, illustrated in Fig. 10.1, is proposed. In this setup, there are distinct synchronous regions with no direct power line connection within the onshore systems, as shown in Fig. 10.1. The ADDC scheme offers rapid frequency support and comprises three components.

- **Onshore Adaptive Dual Droop Control**: The disturbed system detection method is proposed, where undisturbed systems will offer more frequency support power to the disturbed systems by increasing the power-voltage droop coefficients during frequency events, compared with the approach in [1].

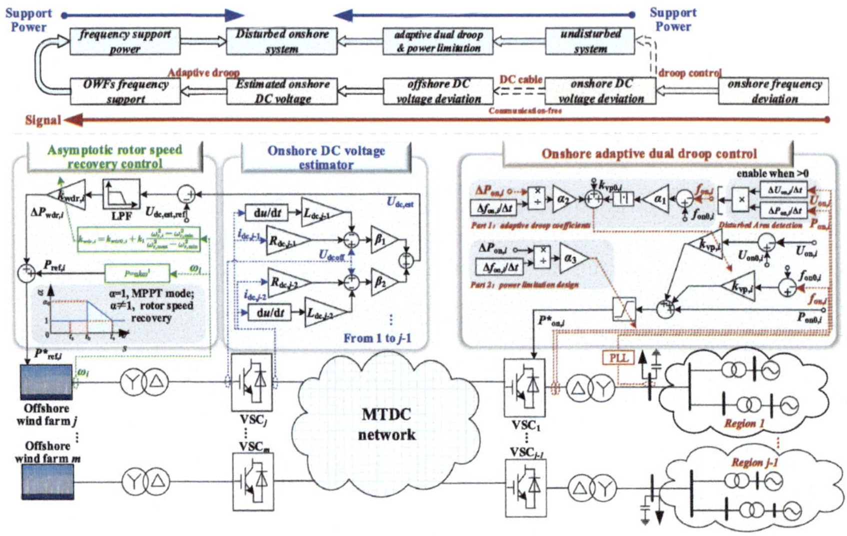

Fig. 10.1 The block diagram of the proposed adaptive dual droop control scheme

- **Support Power Limitation for Undisturbed System**: To keep the frequency of the undisturbed systems within a safe range, the support power limitation is determined by considering the different inertia statuses of the undisturbed onshore converters while providing frequency support power.
- **Offshore Communication-Free Estimator and Asymptotic Recovery Control**: A communication-free estimator is proposed to evaluate the onshore DC voltage for frequency support using locally acquired signals. Simultaneously, the recovery profiles of wind turbines (WTs) are redesigned, and the recovery duration is extended to smoothen the recovery process and minimize the SFD. Notably, the proposed ADDC scheme employs Kirchhoff's law and can be adapted to various configurations.

The main steps of the ADDC scheme are illustrated in Fig. 10.2. First, the disturbed and undisturbed systems are identified, and the droop coefficients of stations linked to the undisturbed systems are adjusted to provide more frequency support power. Concurrently, to keep the frequency of the undisturbed systems within a safe range, the support power limit is set considering the different inertia conditions of the undisturbed onshore converters. A communication-free estimator is introduced to determine the onshore DC voltage, avoiding communication delays for rapid frequency support. After frequency support, offshore wind turbines will restore their rotor speed using asymptotic control to reduce SFD. The proposed ADDC scheme uses local signals as controller inputs at each stage, minimizing the impact of remote communication delays.

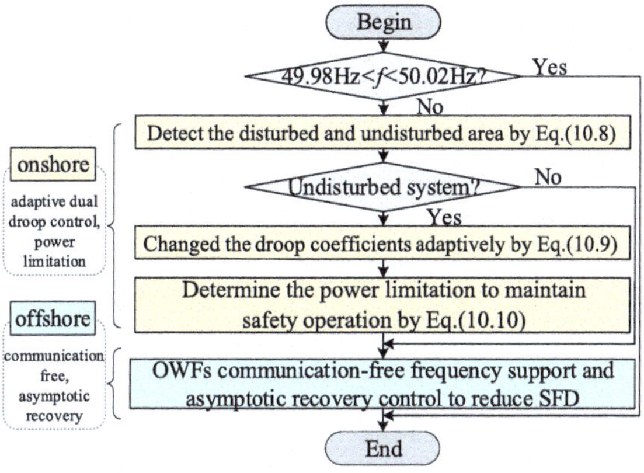

Fig. 10.2 The flow chart of the proposed ADDC scheme

10.2.2 Onshore Adaptive Dual Droop Control

10.2.2.1 Disturbed Area Detection

The initial step of the ADDC scheme involves identifying the disturbed and undisturbed areas when the measured frequency surpasses the threshold (49.98 Hz $< f <$ 50.02 Hz in this study). It is evident from (10.2) that the frequency and DC voltage of converters linked to disturbed systems will drop, while the injected power in the disturbed area will rise. Conversely, the injected power of the undisturbed system will decline. The differences in power and DC voltage between disturbed and undisturbed systems are detailed in Table 10.1.

Thus, the equation $\mathrm{sgn}(\frac{\Delta U_{\mathrm{on},i}}{\Delta t} \cdot \frac{\Delta P_{\mathrm{on},i}}{\Delta t})$ can be utilized to identify disturbed and undisturbed areas. If this index is greater than 0, the connected area is undisturbed. Conversely, if the index is less than 0, the connected area is disturbed.

$$\mathrm{sgn}\left(\frac{\Delta U_{\mathrm{on},i}}{\Delta t} \cdot \frac{\Delta P_{\mathrm{on},i}}{\Delta t}\right) \begin{cases} > 0, \; undisturbed \\ < 0, \; disturbed \end{cases} \tag{10.8}$$

Table 10.1 The power and DC voltage difference of different VSC stations during the frequency events

Index	Load increase		Load decrease	
	Disturbed system	Undisturbed system	Disturbed system	Undisturbed system
$\Delta U_{\mathrm{on},i}/\Delta t$	–	–	+	+
$\Delta P_{\mathrm{on},i}/\Delta t$	+	–	–	+

10.2.2.2 Adaptive Droop Coefficients

After identifying the disturbed area, the voltage droop coefficients of the undisturbed areas in Fig. 10.1 will be increased to offer more frequency support power to the disturbed system. Unlike the existing scheme in (10.1), the calculation formula for $k_{\mathrm{vp},i}$ with the proposed ADDC scheme can be expressed as follows:

$$k_{\mathrm{vp},i} = k_{\mathrm{vp0},i} + \underbrace{\alpha_1 \cdot |\frac{f_{\mathrm{on0},i} - f_{\mathrm{on},i}}{f_{\mathrm{on0},i} - f_{\mathrm{onlim},i}}|}_{\text{frequency compensation}} + \underbrace{\alpha_2 \cdot \frac{\Delta P_{\mathrm{on},i}}{\Delta f_{\mathrm{on},i}/\Delta t}}_{\text{inertia compensation}} \tag{10.9}$$

where $k_{\mathrm{vp0},i}$ represents the initial value of the voltage droop coefficient, α_1 and α_2 are constants, $f_{\mathrm{on0},i}$ and $f_{\mathrm{on},i}$ represent the frequency reference and measurement of onshore converter VSCi, and $f$onlim, i denotes the frequency limit for safe operation. $\Delta P_{\mathrm{on},i}$ represents the measured power of VSC$_i$. The frequency compensation and inertia compensation are designed to adjust the droop coefficients based on the frequency deviations and the inertia statuses of the undisturbed systems, respectively. It is important to note that $k_{\mathrm{vp},i}$ can be safely adjusted by appropriately setting α_1 and α_2.

Regarding α_1, frequency compensation is used to increase the droop control coefficients when the frequency exceeds the dead-band, with $f_{\mathrm{onlim},i}$ set to 49.5 Hz or 50.5 Hz during load increasing and decreasing events. With a base frequency of 50 Hz, $f_{\mathrm{onlim},i}$ is equal to 0.99 or 1.01 pu, and $\left|f_{\mathrm{on0},i} - f_{\mathrm{on},i}\right| / \left|f_{\mathrm{on0},i} - f_{\mathrm{onlim},i}\right| \in [0, 1]$. If $f_{\mathrm{on},i}$ does not exceed the limit, to make the U–P droop coefficients $k_{\mathrm{vp},i}$ change more significantly, α_1 can be set to [1, 10] to ensure safety. In this chapter, α_1 is set to 10 for a more pronounced change in the U–P droop coefficients $k_{\mathrm{vp},i}$.

Regarding α_2, the result of $\Delta P_{\mathrm{sys}}/\Delta f_{\mathrm{co}}/\Delta t$ is related to the equivalent inertia time constant. To make the U–P droop coefficients $k_{\mathrm{vp},i}$ change more significantly while ensuring a safe range ($k_{\mathrm{vp},i} \in [0, 40]$), α_2 is set to [0, 2].

10.2.3 Support Power Limitation for Undisturbed System

The frequency of an undisturbed system will drop when providing support power to a disturbed system. To ensure safe operation, the provided frequency support power must be kept within a safe range. However, with different operational states of synchronous generators, the real-time inertia statuses of undisturbed systems may vary. The greater the inertia, the more frequency support power undisturbed systems can offer, and thus, the support power limitation should be larger. Therefore, to maintain the safe frequency range of the undisturbed system while offering more frequency support power, inertia compensation should be factored into the power limitation design considering the various operational states. The greater the inertia, the more power can be provided. Hence, the power limitation of the undisturbed system can be formulated as follows:

$$P_{\text{onlim},i} = P_{\text{on0},i} + \alpha_3 \cdot \text{sgn}(f_{\text{on0},i} - f_{\text{on},i}) \cdot \frac{\Delta P_{\text{on},i}}{\Delta f_{\text{on},i}/\Delta t} \tag{10.10}$$

where $\text{sgn}(f_{\text{on0},i} - f_{\text{on},i})$ is used to detect overload or underload frequency events, and α_3 is a constant. The power limitation of the undisturbed systems can be determined considering the different states of onshore systems. The greater the inertia, the larger the index $\frac{\Delta P_{\text{on},i}}{\Delta f_{\text{on},i}/\Delta t}$, and thus a larger support power limitation will be set. $P_{\text{VSClim},i}$ represents the power limit of VSCi. Due to the power limitation design, the undisturbed systems can provide more frequency support power to the disturbed system while maintaining their own safe operation, during both overload and underload frequency events. It is important to note that if $P_{\text{onlim},i}$ exceeds the transmission power limit of converters, it will be set to the limit value to avoid overloading. The function of Δt is to create a time interval to get the value of $\Delta P_{\text{on},i}$, $\Delta U_{\text{on},i}$ and $\Delta f_{\text{on},i}$ in (10.8)–(10.10). The value of Δt is related to the signal sampling frequency of the power system. Referred to [2], the sampling frequency of SCADA and PMU in practical power systems are about 20 and 100 Hz, to ensure the effectiveness of the proposed ADDC control scheme, the time interval Δt is set to 0.1 s in this chapter.

10.2.4 Offshore Communication-Free Estimator and Asymptotic Recovery Control

The offshore stations are generally controlled to maintain the offshore AC voltage [3]. When OWFs participate in frequency support and increase their output power, the offshore DC voltage will also rise. Therefore, during frequency support, the DC voltage cannot accurately reflect the onshore frequency deviation with the existing communication-free schemes in [1] (as referred to in (10.7)). To address this issue, an offshore communication-free DC voltage estimator is proposed, which can estimate the onshore average DC voltage for frequency support using local signals instead of directly using the offshore DC voltage. The structure of the proposed offshore communication-free estimator is depicted in Fig. 10.1.

In Fig. 10.1, there are $j - 1$ onshore converters (VSC$_1$ to VSC$_{j-1}$) and $m - j + 1$ offshore converters (VSC$_j$ to VSC$_m$). Assuming the OWFs integrated by VSC$_j$ are participating in frequency support, the DC line currents from VSC$_j$ to other onshore converters can be measured at VSC$_j$. Afterward, the average onshore DC voltage can be estimated using Kirchhoff's law, knowing the DC line parameters. The detailed calculation process is shown in (10.11). where $L_{\text{dc},ij}$ and $R_{\text{dc},ij}$ represent the DC line reactance and resistance for DC Line $i - j$, respectively. These DC line parameters can be obtained for a specific practical project. Therefore, the onshore DC voltage can be estimated using only locally measured electrical quantities in the offshore station. Furthermore, the estimated onshore DC voltage $U_{\mathrm{de},j}$ is calculated as an average value at VSC$_j$ using the weighted factor β_i. The factor β_i is calculated based on the proportion of the DC line current, which can be determined using (10.12).

$$U_{\mathrm{de},j} = \sum_{i=1}^{j-1}\left(\beta_i \cdot \left(U_{\mathrm{d},j} - I_{\mathrm{dc},ij} \cdot R_{\mathrm{dc},ij} - \frac{\mathrm{d}I_{\mathrm{dc},ij}}{\mathrm{d}t} \cdot L_{\mathrm{dc},ij}\right)\right) \tag{10.11}$$

$$\beta_i = \frac{I_{\mathrm{dc},ij}}{\sum_{i=1}^{j-1} I_{\mathrm{dc},ij}} \tag{10.12}$$

To enable OWFs to provide frequency support, additional droop and inertial control must be incorporated into the MPPT control scheme to release kinetic energy. As referred to in [4], the calculation formulas for $\Delta P_{\mathrm{wdr},i}$ and $k_{\mathrm{wdr},i}$ are shown below.

$$\Delta P_{\mathrm{wdr},i} = -k_{\mathrm{wdr},i} \cdot (U_{\mathrm{deref},j} - U_{\mathrm{de},j}) \tag{10.13}$$

$$k_{\mathrm{wdr},i} = k_{\mathrm{wdr0},i} + k_1 \cdot \frac{\omega_{\mathrm{r},i}^2 - \omega_{\mathrm{r,min}}^2}{\omega_{\mathrm{r,max}}^2 - \omega_{\mathrm{r,min}}^2} \tag{10.14}$$

After providing frequency support, the WTs need to restore their rotor speed. However, an unsuitable recovery strategy may cause the SFD during the recovery period [5]. To minimize the SFD, a new P-ω curve is designed as shown in (10.15), and the value of α at different times is depicted in (10.16). During the recovery period $(t_b - t_e)$, α will gradually change from its initial value to 1 to avoid large active power disturbances. Thus, adopting the proposed control strategy can reduce the SFD.

$$P_{\mathrm{ref}} = \alpha f_{\mathrm{MPPT}}(\omega) \tag{10.15}$$

$$\alpha = \begin{cases} 1 & t < t_b \\ \frac{1-\alpha_0}{\Delta t_r} \cdot (t - t_b) + \alpha_0 \ (\alpha_0 = \frac{P_{op1}}{\omega_{op1}}) & t_b \le t \le t_e \\ 1 & t_e < t \end{cases} \tag{10.16}$$

where α_0 represents the initial value at the start of the recovery process, t_{s} is the start time of the frequency events, and t_{b} and t_{e} represent the beginning and ending times of the recovery process, respectively. $\Delta t_{\mathrm{r}} = t_{\mathrm{e}} - t_{\mathrm{b}}$ denotes the duration of the recovery process. To enable rapid rotor speed recovery and minimize the SFD, Δt_{r} is set to 7 s in this chapter.

10.3 Case Study I: 3-Area 3-Terminal System

10.3.1 Benchmark Description

To demonstrate the performance of the proposed ADDC scheme, the 3-area 3-terminal VSC-MTDC based OWFs depicted in Fig. 10.3 are established in the MATLAB/Simulink platform, with parameters shown in Table 10.2 [6, 7]. The sixth-order dynamic model, the IEEE Type 2 speed-governor model, and the Type I excitation system model as referred to in PSAT in [8] are used in the test system. Additionally, the Type 3 wind turbine model is used in this study as referenced in [9]. Different control schemes and their abbreviations are depicted in Table 10.3.

10.3.2 Controller Parameters

The comprehensive control coefficients of the testing system and their values are displayed in Table 10.2. To conduct the minor-signal examination of the test setup, the robust Model Linearization Toolbox included in MATLAB can be extensively employed. By designating input and output locations at various points in the system, linearization can be executed, and the test setup will be assessed at these designated points. Following linearization, the pole-zero diagram can be generated in the Simulink environment. By altering different coefficients, the pole-zero variation curves with different droop coefficients can also be plotted, which are instrumental in parameter design. To assess the impact of onshore droop coefficients on system stability, the small-signal analysis is designed with the $k_{vp,i}$ varying from 0 to 20 and $k_{fp,i}$ varying from -100 to -50. The pole-zero diagrams are shown in Fig. 10.4. It can be observed that the damping ratios of the most critical eigenvalues remain above 0.05 when the droop control parameters of onshore stations change within the specified ranges in Table 10.2, confirming that small-signal stability can be maintained during parameter variations.

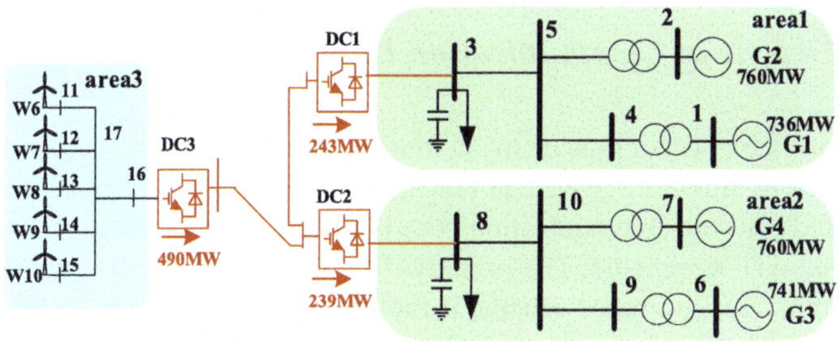

Fig. 10.3 The 3-area 3-terminal test systems

Table 10.2 Parameters of different controllers in the 3-area system

Parameters	Description	Value (unit)
$k_{vp0,i}$	Initial value of $U-P$ droop coefficients	10
$k_{vp,i}$	The $U-P$ droop coefficient	0–40
k_{fpi}	The $f-P$ droop coefficient	100
α_1	The frequency compensation gain	10
α_2 for VSC_1	The inertia compensation gain	1
α_2 for VSC_2	The inertia compensation gain	1
α_3	The gain for support power limitation	–0.12
Δt_s	The RoCoF integration time	0.01 s
$f_{Vi,\text{ref}}$	The reference value of VSC_i	50 Hz
L_{dc31}	The inductance of DC line 3–1	240.34 mH
L_{dc32}	The inductance of DC line 3–2	287.12 mH
R_{dc31}	The resistance of DC line 3–1	1.71 Ω
R_{dc32}	The resistance of DC line 3–2	2.28 Ω
β_1	The estimated weight of VSC_1 DC voltage	0–1
β_2	The estimated weight of VSC_2 DC voltage	0–1
$\omega_{r,i}$	The rotor speed of WT_i	0.7–1.2 pu
$\omega_{r,\max}$	Maximum rotor speed limitation of WTs	1.2 pu
$\omega_{r,\min}$	Minimum rotor speed limitation of WTs	0.7 pu
$k_{\text{wdr},i,0}$	The initial value of droop coefficient	–10
$k_{\text{wdr},i}$	The value of droop coefficient	–20 to 0
k_1	The slope of the inertial coefficient	–10

Table 10.3 The compared control strategies and their abbreviations

Abbr.	Description
ADDC	The proposed adaptive dual droop control scheme
CDC	Droop control in [1] and OWF control in [5]
NOWT	Without offshore WTs participation

10.3.3 Defects of Communication Delay

To investigate the effect of time delay on frequency support performance, a 300 MW load is added to Bus 5 of Area 1 at $t = 10s$ in this scenario. The control effects are compared between communication-based methods with varying time delays and the proposed ADDC scheme. The primary distinction between the ADDC scheme and the communication-based scheme is the control signals of OWFs. The ADDC scheme uses the estimated onshore DC voltage as the frequency response signal, while the communication-based scheme directly employs the onshore frequency signal. The frequency response is illustrated in Fig. 10.5. It is evident that the ADDC

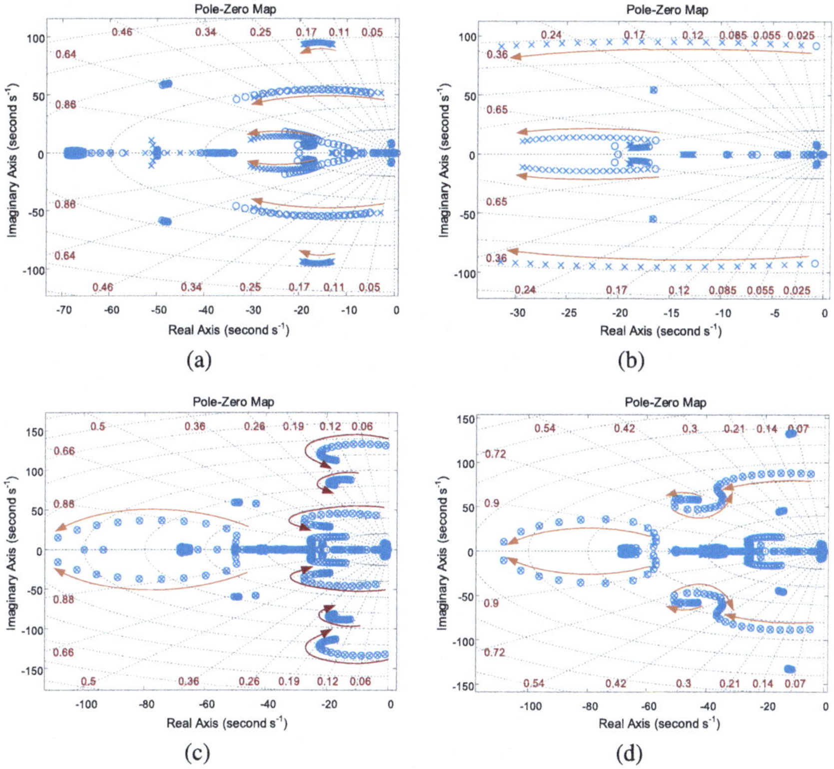

Fig. 10.4 The small-signal analysis for 3-terminal test system with droop coefficients change: **a** k_{vp1} from 0 to 20; **b** k_{vp2} from 0 to 20; **c** k_{fp2} from –100 to –50; **d** k_{fp2} from –100 to –50

scheme achieves comparable performance to the communication-based scheme without time delay (the frequency nadirs are 48.8 Hz for ADDC and 48.78 Hz for the communication-based scheme). As the time delay increases from 0 to 2 s, the control performance of the communication-based schemes deteriorates (the frequency nadir drops to 48.6 Hz with a 2 s delay). Therefore, the ADDC scheme offers superior frequency support performance compared to communication-based schemes, especially when the communication delay is significant. Additionally, the output power of OWF under different time delays is depicted in Fig. 10.5. It is observed that the OWF output power varies with different time delays; the longer the delay, the later the frequency support power is provided.

To further analyze the impact of time delay, the rotor speed of wind turbines under various time delays is shown in Fig. 10.6. It is noted that the rotor speed of wind turbines changes to release kinetic energy at different times with different delays, which explains why the output power of offshore wind farms varies with different time delays.

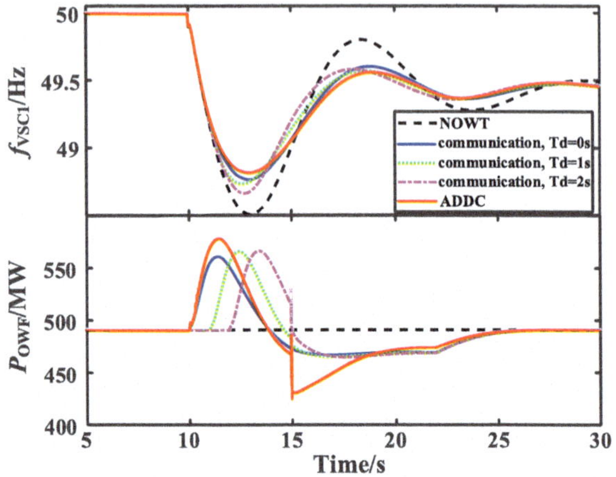

Fig. 10.5 The frequency response with communication-based schemes and ADDC control schemes with load increase 300 MW

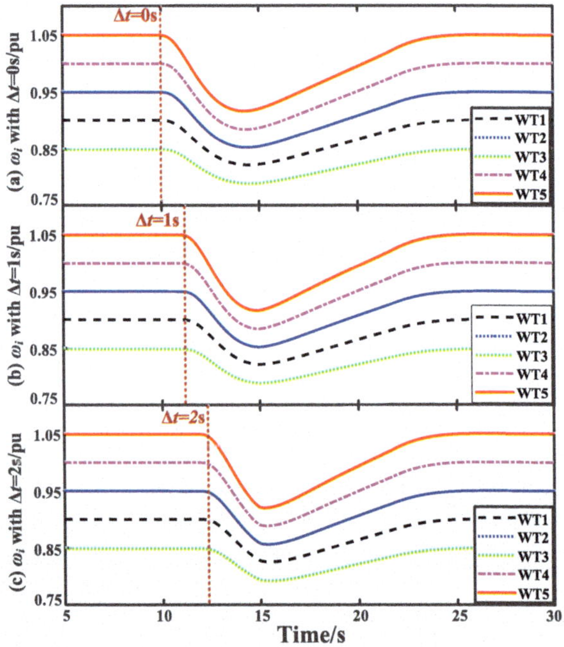

Fig. 10.6 The rotor speed of wind turbines with different time delay

10.3.4 Defects of Signal Sampling Time Interval

To explain the sensitivity of Δt in (10.8)–(10.10) on the overall performance of the control scheme, the The case study in the 3-terminal system with varying time intervals Δt is presented. A 200 MW load increase event in Bus 5 of area 1 occurs at $t = 10s$, and the simulation results are depicted in Fig. 10.7. As seen in Fig. 10.7a, b, with different time intervals ranging from 0.05 to 0.5 s, the disturbance detection

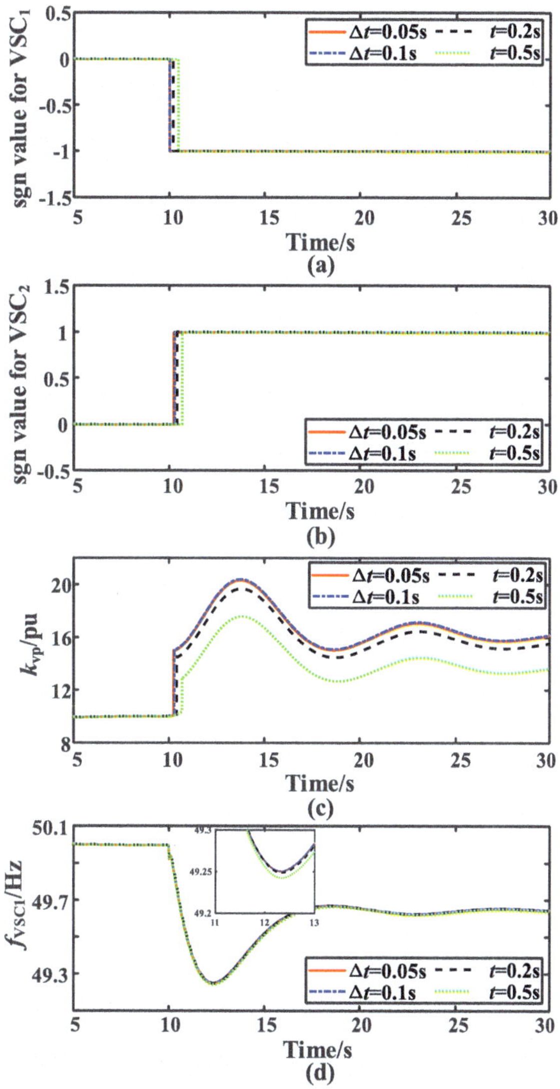

Fig. 10.7 The change of parameters in the 3-terminal test system with different time intervals: **a** the sgn value in (10.8) for VSC_1; **b** the sgn value in (10.8) for VSC_2; **c** the U–P droop control coefficient for VSC_2; **d** the frequency response for VSC_1

using (10.8) is accurate in all scenarios. However, as shown in Fig. 10.7c, the U–P droop control coefficient for VSC_2, $k_{vp,2}$, decreases as the time intervals increase, but this change has only a slight effect on the frequency response of Area 1 according to Fig. 10.7d. Thus, with the time interval Δt varying between 0.05 and 0.5 s, the frequency support performance remains assured.

10.3.5 Comparison with Different Schemes

To demonstrate the effectiveness of the proposed ADDC control compared to the existing schemes in [1], various frequency events in Bus 5 of area 1 are tested at $t = 10s$ in this scenario, and the frequency responses are shown in Figs. 10.8, 10.9 and 10.10. As seen in Fig. 10.8, with a 200 MW load increase, the frequency nadirs are

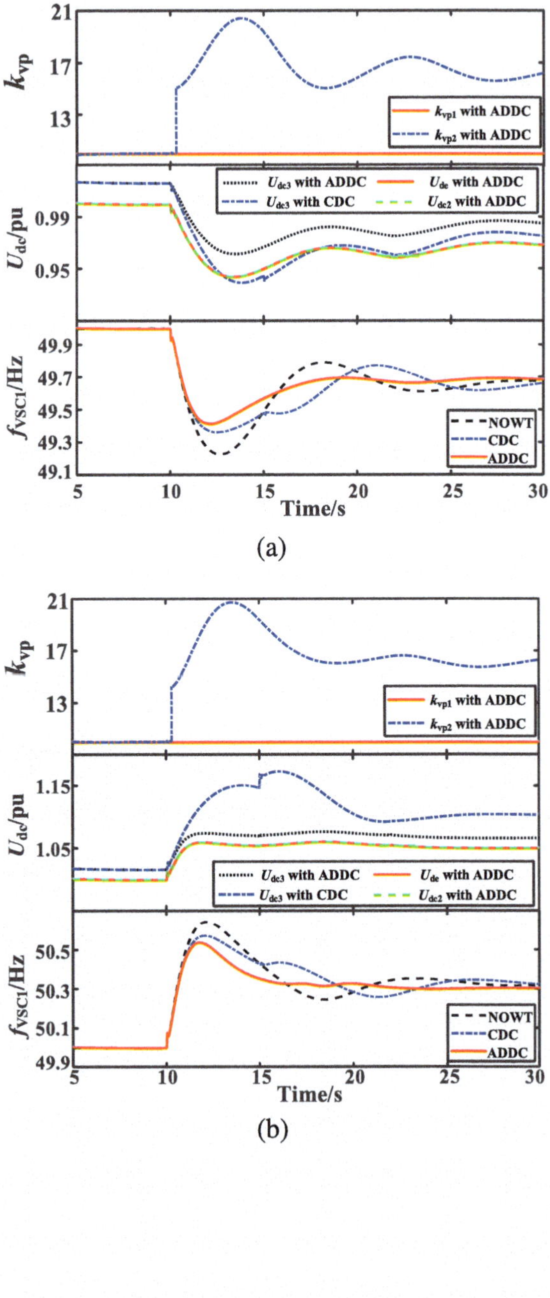

Fig. 10.8 The simulation results with different control schemes in the 3-terminal system: **a** with load increasing 200 MW; **b** with load decreasing 200 MW

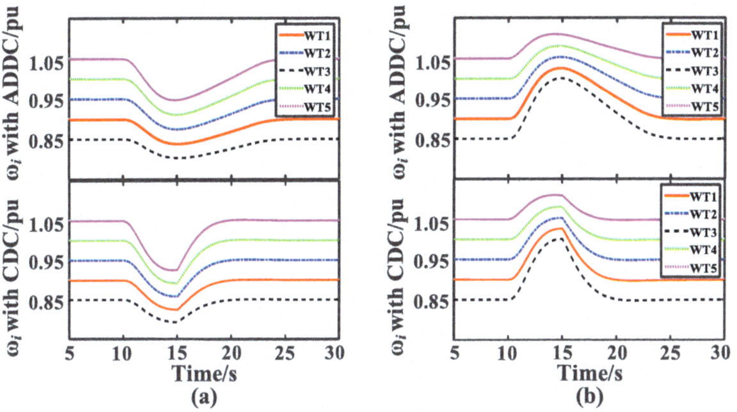

Fig. 10.9 The turbine speed of the WF for different test cases with different control scheme: **a** with 200 MW load increasing; **b** with 200 MW load decreasing

49.45 Hz with the ADDC scheme, 49.35 Hz with the CDC scheme, and 49.2 Hz with the NOWT scheme. Conversely, with a 200 MW load decrease, the frequency peak values reach 50.5 Hz with the ADDC scheme, 50.55 Hz with the CDC scheme, and 50.65 Hz with NOWT. Additionally, it is observed that the power-frequency droop coefficient of VSC_2, connected to the undisturbed system, k_{vp2}, can adaptively change based on the operating condition. Moreover, the offshore DC voltage U_{dc3} with the ADDC scheme (black dot curves in Fig. 10.8) exhibits less fluctuation compared to the CDC scheme. The estimated U_{de} (red curves in Fig. 10.8) closely matches the onshore DC voltage U_{dc2}, making it suitable as a control signal for frequency support. Furthermore, as shown in Fig. 10.10, the proposed ADDC scheme provides superior frequency support performance in reducing frequency deviations and RoCoF compared to the CDC scheme under both load increase and decrease conditions. With the ADDC scheme, the rotor speed recovers more smoothly than with the CDC scheme, aiding in the recovery of the OWF and reducing the SFD.

10.3.6 The Power Pulse Changing Situation

To evaluate the control performance of the proposed ADDC scheme under a power pulse changing situation, a 200 MW load increase occurs in Bus 5 of Area 1 at $t = 10s$, followed by a 200 MW load decrease in Bus 5 of Area 1 at $t = 50\ s$. The comparison results between the proposed ADDC scheme and the conventional CDC scheme are shown in Fig. 10.11.

As illustrated in Fig. 10.11, under a 200 MW power pulse change, the proposed ADDC scheme demonstrates superior frequency support performance compared to

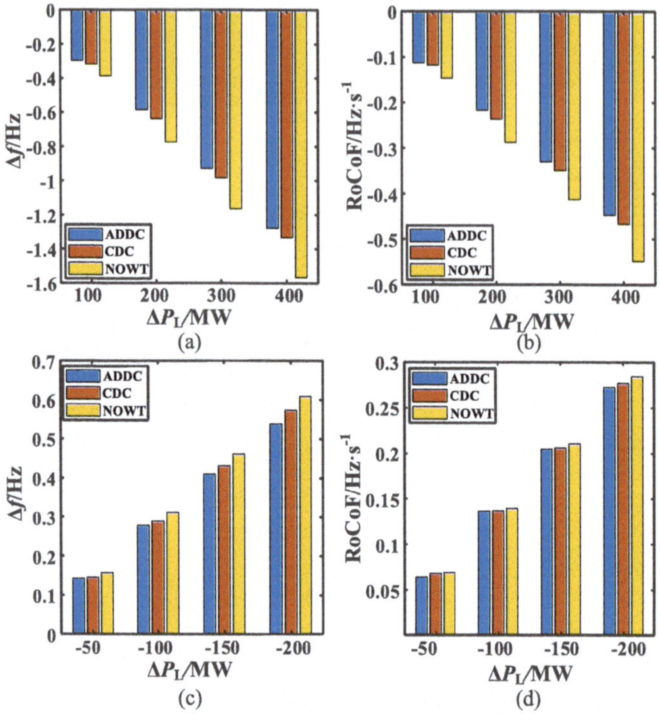

Fig. 10.10 The simulation results with different control schemes in the 3-terminal system: **a** Δf with load increasing; **b** RoCoF with load increasing; **c** Δf with load decreasing; **d** RoCoF with load decreasing

the CDC scheme. Additionally, with the proposed ADDC scheme, wind turbines (WTs) with higher rotor speeds can release more kinetic energy, thereby maximizing the utilization of stored energy.

10.3.7 Comparison with Different Inertia

The inertia of onshore AC systems can vary under different operational conditions, and undisturbed AC systems with larger inertia can provide more frequency support power. The proposed ADDC scheme addresses the support power limitation for undisturbed systems, ensuring that frequency remains within a safe range by considering the real-time inertia and operational conditions of undisturbed systems. To demonstrate the effectiveness of the proposed ADDC scheme compared to existing CDC schemes under various inertia conditions, different frequency events in Bus 5 of area 1 are tested at $t = 10$ s, and the control effect is compared with the CDC scheme. The frequency response is shown in Fig. 10.12.

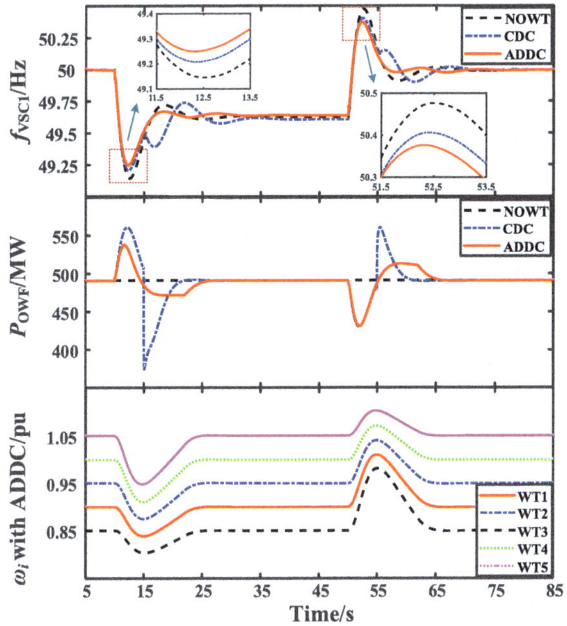

Fig. 10.11 The frequency response with a power pulse changing situation

Figure 10.12a, b depict the frequency response curves of VSC_1 (connected to the disturbed system) and VSC_2 (connected to the undisturbed system), respectively. As shown in Fig. 10.12a, the proposed ADDC scheme achieves better frequency support performance than the CDC and NOWT schemes under different inertia conditions. Furthermore, in Fig. 10.12b, the frequency of VSC_2 drops below 49.5 Hz with the NOWT scheme due to coupling caused by droop control. However, with the proposed ADDC scheme, the frequency of VSC_2 is maintained within a safe range (above 49.5 Hz in this study).

10.3.8 Gaussian Noise

In practical power systems, noise in control signals can impact frequency support performance. To study this influence on the performance of the proposed ADDC scheme, different levels of Gaussian noise are added to the offshore DC voltage and current measurements in this scenario: (i) with 1% Gaussian noise (signal-to-noise ratio (SNR) is 40 dB) and (ii) with 5% Gaussian noise (SNR is 26 dB). The estimated DC voltage is shown in Fig. 10.13. A 300 MW load increase occurs in Bus 7 at $t = 10s$.

As shown in Fig. 10.14, the frequency response curves with 1 and 5% noise are nearly identical to the curve without noise, although the frequency deviation increases

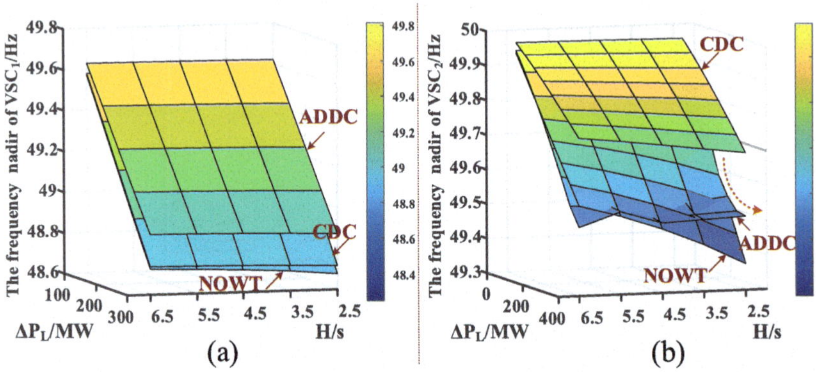

Fig. 10.12 The simulation results with different control schemes: **a** f_{VSC1}; **b** f_{VSC2}

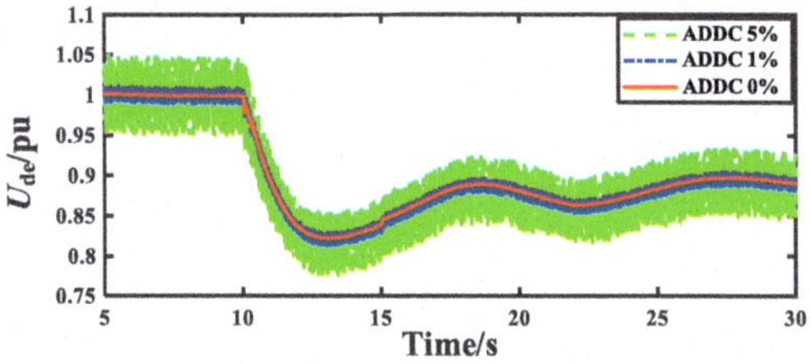

Fig. 10.13 The estimated DC voltage with different Gaussian noise

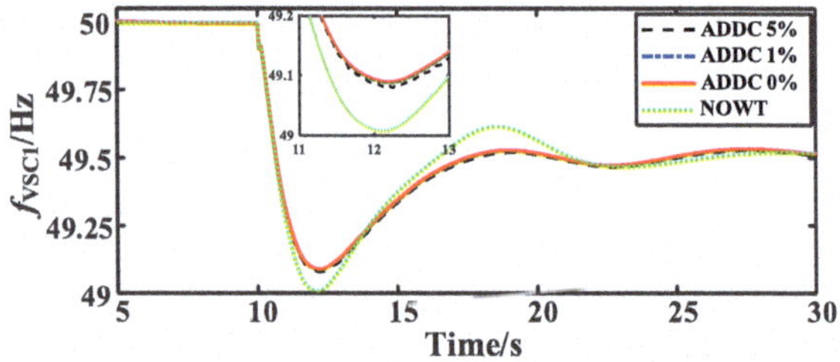

Fig. 10.14 The frequency response with different Gaussian noise

with higher noise levels. Overall, the proposed ADDC scheme maintains good frequency support performance with noise disturbances up to 5%, demonstrating the robustness of the ADDC scheme to Gaussian noise.

10.3.9 Parameter Uncertainties

To study the influence of the parameter uncertainty, positive and negative parameter uncertainties are added on DC line parameters L_{dc31}, L_{dc32}, R_{dc31}, R_{dc32} under this scene. A 300 MW load increase event occurs at $t = 10\ s$, the frequency response with different parameter uncertainties are shown as Fig. 10.15. It could be found in Fig. 10.15a that the frequency nadir with +25 and –25% parameter uncertainty is

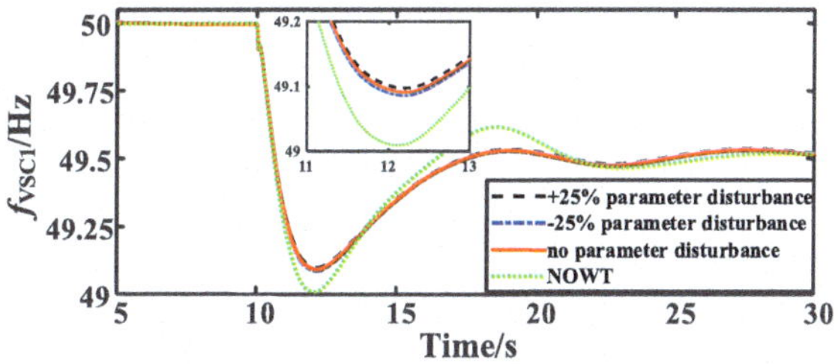

Fig. 10.15 The frequency response with different parameter disturbance

almost the same as that without parameter uncertainty (49.76 Hz), which illustrates the robustness of the proposed ADDC scheme to the parameter uncertainties.

To investigate the impact of parameter uncertainty, both positive and negative uncertainties are introduced to the DC line parameters L_{dc31}, L_{dc32}, R_{dc31}, R_{dc32} in this scenario. At $t = 10$ s, a 300 MW load increase event occurs. The frequency response under different parameter uncertainties is depicted in Fig. 10.15. It can be observed from Fig. 10.15a that the lowest frequency achieved with +25 and –25% parameter uncertainty remains nearly the same as that without any parameter uncertainty (49.76 Hz), demonstrating the resilience of the proposed ADDC scheme against parameter variations.

10.4 Case Study II: 4-Area 5-Terminal System

To demonstrate the effectiveness of the proposed ADDC scheme across various network configurations, we analyze a 5-terminal system as referenced in [10]. This section focuses on onshore station parameters utilizing the proposed ADDC control. The parameters for the MTDC network are sourced from [7, 10], detailed in Table 10.4.

10.4.1 Comparison with Different Schemes

To evaluate the support capabilities of the proposed ADDC scheme in comparison to the existing communication-based CDC scheme, various frequency events in Bus 3 of area 1 within a 4-area 5-terminal system are analyzed at $t = 10s$. Figures 10.17 and 10.19 depict the frequency responses.

In Fig. 10.17, during a 500 MW load increase event, the frequency nadirs are observed to be 49.56 Hz with the ADDC scheme, 49.48 Hz with the CDC scheme, and 49.38 Hz with NOWT. Conversely, during a 500 MW load decrease event, the frequency peaks reach 50.4 Hz with the proposed ADDC scheme, 50.51 Hz with the CDC scheme, and 50.55 Hz with NOWT. Notably, Fig. 10.19 illustrates that across

Table 10.4 Parameters of the 4-area 5-terminal system

Parameters	Description	Value (unit)
$k_{vp0,i}$	Initial value of $U-P$ droop coefficient	10
$k_{vp,i}$	The $U-P$ droop coefficient	0–40
k_{fpi}	The $f-P$ droop coefficient	150
α_1	The frequency compensation gain	10
α_2 for VSC_1	The inertia compensation gain	0.05
α_2 for VSC_2	The inertia compensation gain	0.05
α_2 for VSC_3	The inertia compensation gain	0.15
α_3	The gain for support power limitation	–0.072
L_{dc51}	The inductance of DC line 5–1	240.34 mH
L_{dc52}	The inductance of DC line 5–2	240.34 mH
L_{dc43}	The inductance of DC line 4–3	240.34 mH
L_{dc12}	The inductance of DC line 1–2	120.17 mH
L_{dc23}	The inductance of DC line 2–3	120.17 mH
R_{dc51}	The resistance of DC line 5–1	1.71 Ω
R_{dc52}	The resistance of DC line 5–2	1.71 Ω
R_{dc43}	The resistance of DC line 4–3	1.71 Ω
R_{dc12}	The resistance of DC line 1–2	0.855 Ω
R_{dc23}	The resistance of DC line 2–3	0.855 Ω

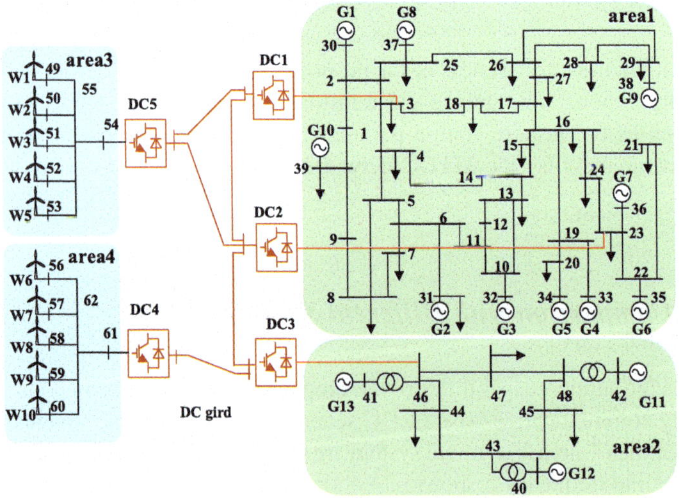

Fig. 10.16 The 4-area 5-terminal test systems

Fig. 10.17 The simulation results with different control schemes in the 5-terminal system: **a** with load increasing 500 MW; **b** with load decreasing 500 MW

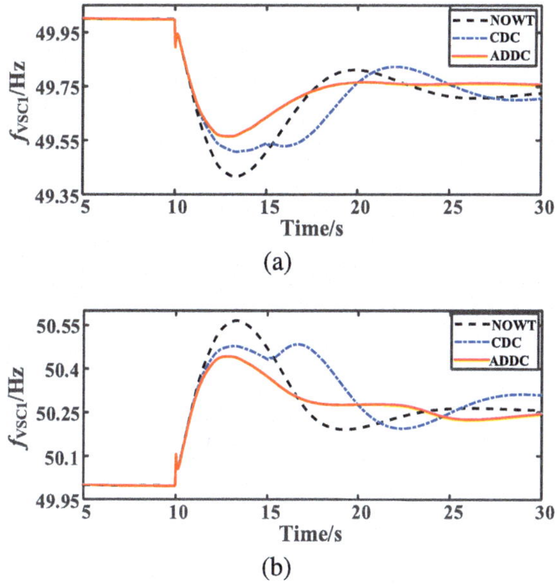

Fig. 10.18 The second frequency drop nadir with different recovery times

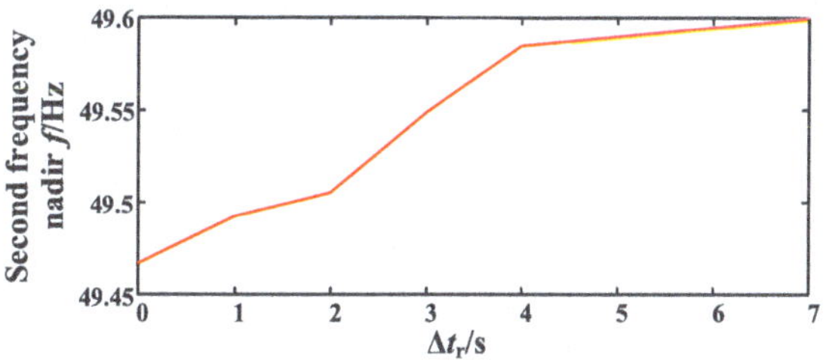

various load change events, the proposed ADDC scheme demonstrates superior frequency support performance compared to both the CDC and NOWT schemes.

Moreover, in Fig. 10.17, it is evident that the CDC scheme experiences a second frequency drop or rise during the OWFs recovery processes, whereas the frequency curve stabilizes quickly and smoothly with the proposed ADDC scheme.

Furthermore, to determine the optimal recovery time for rotor speed and minimize the second frequency drop, different recovery times are investigated in the 4-area 5-terminal system following a 500 MW load increase event at $t = 10$ s. Figure 10.18 illustrates that as recovery time increases, the nadir of the second frequency drop decreases. However, beyond a recovery time of 5 s, the impact on the second frequency nadir diminishes. To facilitate rapid rotor speed recovery and reduce the second frequency drop, a recovery time Δt_r of 7 s is adopted in this study.

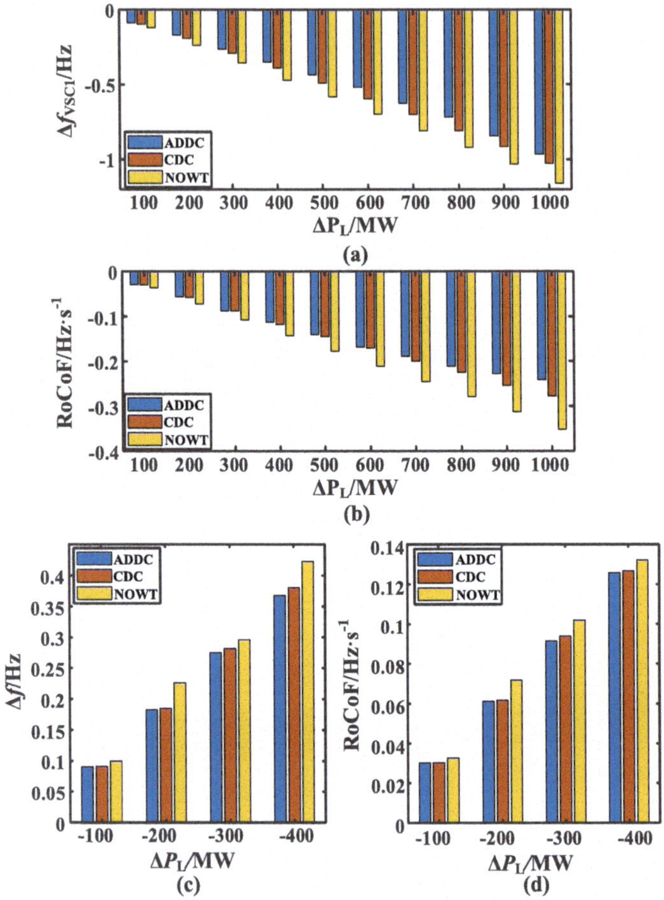

Fig. 10.19 The simulation results with different control schemes in the 5-terminal system: **a** Δf with load increasing; **b** RoCoF with load increasing; **c** Δf with load decreasing; **d** RoCoF with load decreasing

10.4.2 Comparison with Different Inertia

To investigate the impact of system inertia, various frequency events at Bus 3 of area 1 are examined at $t = 10\,s$. The frequency responses are depicted in Fig. 10.20, where Fig. 10.20a, b represent the frequency response curves of VSC_1 and VSC_3, respectively. It is evident from Fig. 10.20a that the proposed ADDC scheme achieves superior frequency support performance compared to the CDC and NOWT schemes across different inertia settings.

In Fig. 10.20b, the frequency of the undisturbed area is effectively controlled within a safe range (above 49.5 Hz) while actively providing frequency support power.

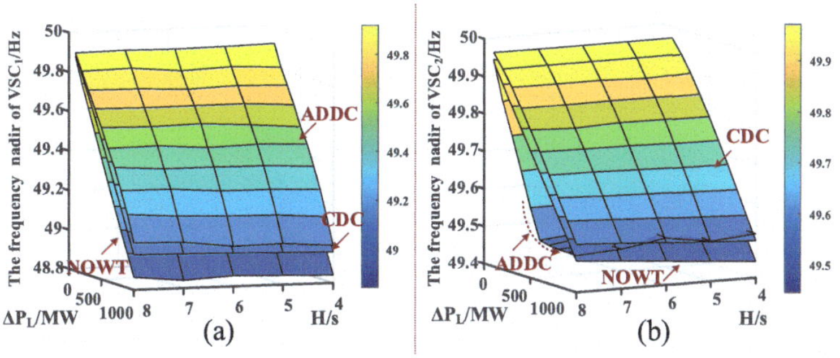

Fig. 10.20 The simulation results with different control schemes in the 4-area 5-terminal system: **a** f_{VSC1}; **b** f_{VSC3}

10.4.3 Opal-RT Real-Time Simulation

To further demonstrate the effectiveness of the proposed ADDC scheme, we conducted studies on the 4-area 5-terminal system illustrated in Fig. 10.16 using the Opal-RT real-time simulation platform. The configuration diagram and physical setup of the experimental platform are depicted in Fig. 10.21. The simulation steps for both onshore systems in Opal-RT are set to 0.0001 s.

The frequency response curves obtained from the Opal-RT platform with different control schemes are presented in Fig. 10.22. It is evident that the proposed ADDC scheme outperforms the other two control schemes in terms of frequency regulation.

Additionally, the frequency output signals from the Opal-RT real-time platform were observed using an oscilloscope. It was observed that under a 600 MW load increase, the ADDC scheme elevated the frequency nadir to 49.45 Hz. Based

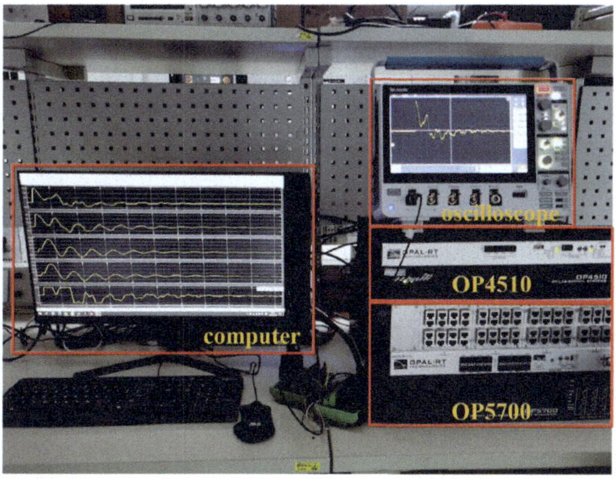

Fig. 10.21 The Opal-RT real time simulation test system

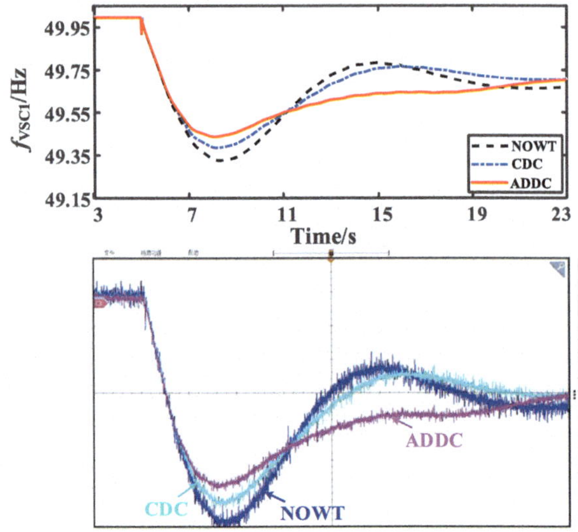

Fig. 10.22 Experimental result on Simulink and Opal-RT with load increase

on these real-time simulation results from Opal-RT, the proposed ADDC scheme demonstrates superior frequency support performance compared to the alternative schemes.

10.5 Summary

This chapter introduces an Adaptive Dual Droop Control (ADDC) scheme for integrating MTDC-VSC offshore wind farms (OWFs) with onshore systems to enhance fast frequency support. The ADDC scheme optimizes the voltage-power droop coefficients of onshore converters in undisturbed systems to increase frequency support power for affected systems, while ensuring operational safety within support power limits through inertia compensation. Unlike other control schemes, ADDC achieves superior performance in both over-frequency and under-frequency scenarios without requiring remote communication between onshore and offshore systems.

Moreover, employing asymptotic recovery control, ADDC mitigates frequency deviations caused by wind turbine (WT) recovery operations. Case studies conducted on two different test systems and the Opal-RT real-time simulation platform validate the effectiveness of ADDC in improving frequency resilience and reducing frequency deviations across various operational conditions. The scheme's robustness against parameter uncertainties (–25 to +25%) and noise disturbances under 5% is demonstrated. Oscilloscope observations from the Opal-RT platform further underscore the efficacy of ADDC in practical applications.

References

1. Akkari S, Dai J, Petit M, Guillaud X (2016) Interaction between the voltage-droop and the frequency droop control for multi-terminal HVDC systems. IET Gener Transm Distrib 10(6):1345–1352. https://doi.org/10.1049/iet-gtd.2015.0814
2. Xiong Y, Yao W, Wen J, Lin S, Ai X, Fang J, Wen J, Cheng S (2021) Autonomous synchronizing and frequency response control of multi-terminal DC systems with wind farm integration. IEEE Trans Power Syst 36(1):781–792. https://doi.org/10.1109/TSTE.2020.2964145
3. Kou P, Liang D, Wu Z, Ze Q, Gao L (2018) Frequency support from a DC-grid offshore wind farm connected through an HVDC link: a communication-free approach. IEEE Trans Energy Convers 33(3):1297–1310. https://doi.org/10.1109/TEC.2018.2814604
4. Xiong Y, Yao W, Wen J, Lin S, Ai X, Fang J, Wen J, Cheng S (2021) Two-level combined control scheme of VSC-MTDC integrated offshore wind farms for onshore system frequency support. IEEE Trans Power Syst 36(1):781–792. https://doi.org/10.1109/TPWRS.2020.2998579
5. Lu Z, Ye Y, Qiao Y (2019) An adaptive frequency regulation method with grid-friendly restoration for VSC-HVDC integrated offshore wind farms. IEEE Trans Power Syst 34(5):3582–3593. https://doi.org/10.1109/TPWRS.2019.2901986
6. Adeuyi OD, Cheah-Mane M, Liang J, Jenkins N (2017) Fast frequency response from offshore multiterminal VSC-HVDC schemes. IEEE Trans Power Deliv 32(6):2442–2452. https://doi.org/10.1109/TPWRD.2016.2632860
7. Xiao L, Xu Z, Xiao H, Zhang Z, Wang G, Xu Y (2019) Electro-mechanical transient modeling of mmc based multi-terminal HVDC system with DC faults considered. Int J Electr Power Energy Syst 113:1002–1013. https://doi.org/10.1016/j.ijepes.2019.06.003
8. Milano F (2005) An open source power system analysis toolbox. IEEE Trans Power Syst 20(3):1199–1206. https://doi.org/10.1109/TPWRS.2005.851911
9. Wu Q, Sun Y (2018) Modeling and modern control of wind power. Wiley, Hoboken, NJ
10. Wang Z, Wu W (2018) Coordinated control method for DFIG-based wind farm to provide primary frequency regulation service. IEEE Trans Power Syst 33(3):2644–2659. https://doi.org/10.1109/TPWRS.2017.2755685

Chapter 11
Distributed Cooperative Control of Offshore Wind Farms Integrated via MTDC System for Fast Frequency Support

Abstract This chapter introduces a Distributed Coordinated Control (DCC) scheme designed for offshore wind farms (OWF) connected through Multi-Terminal Direct Current (MTDC) systems, aimed at providing rapid frequency support. By employing a consensus algorithm, the DCC strategy effectively distributes frequency support power among the offshore wind turbines (WT). The proposed DCC framework ensures the maximum use of kinetic energy from all WTs and maintains safety by utilizing a consensus state index and adaptively modifying droop control coefficients. Following the frequency support, the strategy directs the WTs to begin an asymptotic recovery process, promoting smooth re-storage of rotor speed, with other WTs following suit to prevent secondary frequency drops. Additionally, to facilitate rapid frequency support, a communication-free estimator is employed to estimate the onshore DC voltage using locally measured offshore signals. The performance of the proposed scheme is evaluated through case studies conducted on MATLAB and OPAL-RT real-time simulation platforms. Various control schemes are compared, taking into account parameter uncertainties and noise interference, showcasing the effectiveness of the proposed DCC scheme.

11.1 Structure of The Proposed DCC Scheme

The proposed DCC framework comprises three components:

- The consensus protocol is utilized among all WTs to equitably share the frequency support load, taking into account the wake effect and varying operating conditions. Using the consensus protocol, the WT closest to the offshore station is designated as the leader WT, with other WTs following the leader to distribute the frequency support power fairly.
- The cause of the SFD is analyzed mathematically. To minimize the SFD, a new asymptotic recovery control strategy is proposed for the leader WT to restore its rotor speed smoothly, with other WTs merely needing to follow the leader for smooth recovery.

W. Yao et al., *Active Control of Large-Scale Offshore Wind Farms Connected Via VSC-HVDC*, Power Systems, https://doi.org/10.1007/978-981-97-9346-4_11

- To achieve a rapid frequency response, a communication-free estimator is proposed at the offshore converters to estimate the onshore DC voltage, which serves as the control signal for OWFs to provide frequency support.

The block diagram of the proposed DCC framework is illustrated in Fig. 10.1. With the proposed DCC scheme, the leader WT can provide rapid frequency response by responding to the change of the estimated DC voltage. After that, by exchanging the consensus index x_i, other WTs may provide different frequency support power regarding their wording conditions.

11.1.1 Distributed Consensus Framework

The consensus protocol has been extensively investigated in [1–4]. To decrease communication traffic and improve control reliability, a distributed consensus framework is introduced, based on the existing protocol, to coordinate WTs' participation in frequency regulation. This framework equitably distributes the frequency support burden, allowing WTs with higher rotor speeds and kinetic energy to contribute more power.

Assuming there are 25 WTs connected by 5 strings in an OWF, as illustrated in Fig. 11.1, each string contains 5 WTs. WT_1 is designated as the leader, and the other WTs act as follower agents to share the frequency support burden. As shown in Fig. 11.1, to ensure reliable communication, each WT can communicate with at least two neighboring WTs and exchange the consensus index x_i. The available communication between WTs depends on their geographical location within the practical power system, and the communication network can be redesigned based on different scenarios. The detailed control processes for the leader and follower WTs are shown in Fig. 11.2. To ensure that all WTs equitably share the frequency support regarding their conditions, the consensus index x_i is predefined as

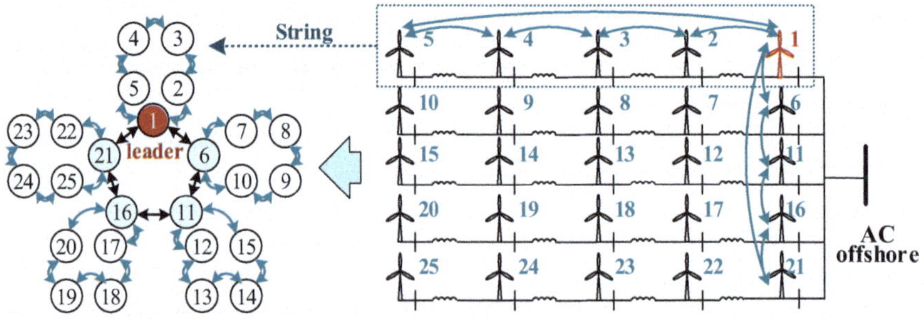

Fig. 11.1 Communication network for OWFs with consensus protocol

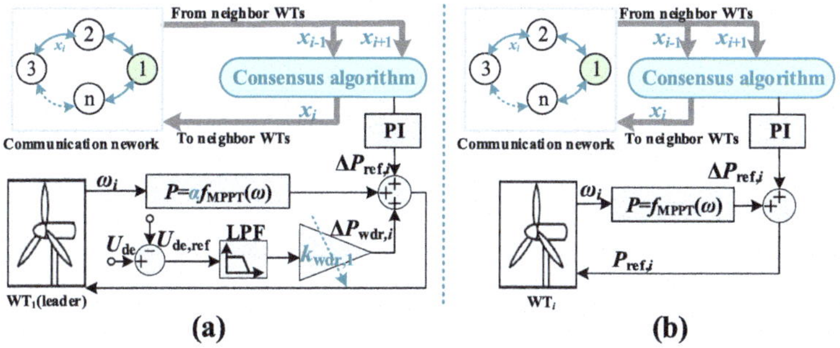

Fig. 11.2 Consensus framework with leaders for OWFs: **a** control diagram for the leader WT, **b** control diagram for follower WTs

$$x_i = \frac{\omega_{r,i}^2 - \omega_{r,\min}^2}{\omega_{r,0}^2 - \omega_{r,\min}^2} \tag{11.1}$$

where, $\omega_{r,i}$ denotes the rotor speed of WTi, ωr, 0 denotes the initial value of the rotor speed when the frequency event starts, and $\omega_{r,\min}$ denotes the minimum limit of rotor speed. x_i is employed as the variable consensus index, thus, the power reference deviation $\Delta P_{\text{ref},i}$ can be written as:

$$\Delta P_{\text{ref},i} = -k_{P,i} \sum_{j=1}^{N} a_{ij}(x_i - x_j) - k_{I,i} \int \sum_{j=1}^{N} a_{ij}(x_i - x_j) \tag{11.2}$$

where, $k_{P,i}$ and $k_{I,i}$ denote the proportional and integral coefficients for the consensus power controller in Fig. 11.2, respectively, while a_{ij} denotes the communication weight for information exchange between WT$_j$ and WT$_i$. $a_{ij} = 1$ once WT$_j$ is directly connected to WT$_i$; otherwise, $a_{ij} = 0$. Using this consensus index, more frequency support power can be provided by releasing the kinetic energy from that WT with higher the rotor speed, ensuring the frequency support burden is equitably distributed among WTs.

11.1.1.1 Control for the Leader WT

The leader WT is regulated as the reference during both the frequency support and rotor recovery processes. To achieve these objectives, the estimated DC voltage depicted in Fig. 11.2 are transmitted to the leader WT as the control signal. Subsequently, adaptive droop controller is integrated into the MPPT controller, enabling the leader WT to provide frequency support power by releasing the KE while maintaining the rotor speed within a safe range, as referenced in [5]. $\Delta P_{\text{wdr},i}$ and $k_{\text{wdr},i}$ can be represented as follows:

$$\Delta P_{\mathrm{wdr},i} = -k_{\mathrm{wdr},i} \cdot (U_{\mathrm{deref},j} - U_{\mathrm{de},j}) \tag{11.3}$$

$$k_{\mathrm{wdr},i} = k_{\mathrm{wdr0},i} + k_0 \cdot \frac{\omega_{\mathrm{r},i}^2 - \omega_{\mathrm{r,min}}^2}{\omega_{\mathrm{r,max}}^2 - \omega_{\mathrm{r,min}}^2} \tag{11.4}$$

where, k_0 defines the slope of the droop control coefficient. $k_{\mathrm{wdr},i}$ and $k_{\mathrm{wdr0},i}$ denote the adaptive droop control coefficients and their initial values for WT_i, respectively. $U_{\mathrm{de},j}$ and $U_{\mathrm{deref},j}$ denote the estimated onshore DC voltage and its reference for VSC_j. The value of $kwdr,i$ can be adjusted adaptively during the frequency support process to keep the rotor speed within the safe range, i.e., 0.7–1.2 pu. The parameter design and stability analysis are detailed in Sect. III. Subsequently, $\Delta P_{\mathrm{ref},i}$ and $\Delta P_{\mathrm{wdr},i}$ are incorporated into the MPPT control, and the modified power reference for the leader WT can be expressed as follows:

$$P_{\mathrm{ref},i} = \alpha f_{MPPT}(\omega) + \Delta P_{\mathrm{ref},i} + \Delta P_{\mathrm{wdr},i} \tag{11.5}$$

where, α is time-varying and can help the leader WT recover smoothly after the frequency support, which is investigated in following sections detailedly.

11.1.1.2 Control for Follower WTs

The follower WTs in different strings are regulated by exchanging the information, i.e., the index x_i, with neighboring WTs according to (11.2). The output power reference for the follower WT_j, i.e., $P_{\mathrm{ref},j}$ can be expressed as:

$$P_{\mathrm{ref},j} = f_{MPPT}(\omega) + \Delta P_{\mathrm{ref},j} \tag{11.6}$$

11.1.2 Smooth Recovery Control for WTs

After providing frequency support, WTs must restore their kinetic energy. However, improper recovery strategies may lead to the SFD [6], as illustrated in Fig. 11.3a. f_{MPPT} (ω) denotes the MPPT curve for normal operation. When frequency events begin, the WT releases its kinetic energy, causing its operating point to shift from the MPPT point to P_{op1}. After frequency support, the WT must return to the MPPT status, changing its operating point from P_{op1} to P_{op2}. This abrupt change creates a significant active power deficit (see ΔP in Fig. 11.3a), potentially causing the SFD. Inadequate recovery schemes could result in an SFD greater than the initial frequency drop, leading to severe consequences. To elucidate the cause of the SFD and simplify the analysis, the swing equation for the onshore system and the OWF can be expressed as:

$$\frac{\mathrm{d}f_{\mathrm{on}}}{\mathrm{d}t} = \frac{1}{H_{\mathrm{g}}}(P_{\mathrm{G}} + P_{\mathrm{VSC}} - P_{\mathrm{L}}) \approx \frac{1}{H_{\mathrm{g}}}(P_{\mathrm{G}} + P_{\mathrm{we}} - P_{L}) \tag{11.7}$$

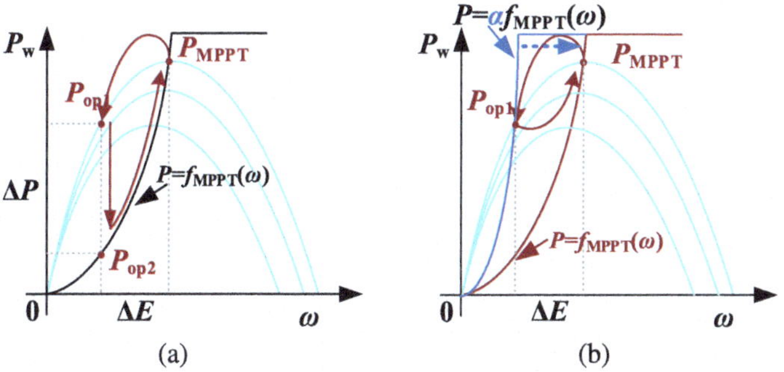

Fig. 11.3 The output power curve for OWFs: **a** conventional control scheme; **b** proposed asymptotic recovery control strategy

$$\omega_r \frac{\mathrm{d}\omega_r}{\mathrm{d}t} = \frac{1}{H_\mathrm{w}}(P_\mathrm{wm} - P_\mathrm{we}) \tag{11.8}$$

where, P_G and P_L denote the power supplied by synchronous generators and the total load, respectively. P_wm and P_we represent the mechanical and electrical power of OWFs, respectively. Ignoring the power loss in DC cables, the integrated power via VSC converters P_VSC is approximately equal to the output power of OWFs P_we. Assuming t_b is the recovery start time, during the recovery process, (11.8) can be rewritten as follows:

$$\frac{\mathrm{d}f_\mathrm{on}(t)}{\mathrm{d}t} - \left.\frac{\mathrm{d}f_\mathrm{on}(t)}{\mathrm{d}t}\right|_{t=t_b} = \frac{1}{H_\mathrm{g}}\left(-\frac{1}{R}\Delta f_\mathrm{on}(t) + \Delta P_\mathrm{we}(t)\right) \tag{11.9}$$

where, $-\frac{1}{R}\Delta f(t)$ represents the primary frequency support for generators, and ΔP_we denotes the output power deviation of OWFs during the recovery process. When the frequency drops to its minimum value, the derivative of the onshore frequency $\frac{\mathrm{d}f_\mathrm{on}(t)}{\mathrm{d}t} = 0$. Therefore, (11.9) can be expressed as follows:

$$\Delta f_\mathrm{on}(t) = R \cdot [H_\mathrm{g} \cdot \left.\frac{df_\mathrm{on}(t)}{dt}\right|_{t=t_b} + \Delta P_\mathrm{we}(t)] \tag{11.10}$$

To minimize the SFD, $\left.\frac{\mathrm{d}f_\mathrm{on}(t)}{\mathrm{d}t}\right|_{t=t_b}$ should be kept low, which can be accomplished by extending the recovery duration. Thus, a new smooth recovery control scheme for WTs is introduced. As the WT commences rotor speed restoration, a modified P–ω curve is designed as shown in (11.11). The time-varying coefficient α changes over time as indicated in (11.12). During the recovery interval $(t_b - t_e)$, α transitions gradually from its initial value to 1, preventing large active power disturbances and thereby reducing the SFD through the proposed control strategy.

$$f^*_\mathrm{MPPT}(\omega) = \alpha f_\mathrm{MPPT}(\omega) \tag{11.11}$$

$$\alpha = \begin{cases} 1 & t < t_b \\ \frac{1-\alpha_0}{\Delta t_r} \cdot (t - t_b) + \alpha_0 \ (\alpha_0 = \frac{P_{\mathrm{we}}|_{t=t_b}}{\omega_{\mathrm{r}}|_{t=t_b}}) & t_b \leq t \leq t_e \\ 1 & t_e < t \end{cases} \tag{11.12}$$

where, α_0 denotes the initial value at the onset of rotor speed recovery, determined by the real-time operating point of various wind turbines at the start of the speed recovery. $P_{\mathrm{we}}|t = t_b$ and $\omega\mathrm{r}|_{t=t_b}$ represent the measured output power and rotor speed, respectively. t_s indicates the beginning time of frequency events, while t_b and t_e mark the start and end times of rotor speed re-storage, with $\Delta t_r = t_e - t_b$ being the duration of re-storage ($\Delta t_r = 7s$ in this study). Using the smooth recovery control, the leader WT can restore its rotor speed gradually, and the consensus index x_i of the leader WT will also recover steadily. Consequently, other WTs can restore their rotor speeds by following the consensus index x_i of leader WT, enabling a smooth recovery for the entire OWF.

11.1.3 DC Voltage Estimator

In this section, to estimate the onshore average DC voltage for frequency support using offshore local signals, a communication-free DC voltage estimator is proposed, whose control structure can be depicted in Fig. 11.4.

With the frequency-voltage-power droop control employed for all onshore converters as referred to [7], the power reference can be write as:

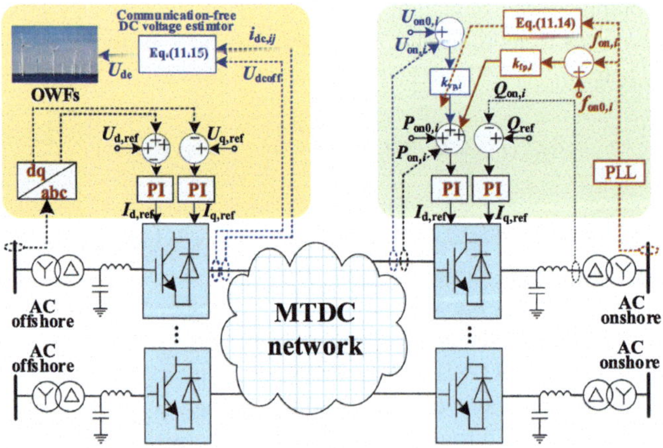

Fig. 11.4 Communication-free estimator for MTDC system

$$P^*_{\mathrm{on},i} = P_{\mathrm{on0},i} - k_{\mathrm{vp},i} \cdot (U_{\mathrm{on0},i} - U_{\mathrm{on},i}) + k_{\mathrm{fp},i} \cdot (f_{\mathrm{on0},i} - f_{\mathrm{on},i}) \tag{11.13}$$

$$k_{\mathrm{vp},i} = k_{\mathrm{vp},0} + \int_{t_1}^{t_1+\Delta t_s} k_1 \cdot (f_{\mathrm{on0},i}(t) - f_{\mathrm{on},i}(t))\,dt \tag{11.14}$$

where $k_{\mathrm{fp},i}$ and $k_{\mathrm{vp},i}$ denote the frequency and voltage droop parameters of VSC$_i$, respectively. $U_{\mathrm{on},i}$ and $U_{\mathrm{on0},i}$ denote the measured DC voltage its reference, respectively. t_1 represents the time point when the onshore frequency deviation surpasses the dead-band threshold. Δt_s represents the integration time (0.01 s in this study). Once the onshore converter is closer to the disturbance bus, a higher $k_{\mathrm{vp},i}$ will result, allowing more frequency support power to be allocated and transmitted to onshore systems through this converter. This approach can reduce power flow changes, result in better support performance.

Referred to Fig. 10.1, assuming $j-1$ onshore converters (VSC$_1$ to VSC$_{j-1}$) and $m-j+1$ offshore converters (VSC$_j$ to VSC$_m$) are included in this system, the OWFs integrated by VSC$_j$ participate in frequency support. The DC line currents from VSC$_j$ to other onshore converters can be measured at VSC$_j$. Using Kirchhoff's law and knowledge of DC line parameters, the average onshore DC voltage can be estimated. The detailed calculation process is shown in (11.15).

$$U_{\mathrm{de},j} = \sum_{i=1}^{j-1}\left(\beta_i \cdot \left(U_{\mathrm{d},j} - I_{\mathrm{dc},ij} \cdot R_{\mathrm{dc},ij} - \frac{\mathrm{d}I_{\mathrm{dc},ij}}{\mathrm{d}t} \cdot L_{\mathrm{dc},ij}\right)\right) \tag{11.15}$$

$$\beta_i = \frac{I_{\mathrm{dc},ij}}{\sum_{i=1}^{j-1} I_{\mathrm{dc},ij}} \quad \left(\sum_{i=1}^{j-1}\beta_i = 1\right) \tag{11.16}$$

where $L_{\mathrm{dc},ij}$ and $R_{\mathrm{dc},ij}$ denote the reactance and resistance for DC Line $i-j$, respectively. With specific practical projects, these parameters can be given directly. Consequently, the onshore DC voltage can be estimated using locally measured electrical quantities in the offshore converters. The estimated onshore DC voltage $U_{\mathrm{de},j}$ is computed at VSC$_j$ and serves as the input control signal for OWFs. This voltage is an average value, calculated with the weighted factor β_i determined by the proportion of DC line currents in (11.16).

To enable communication-free processes between onshore and offshore systems, the frequency deviation in onshore systems will be reflected in the DC voltage change once the frequency surpasses the dead-band threshold. Under the proposed DCC control scheme, if frequency events occur in the onshore systems, the DC voltage will fluctuate. The leader WT will release its kinetic energy to provide frequency support power. Other WTs will support the frequency by communicating with neighboring WTs and exchanging the consensus index x_i. The higher the rotor speed, the more frequency support power will be provided by releasing kinetic energy from that WT.

11.2 Test System and Control Parameter Determination

11.2.1 Test System and Parameters

To demonstrate the effectiveness of the proposed DCC scheme, a general 5-terminal test system, as described in [8], is analyzed. Parameters of this test system are presented in Table 11.1, with references to [5, 8].

11.2.2 Small-Signal Stability Analysis

To investigate the impact of onshore droop coefficients on system stability, small signal analysis is conducted with the P-U droop coefficients (k_{vp1}, k_{vp2}, k_{vp3}) varying from 10 to 30. The powerful Model Linearizer embedded in MATLAB, as referenced in [9], is utilized to perform the small-signal analysis of the test system, eliminating the need to derive the state-space model manually. The entire test system comprises 212 eigenvalues, with Table 11.2 displaying only the dominant eigenvalues. The dominant eigenvalues and their states are listed in Table 11.2. As shown in Fig. 11.6, the damping ratios of most eigenvalues remain above 0.05 when the onshore droop coefficients vary within the specified ranges in Table 11.1.

Table 11.1 Parameters of the 5-terminal test system

Parameters	Description	Value
$k_{vp0,i}$	The initial value of U–P droop	10
$k_{vp,i}$	The U–P droop coefficient	10–30
$k_{fp,i}$	The f–P droop coefficient	-100
$f_{Vi,\text{ref}}$	The reference value of VSC$_i$	50 Hz
L_{dc51}, L_{dc52}, L_{dc43}	The DC line inductance 5–1, 5–2, 4–3	240.34 mH
L_{dc12}, L_{dc23}	The DC line inductance 1–2, 2–3	120.17 mH
R_{dc51}, R_{dc52}, R_{dc43}	The DC line resistance 5–1, 5–2, 4–3,	1.71 Ω
R_{dc12}, R_{dc23}	The DC line resistance 1–2, 2–3	0.855 Ω
β_i	The estimated weight of VSC$_i$	0–1
$\omega_{\text{r},i}$	The rotor speed of WT$_i$	0.7–1.2 pu
$\omega_{\text{r,max}}$	The maximum rotor speed limitation	1.2 pu
$\omega_{\text{r,min}}$	The minimum rotor speed limitation	0.7 pu
$k_{\text{wdr},i,0}$	The initial value of droop coefficient	–10
$k_{\text{wdr},i}$	The value of droop coefficient	–20 to 0
k_0	The slope of the inertial coefficient	–10

Table 11.2 Parts of dominant eigenvalues and their state variables

λ	Eigenvalue	Dominant state
$\lambda_{20}, \lambda_{21}$	$-43.08 \pm 56.56\,i$	PLL control of VSC_3
$\lambda_{22}, \lambda_{23}$	$-65.81 \pm 6.61\,i$	Outloop control of VSC_3
$\lambda_{28}, \lambda_{29}$	$-8.81 \pm 34.09\,i$	ΔU_{dc2}
$\lambda_{46}, \lambda_{47}$	$-19.28 \pm 27.50\,i$	ΔU_{dc1}
$\lambda_{48}, \lambda_{49}$	$-17.49 \pm 26.29\,i$	ΔU_{dc3}

11.3 Simulation Verification

To examine the effectiveness of the proposed DCC scheme, case studies are performed on the MATLAB platform. The test setup is shown in Fig. 11.5, with parameters listed in Table 11.1. The system includes two OWFs, each with 5 × 5 WTs (five strings). The communication topology of each OWF is illustrated in Fig. 11.1; it is designed as a ring shape in every WT string to ensure connectivity in case of a single-link failure. Different scenarios compare the centralized control scheme (CC) and the proposed DCC scheme. The CC scheme, as referenced in [6], employs power-voltage-frequency droop control at the onshore VSC stations and adaptive inertia emulation control at the offshore WTs.

The compared control strategies and their abbreviations are shown in Table 11.3.

With the centralized control, all WTs are required to communicate with the central controller, and the frequency support signal should be transmitted to every WT. With the DCC scheme, the leader WT exchanges information with the offshore VSC station and provide frequency support, other WTs only communicate with neighbor WTs with the consensus algorithm to share the frequency support burden fairly.

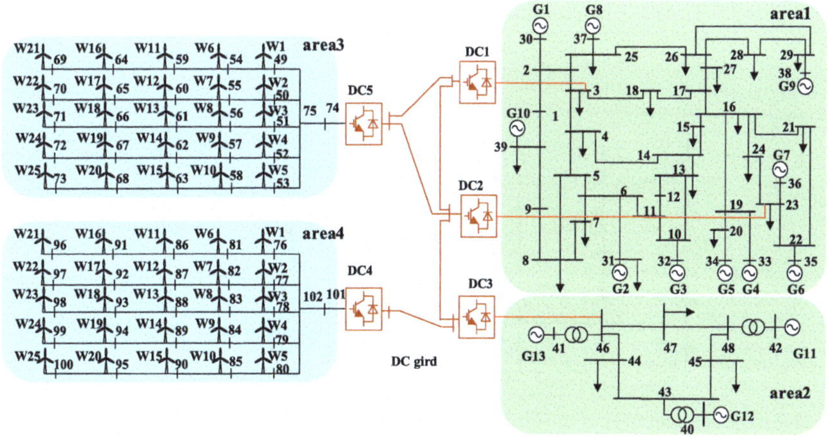

Fig. 11.5 The 4-area 5-terminal test systems

Table 11.3 The compared control strategies and their abbreviations

Abbr.	Description
DCC	The proposed distributed cooperative control
CC	The existing centralized control scheme referred to [6]
NOWT	Without offshore WTs participation

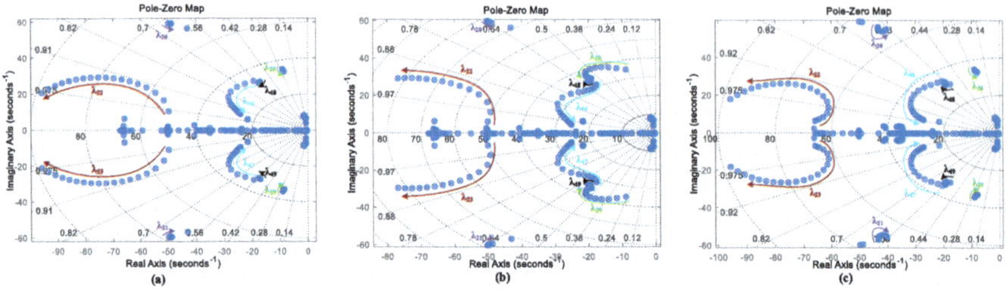

Fig. 11.6 The small signal analysis results with: **a** $k_{vp,1}$ changing from 10 to 30, **b** $k_{vp,2}$ changing from 10 to 30, **c** $k_{vp,3}$ changing from 10 to 30

11.3.1 Comparison Considering the Wake Effect

In this scenario, the wake effect is taken into account, and the wind speeds of WTs are calculated utilizing the wake model referenced in [10]. Figure 11.8 displays the wind speeds under four wind directions (0, 30, 60, and 90). As shown in Fig. 11.9, the frequency deviation is reduced under different wind directions when compared to the CC scheme. Moreover, the rotor speeds of different WTs are shown in Fig. 11.10. With the CC scheme, rotor speeds decrease uniformly, which is unfair in distributing the frequency support burden and increases the risk for WTs with low rotor speeds. However, with the proposed DCC scheme, WTs with higher rotor speeds release more KE, while those with lower rotor speeds release less KE, ensuring safer operation.

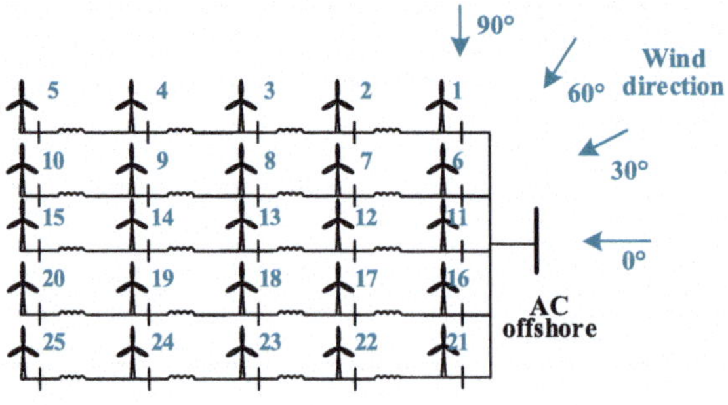

Fig. 11.7 The configuration of the offshore wind farm with different wind directions

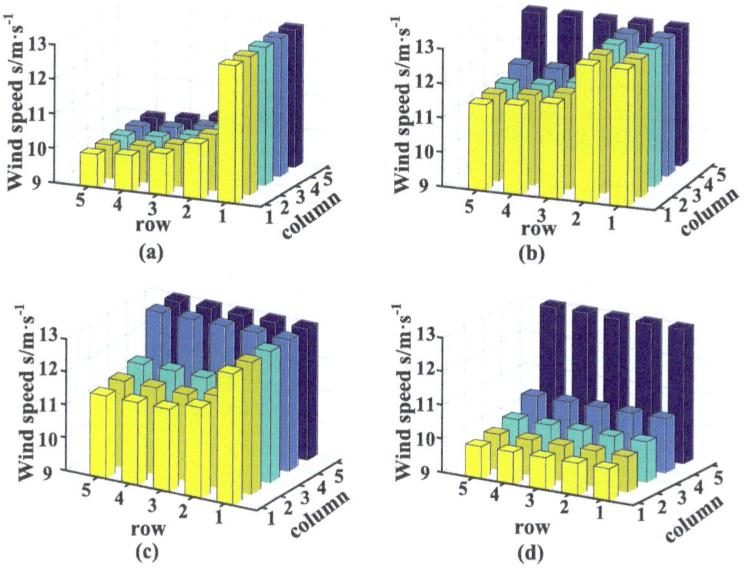

Fig. 11.8 The different wind speeds under different wind directions: **a** 0°, **b** 30°, **c** 60°, **d** 90°

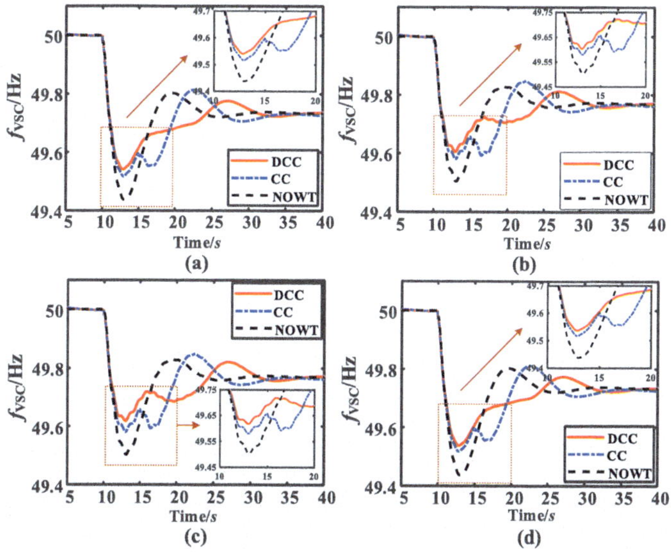

Fig. 11.9 The frequency response results under different wind directions: **a** 0°, **b** 30°, **c** 60°, **d** 90°

The four different wind directions, 0, 30, 60, and 90, are illustrated in Fig. 11.7, with both wind directions and the wake effect model referenced from [10].

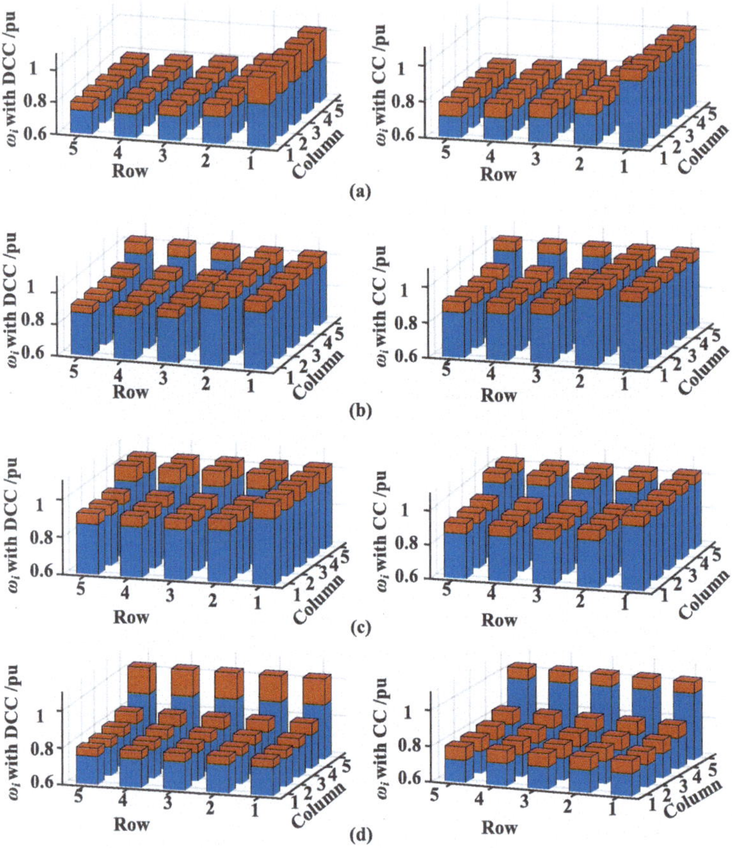

Fig. 11.10 The rotor speed of different WTs under different wind directions: **a** 0°, **b** 30°, **c** 60°, **d** 90°

11.3.2 Load Increase and Decrease

To highlight the effectiveness of the proposed DCC scheme during load increase and decrease events, a 500 MW load increase occurs in Bus 3 of Area 1 at $t = 10\ s$, followed by a 500MW load decrease in the same bus at $t = 50\ s$. The comparison results between the DCC and CC schemes are illustrated in Fig. 11.11.

In Fig. 11.11a, the proposed DCC scheme demonstrates superior frequency support performance compared to the CC scheme during both the 500 MW load increase and decrease events. Additionally, the SFD is also reduced with the proposed DCC scheme. Figure 11.11b, c present the rotor speed ω_i and the index x_i of different WTs under the DCC scheme, while Fig. 11.11d, e show these two indexes under the CC scheme. The proposed DCC scheme achieves consensus of x_i among different WTs,

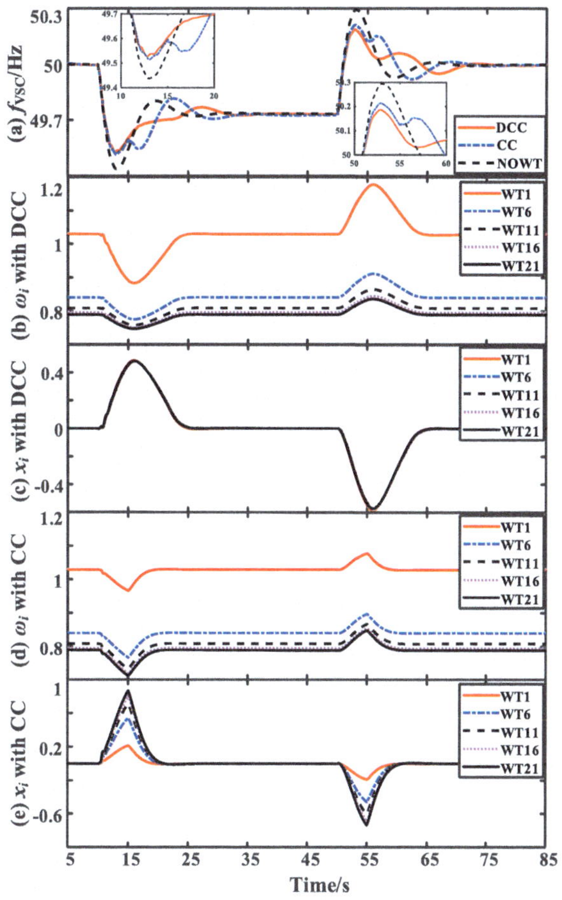

Fig. 11.11 The comparison results between the CC and DCC schemes

ensuring that WTs with higher rotor speeds release more KE, while those with lower rotor speeds release less KE, thereby enhancing operational security.

11.3.3 Communication Failure

To assess the frequency support performance during communication failures, scenarios including N-1 communication failure between WT_3 and WT_4 and N-2 communication failure among WT_3, WT_4, and WT_5 are compared against a normal communication scenario. A 500 MW load increase event occurs in Bus 3 of Area 1 at $t = 10\ s$. The frequency response curves for various communication scenarios are shown in Fig. 11.12, and the x_i index for different conditions is displayed in Fig. 11.13.

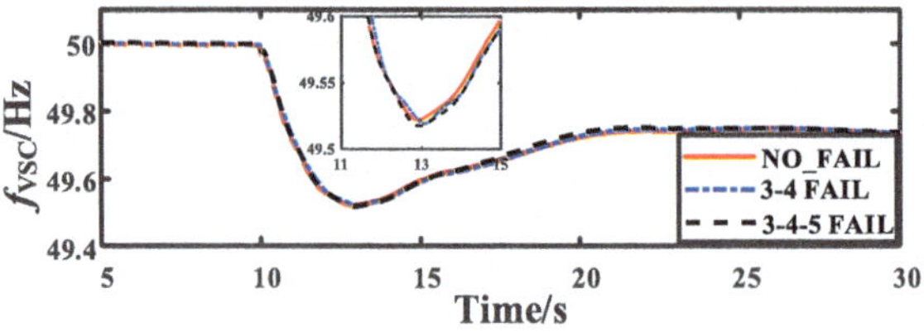

Fig. 11.12 The frequency response under different communication failure situations

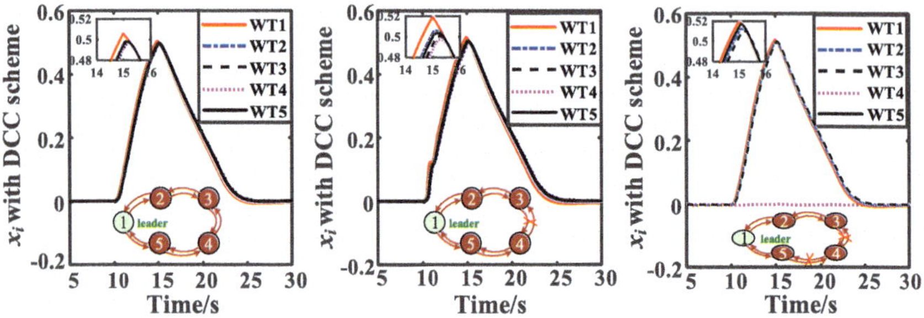

Fig. 11.13 The x_i index under different communication failure situations

It is observed that, with an N-1 communication failure, WTs can still receive consensus information from other WTs, enabling the x_i index to reach consensus. However, in the event of an N-2 communication failure, WT_4 cannot receive the consensus information and therefore does not participate in the frequency support.

11.3.4 Gaussian Noise

In real-world power systems, noise in control signals can impact frequency support performance. To investigate the influence of Gaussian noise, varying levels of Gaussian noise are introduced to the offshore DC voltage and current measurements in this scenario: (i) with 1% Gaussian noise and (ii) with 5% Gaussian noise. A 500 MW load increase occurs in Bus 3 at $t = 10\ s$. The Gaussian noise is added to the estimated onshore DC voltage using the Uniform Random Number block in the Simulink platform. In (11.17), the expression $-a + 2a \cdot rand(1)$ generates random values within the range [–a, a].

$$U^*_{\text{de},j} = U_{\text{de},j} \cdot [1 + (-a + 2a \cdot rand(1))] \tag{11.17}$$

where, $U^*_{\text{de},j}$ indicates the revised estimated onshore DC voltage with Gaussian noise. The constant a in $-a + 2a \cdot rand(1)$ generates random values within $[-a, a]$, simu-

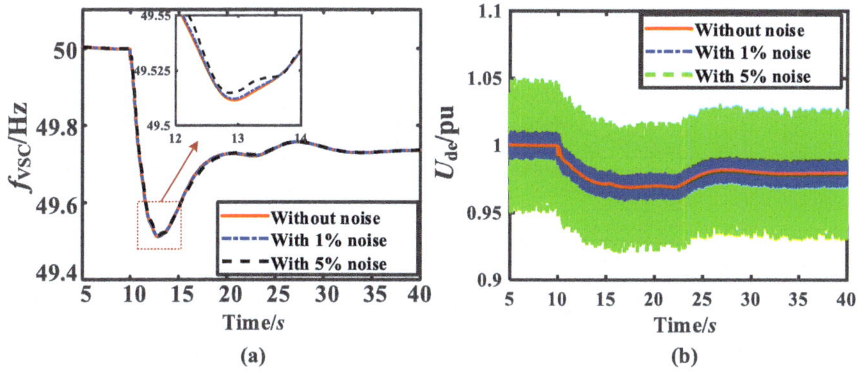

Fig. 11.14 The simulation results with different noise: **a** f_{VSC} **b** U_{de}

lating Gaussian noise. In this study, a is set to 0.01 and 0.05 for 1% and 5% Gaussian noise, respectively.

As illustrated in Fig. 11.14, the frequency response curves with 1% and 5% noise are nearly identical to the curve without noise. However, the frequency deviation increases with higher noise levels. Overall, the proposed DCC scheme maintains effective frequency support performance with noise disturbances below 5%, demonstrating the robustness of the proposed DCC scheme against Gaussian noise.

11.3.5 Parameter Uncertainties

To assess the impact of parameter uncertainty, positive and negative uncertainties are applied to the DC line parameters L_{dc51}, L_{dc52}, R_{dc51}, R_{dc52} in this scenario. A 500 MW load increase event occurs at $t = 10s$. The frequency responses with varying parameter uncertainties are depicted in Fig. 11.15. In Fig. 11.15a, the frequency nadir with +30% and –30% parameter uncertainty is almost identical to that without parameter uncertainty (49.525 Hz), illustrating the robustness of the proposed DCC scheme against parameter uncertainties.

11.3.6 Communication Time Delays Among WTs

To evaluate the impact of communication time delays among different WTs, various delays (0, 50, and 80 ms) are considered and tested in multiple case studies. A 500 MW load increase event occurs at $t = 10\ s$. The frequency responses under different time delays are presented in Fig. 11.16, and the consensus index x_i for various time delays is shown in Fig. 11.17. It is noted that increasing the communication time delay between neighboring WTs slightly affects the consensus convergence. With

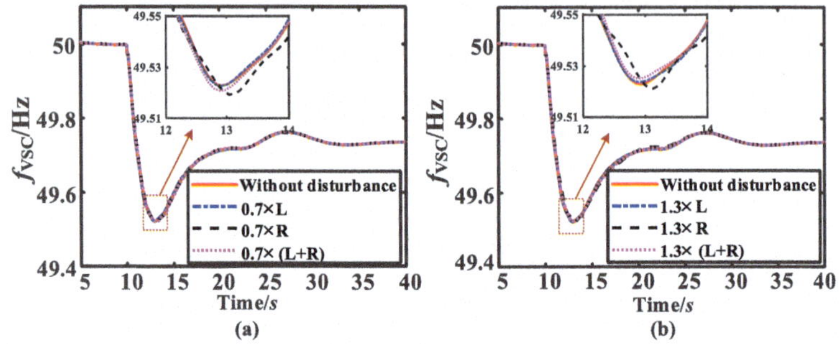

Fig. 11.15 The simulation results with different parameter uncertainties: **a** with parameters decreasing **b** with parameters increasing

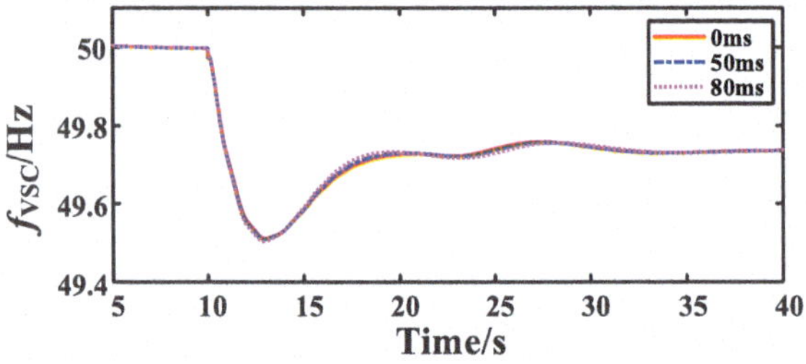

Fig. 11.16 The frequency response with different communication time delays among WTs

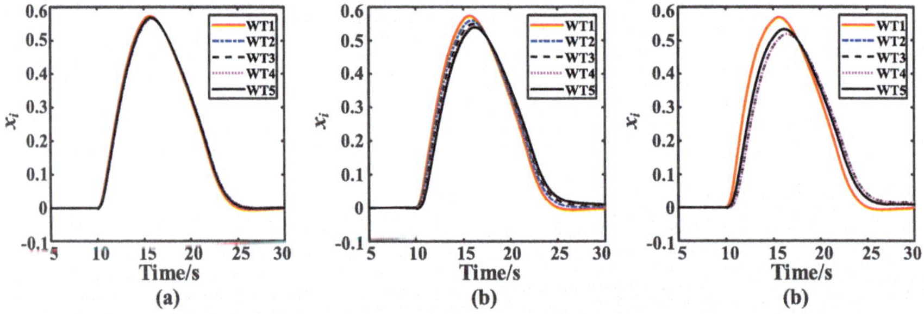

Fig. 11.17 The consensus index x_i with different communication time delays among WTs: **a** with 0 ms time delay, **b** with 50 ms time delay, **c** with 80 ms time delay

an 80 ms delay, the overall consensus convergence might be impacted. However, the frequency support performance with an 80 ms time delay is only slightly affected because the WTs still follow the leader WT to release their kinetic energy and provide frequency support under this condition.

11.4 Experimental Verification

To further validate the effectiveness of the proposed DCC scheme, the 4-area 5-terminal system depicted in Fig. 11.5 is modeled on the OPAL-RT real-time simulation platform. The configuration diagram and physical picture of the experimental setup are shown in Fig. 11.18, with simulation steps of 0.0001 s. In this scenario, a 500 MW load increase event occurs in Bus 3 of Area 1 at $t = 5$ s. The main steps for OPAL-RT real-time simulation are detailed in Fig. 11.19.

First, a system model is created on the Simulink platform. Since real-time simulation supports only a fixed-step dynamic solver, the model must undergo offline simulation with a fixed-step dynamic solver. Next, the AC grid, divided into different areas, is segmented into various parts and placed into three distinct submodules. The OPAL-RT simulator software on the upper computer automatically translates the model into code, assigns, and downloads each submodule to a CPU core in the lower computer. Finally, the model runs on the lower computer and interacts with actual controllers.

The OPAL-RT real-time simulation platform comprises two simulators: OP5700 and OP4510, as shown in Fig. 11.19. To simulate signal transmission under real conditions, the entire 4-area 5-terminal system is modeled in the OP5700 simulator, while the adaptive droop control of the leader WT is implemented in the OP4510 simulator. Consequently, the frequency support signal (the difference in estimated

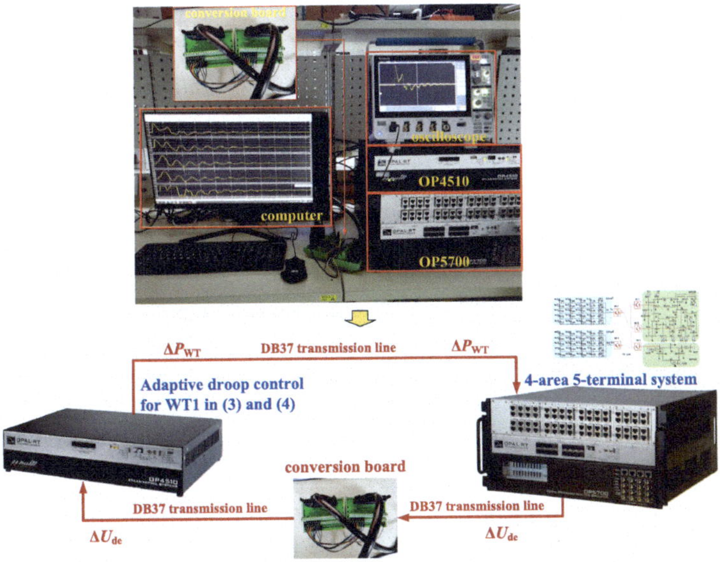

Fig. 11.18 The OPAL-RT real time simulation test system

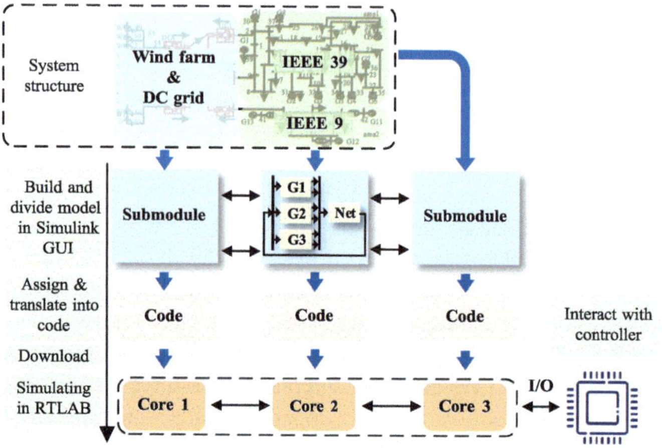

Fig. 11.19 The main steps for OPAL-RT real-time simulation

DC voltage ΔU_{de}) is transmitted from the OP5700 simulator through DB37 lines and a conversion board to the OP4510 simulator. The adaptive droop control, as detailed in (11.3) and (11.4), is executed in the OP4510 simulator. If the input signal ΔU_{de} is not zero, the adaptive droop control outputs the frequency support power increment, which is then transmitted back to WT1 in the OP5700 simulator. This setup separates the adaptive droop controller from the leader WT and simulates the signal transmission process. The signal transmission process and the real-time simulator are depicted in Fig. 11.18. As shown in Fig. 11.21, the transmission signal ΔU_{de} remains nearly identical through the transmission line and conversion board. The frequency support performance with the proposed DCC scheme exceeds that of the CC control scheme, validating the effectiveness of the DCC scheme with signals exchanged through real transmission lines.

The frequency curves for different schemes on the OPAL-RT platform are illustrated in Figs. 11.20, 11.21 and 11.22. The proposed DCC scheme demonstrates superior performance compared to the CC scheme. Frequency output signals from the OPAL-RT real-time platform under different schemes are observed using an oscilloscope. The oscilloscope shows that the DCC scheme raises the frequency nadir to 49.45 Hz under 600 MW load increase conditions. Based on the real-time OPAL-RT simulation results, the proposed DCC achieves better frequency support performance compared to centralized control schemes.

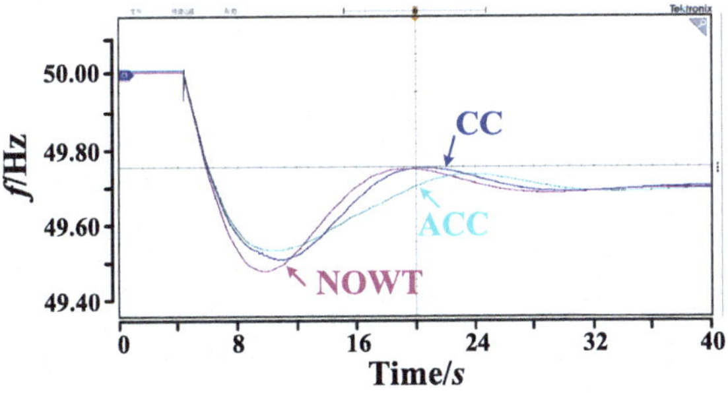

Fig. 11.20 The frequency response curves with different controls schemes in OPAL-RT

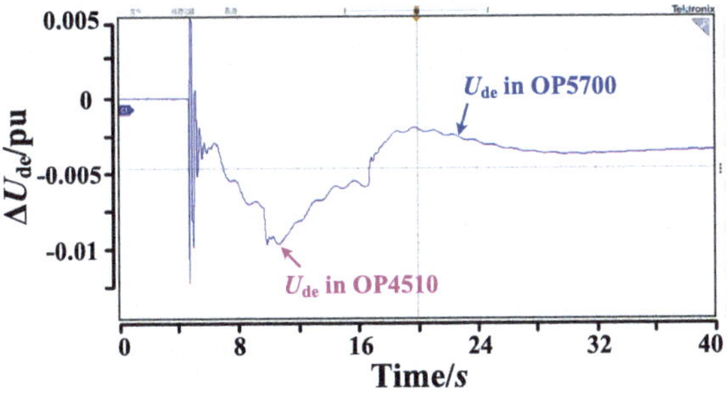

Fig. 11.21 The estimated DC voltage ΔU_{de} measured at different simulators

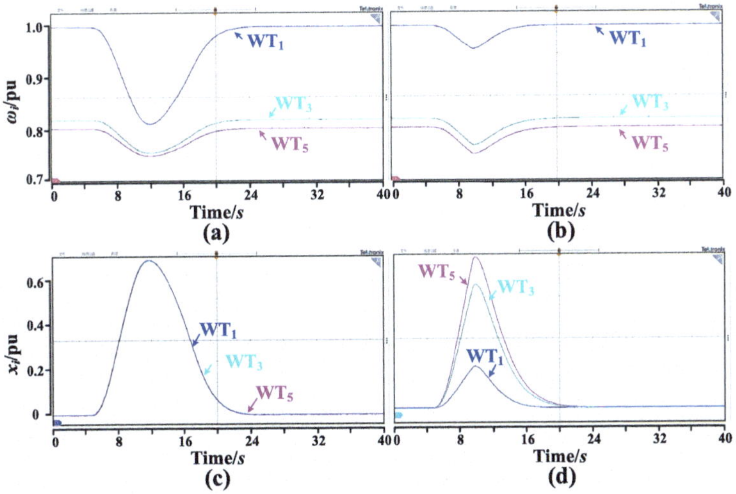

Fig. 11.22 The experimental results with different controls schemes in OPAL-RT: **a** x_i with DCC, **b** x_i with CC, **c** ω_i with DCC, **d** ω_i with CC

11.5 Summary

This chapter introduces a distributed cooperative control (DCC) scheme for MTDC-integrated OWFs, aimed at delivering rapid frequency support to onshore asynchronous systems. The DCC scheme leverages the consensus algorithm to efficiently distribute the frequency support burden among offshore WTs. In contrast to the centralized control scheme, the proposed DCC approach can effectively utilize the kinetic energy (KE) of all WTs by employing the consensus state and adaptively adjusting the droop control coefficients, ensuring safe operation. The DCC scheme exhibits superior performance in both load increase and decrease scenarios without necessitating remote communication between onshore and offshore systems. Additionally, due to the asymptotic recovery control, the proposed DCC scheme can alleviate the SFD caused by the recovery of WTs. Case studies confirm the effectiveness of the proposed DCC scheme in raising the frequency nadir and reducing the SFD under various operating conditions. The robustness of the DCC scheme to 30% parameter uncertainties and noise disturbances below 5% is also demonstrated. Moreover, the proposed DCC scheme is assessed on the OPAL-RT simulation platform, with real-time simulation results emphasizing its performance. Future work includes the flexible selection of the leader WT, taking into account different working conditions, geographical locations of WTs, wake effects, and communication failures.

References

1. Saber RO, Fax JA, Murray RM (2007) Consensus and cooperation in networked multi-agent systems. Proc IEEE 95(1):215–233. https://doi.org/10.1109/JPROC.2006.887293
2. Chen X, Shi M, Sun H, Li Y, He H (2018) Distributed cooperative control and stability analysis of multiple DC electric springs in a dc microgrid. IEEE Trans Ind Electron 65(7):5611–5622. https://doi.org/10.1109/TIE.2017.2779414
3. Chen J, Yao W, Zhang CK, Ren Y, Jiang L (2019) Design of robust mppt controller for grid-connected PMSG-based wind turbine via perturbation observation based nonlinear adaptive control. Renew Energy 134:478–495. https://doi.org/10.1016/j.renene.2018.11.048
4. Chen X, Shi M, Zhou J, Chen Y, Zuo W, Wen J, He H (2020) Distributed cooperative control of multiple hybrid energy storage systems in a DC microgrid using consensus protocol. IEEE Trans Ind Electron 67(3):1968–1979. https://doi.org/10.1109/TIE.2019.2898606
5. Yang R, Shi G, Cai X, Zhang C, Li G, Liang J (2020) Autonomous synchronizing and frequency response control of multi-terminal DC systems with wind farm integration. IEEE Trans Sustain Energy 11(4):2504–2514. https://doi.org/10.1109/TSTE.2020.2964145
6. Lu Z, Ye Y, Qiao Y (2019) An adaptive frequency regulation method with grid-friendly restoration for VSC-HVDC integrated offshore wind farms. IEEE Trans Power Syst 34(5):3582–3593. https://doi.org/10.1109/TPWRS.2019.2901986
7. Akkari S, Dai J, Petit M, Guillaud X (2016) Interaction between the voltage-droop and the frequency droop control for multi-terminal HVDC systems. IET Gener Transm Distrib 10(6):1345–1352. https://doi.org/10.1049/cp.2015.0036
8. Wang Z, Wu W (2018) Coordinated control method for DFIG-based wind farm to provide primary frequency regulation service. IEEE Trans Power Syst 33(3):2644–2659. https://doi.org/10.1109/TPWRS.2017.2755685

9. Mathworks Inc. Linearization (2021) https://ww2.mathworks.cn/help/slcontrol/linearization.html
10. Ma S, Geng H, Yang G, Pal BC (2018) Clustering-based coordinated control of large-scale wind farm for power system frequency support. IEEE Trans Sustain Energy 9(4):1555–1564. https://doi.org/10.1109/TSTE.2018.2794622